RADICALS OF RINGS

F. A. SZÁSZ

MATHEMATICAL INSTITUTE
HUNGARIAN ACADEMY OF SCIENCES

A Wiley–Interscience Publication

JOHN WILEY & SONS
Chichester · New York · Brisbane · Toronto

This Monograph was published in German
as Vol. 6 and in English as Vol. 11 in the series
DISQUISITIONES MATHEMATICAE HUNGARICAE
by Akadémiai Kiadó, Budapest

Published in co-edition with AKADÉMIAI KIADÓ, Budapest

***British Library Cataloguing in Publication Data*:**

Szász, F. A.
Radicals of rings.
1. Radical theory
I. Title
512′.4 QA251.3 79–40509

ISBN 0 471 27583 2

Printed in Hungary

PREFACE

This book is an introduction to the theory of radicals of rings including some results concerning nil rings since historically the nil radical was the very first concrete type of radical. The reader I had in mind when writing this book was the university student. Presumably, the book will be of use for research workers too, who aim at the solution of open problems. To understand this book, no special ring-theoretical knowledge is required except for the notions of ring, ideal, one-sided ideal, homomorphism, the Isomorphism Theorems, etc. Nevertheless, it is necessary to have some practice in mathematical reasoning. We shall deal with associative rings only, but the bibliography also includes items from the field of radical theory of different generalizations and analogues of associative rings.

As is generally known, the primary aim of the theory of rings is to determine the structure of all rings. The main task of a structure theorem is to find a complete system of invariants describing the ring up to isomorphism. The well-known Wedderburn–Artin Theorem is a good example: it says that every ring that has no nilpotent ideals and at the same time satisfies the minimum condition for right ideals, is isomorphic to the (ring-theoretical) direct sum of a finite number of full matrix rings over certain skew fields. Thus, the system of invariants for these special rings consists of finitely many skew fields and of finitely many natural numbers (the dimensions of the matrix rings over the corresponding skew fields). In the case of rings with minimum condition for right ideals, the sum of all nilpotent right ideals is an ideal which is also nilpotent and the factor ring by it does not contain non-zero nilpotent ideals. This same ideal has been termed the radical of the ring with minimum condition for right ideals and can be considered as a measure of the singularity of the ring. Namely, if the radical is small, e.g. it coincides with the zero ideal, then the ring is "well-behaved" inasmuch as one can obtain a structure theorem for it. On the other hand, if the radical is large, e.g. it coincides with the ring itself, then the ring is in general "badly behaved". ("Large" and "small" are used here in a sense different from the usual module-theoretical notions.) Analogously, the nilpotent elements (i.e. the elements x such that $x^n=0$ for some n: roots of zero, radical elements) of a commutative ring form an ideal, the factor ring by it does not contain non-zero nilpotent elements.

Later Baer, Levitzki, Jacobson, Brown and McCoy indicated explicitly constructed ideals in arbitrary rings (without any finiteness condition) that, in a certain sense, also characterize the measure of singularity of the ring, and therefore may be accepted for radicals, too. In all of these cases the factor ring of the ring by its radical has no non-zero radical and obeys some weaker structure theorem.

Amitsur and Kurosh, independently of each other, noticed the common features of the different radicals and developed the elements of a general theory of radicals of rings. As every class of rings can be embedded in a class of radical rings (with a suitable radical), a general Amitsur–Kurosh radical does not need to measure any singularity of the ring. The literature dealing with this general notion is extensive (see, for example, the works of Andrunakievich, Ryabukhin, Leavitt, Suliński, Divinsky, Leeuwen, Gardner, Wiegandt, Szász, etc.). Concerning the generalization of the radical theory for universal algebras we refer to Hoehnke's papers and with respect to its generalization for categories we refer the reader to the papers of MacLane, Shulgeifer, Ryabukhin, Livshits, Dickson, Wiegandt, Szász, etc.

Let us mention that, for associative and for alternative rings, radical and semisimplicity are dual and also equivalent notions which mutually determine one another.

While writing this book I often thought of Divinsky's classic work, "Rings and Radicals". The years that have passed since the publication of Divinsky's book have brought such a large number of new results that it seems justified to publish a new book on radical theory. The manuscript has been completed in 1968. I tried to give a comprehensive analysis of the general radical theory (Chapter I), of the theories of supernilpotent and special radicals (Chapter II), of nil radicals (Chapter III), of the Jacobson radical (Chapter IV), of the Brown–McCoy radical (Chapter V), and of some other types of radicals and zeroid-pseudoradicals (Chapter VI). Furthermore, my intention was to give an extensive (though, naturally not complete) bibliography on the radical theory of rings, on nil rings and related topics. I also endeavoured to discuss in detail some important examples. The open problems listed at the end of each chapter are meant to stimulate the further development of the field.

The book has been translated from German into English by the author except for the Preface which was translated by György Pollák, whom I thank. I also would like to thank Sándor Lajos for his thorough proofreading and for compiling the indexes. However, the English version is not a mere translation of the German original, since it also contains an outline of Tangeman's paper in § 4 and of a paper by Wiegandt and the author. In view of the fact that in § 4 we refer to several results which are more recent than the bibliography at the end of book, there we give full bibliographical data. The problems which were solved after the publication of the German version are indicated by a dagger and listed in an Appendix.

I dedicate this book to the memory of my teacher, the late Professor Tibor Szele (1918–1955), the excellent and unselfish mathematician who acquainted me as a student of the Debrecen University with the research methods of modern algebra.

The author

CONTENTS

CHAPTER I

GENERAL THEORY OF THE RADICALS

In this chapter we begin with the definition and the most important properties of an Amitsur–Kurosh radical of rings. We then show the relative independence of the defining system of axioms for the Amitsur–Kurosh radical property. The upper and lower radicals will also be discussed. We investigate how intersections and unions of radical classes determine a further radical. In connection with the radical theory the decompositions of the class of rings having no non-trivial ideals will also be treated. Furthermore, we examine the radical of an algebra over an operator domain and the radical of a full matrix ring. We then look at a property of hereditary radicals and of the strong semisimplicities. We also give a definition of radicals by means of operator modules and the definition of a Maranda–Michler quasiradical. Finally, we mention some open problems.

§ 1. THE DEFINITION AND THE MOST IMPORTANT PROPERTIES OF RADICALS

For the subject-matter of this section we refer the reader to Amitsur [4, 5, 6], Kurosh [3], and Anderson, Divinsky and Suliński [1, 2].

Let **R** be a property of rings. A ring with property **R** will be called an **R**-*ring*. We suppose that an arbitrary isomorphic image A' of an **R**-ring A is again an **R**-ring. This condition means in other words that **R** is an abstract algebraic property (corresponding to the well-known Noetherian isomorphism principle of algebra by which any two isomorphic structures which are not substructures of a common structure can essentially be identified). A (two-sided or right) ideal I of a ring A is said to be an **R**-*ideal* if the ring I is an **R**-ring. There exist abstract properties **R** for which the sum of two **R**-ideals or the union of an ascending chain of **R**-ideals is not an **R**-ideal.

If there exists an **R**-ideal $\mathbf{R}(A)$ of the ring A which contains every other **R**-ideal of A then $\mathbf{R}(A)$ is called the **R**-*radical* of A. The **R**-radical $\mathbf{R}(A)$ of the ring A need not exist for every abstract property **R** (cf. § 2).

A ring A is said to be **R**-*semisimple* if the zero-ideal (0) is the only **R**-ideal in A. Thus A is **R**-semisimple if and only if (0) is the **R**-radical $\mathbf{R}(A)$ of A.

We now come to the definition of a radical property **R** in the sense of Amitsur and Kurosh:

The property **R** is called a *radical property,* if **R** satisfies the following three axioms:

(i) Every homomorphic image A' of an **R**-ring A is again an **R**-ring.

(ii) Every ring A has an **R**-radical $\mathbf{R}(A)$.

(iii) The factor ring $A/\mathbf{R}(A)$ of A with respect to $\mathbf{R}(A)$ is **R**-semisimple, i.e. $\mathbf{R}(A/\mathbf{R}(A))=(0)$.

The radical property defined by this system of axioms can be called an "outer" property, since $\mathbf{R}(A)$ is defined in a fixed ring A "from outside" with reference to the class of **R**-rings.

If $\mathbf{R}(A)=A$ holds for a radical property **R** and for a ring A, then we say that A is an **R**-*radical ring.*

The property of **R** described by axiom (i) is very important and it is closely related to the definition of a preradical in the sense of Maranda [1] and Michler [2]. We shall discuss this later (cf. § 10). There exist properties for which (i) holds (e.g. the nilpotence of a ring) and there exist properties for which (i) does not hold (e.g. the property of not containing divisors of zero). Axiom (ii) ensures that every ring A has at least one **R**-ideal. Again (ii) implies that the zero ideal (0) of any ring is an **R**-ideal since by (ii) the ring (0) has an **R**-radical which must coincide with (0). Axiom (iii) ensures the existence of **R**-semisimple rings. Obviously, the ring (0) is the only ring which is for a radical property **R** both an **R**-radical ring and an **R**-semisimple ring. Note that (i) does not in general imply that (0) is an **R**-radical ring since the class of rings satisfying axiom (i) can be chosen to be empty. The relative independence of the axioms (i), (ii) and (iii) is shown in § 2.

Now we mention some properties of $\mathbf{R}(A)$ if **R** is a radical property.

THEOREM 1.1. $\mathbf{R}(A)$ is invariant in A with respect to every endomorphism φ of A onto itself.

PROOF. $(\mathbf{R}(A))\varphi$ is an **R**-ideal by (i). Hence by (ii) we have $(\mathbf{R}(A))\varphi\subseteq\mathbf{R}(A)$. □

THEOREM 1.2. If B is an ideal of a ring A and if B and A/B are **R**-radical rings, then A is also an **R**-radical ring.

PROOF. Suppose $\mathbf{R}(A)\not\supseteq B$. We have $(\mathbf{R}(A)+B)/\mathbf{R}(A)\cong B/(B\cap\mathbf{R}(A))$.

The right-hand side of this isomorphism is, by axiom (i), an **R**-ring. Therefore $(\mathbf{R}(A)+B)/\mathbf{R}(A)$ is also an **R**-ring. However this contradicts axiom (iii). Therefore $\mathbf{R}(A)\supseteq B$. But, then $A/\mathbf{R}(A)$ is **R**-semisimple by (iii). The existence of the homomorphism $A/B \to A/\mathbf{R}(A)$ together with the assumption on A/B imply that $A/\mathbf{R}(A)$ is also an $\mathbf{R}$-ring. From this it follows that $A/\mathbf{R}(A)=0$ and thus $\mathbf{R}(A)=A$. □

THEOREM 1.3. In a ring A the union $\bar{A}$ of any ascending chain of $\mathbf{R}$-ideals A_α is again an $\mathbf{R}$-ideal.

PROOF. Let

$$A_0 \subseteqq A_1 \subseteqq A_2 \subseteqq \dots \subseteqq A_\alpha \subseteqq \dots$$

be an ascending chain of $\mathbf{R}$-ideals of the ring A, $\bar{A}$ be the union of this chain, and $\mathbf{R}(\bar{A})$ be the $\mathbf{R}$-radical of $\bar{A}$. Then we have the isomorphism

$$(A_\alpha+\mathbf{R}(\bar{A}))/\mathbf{R}(\bar{A}) \cong A_\alpha/(A_\alpha \cap \mathbf{R}(\bar{A})).$$

The right-hand side is an $\mathbf{R}$-ring by axiom (i). However, since $\bar{A}/\mathbf{R}(\bar{A})$ is $\mathbf{R}$-semisimple by (iii), we obtain $A_\alpha \subseteqq \mathbf{R}(\bar{A})$ for every index α. Thus $\bar{A}=\bigcup_\alpha A_\alpha$ yields $\bar{A} \subseteqq \mathbf{R}(A)$. It follows that $\mathbf{R}(\bar{A})=\bar{A}$ and $\bar{A}$ is an $\mathbf{R}$-radical ring. □

THEOREM 1.4. The sum I_1+I_2 of the $\mathbf{R}$-ideals I_j $(j=1, 2)$ is again an $\mathbf{R}$-ideal of A.

PROOF. We have

$$(I_1+I_2)/I_2 \cong I_1/(I_1 \cap I_2),$$

and $I_1/(I_1 \cap I_2)$ is an $\mathbf{R}$-ideal by axiom (i) and by our assumption. But, since both $(I_1+I_2)/I_2$ and I_2 are $\mathbf{R}$-rings, we deduce from theorem 1.2 that I_1+I_2 is an $\mathbf{R}$-ideal. □

THEOREM 1.5. The $\mathbf{R}$-radical $\mathbf{R}(A)$ of an arbitrary ring A is the sum of all $\mathbf{R}$-ideals.

PROOF. This follows from theorems 1.3, 1.4 and axiom (ii). □

THEOREM 1.6. If B, C and D are ideals of a ring A, such that $B \supseteqq C$ and B/C is an $\mathbf{R}$-radical ring, then $(B+D)/(C+D)$ is also an $\mathbf{R}$-radical ring.

PROOF. Since $C \subseteqq B \cap (C+D)$, we have the following isomorphisms:

$$(B/C)/(B \cap (C+D))/C \cong B/(B \cap (C+D)) \cong (B+C+D)/(C+D) = \\ = (B+D)/(C+D).$$

The left-hand side is the factor ring of the radical ring B/C with respect to the ideal $(B \cap (C+D))/C$; thus, axiom (i) can be applied. □

The following fundamental theorem of Anderson, Divinsky and Suliński [7] which naturally was not yet known at the time when the important papers of

Amitsur and Kurosh on radicals appeared renders possible to establish a duality principle in the radical theory for associative rings (cf. e.g. Szász and Wiegandt [2] or Ryabukhin [2]).

THEOREM 1.7. If $\mathbf{R}$ is a radical property and I an ideal of the ring A then $\mathbf{R}(I)$ is also an ideal of A.

PROOF. If $\mathbf{R}(I)$ is not an ideal of the ring A, then there exist elements x and y in A satisfying $x\mathbf{R}(I)\not\subseteq\mathbf{R}(I)$ or $\mathbf{R}(I)y\not\subseteq\mathbf{R}(I)$. Let us suppose $x\mathbf{R}(I)\not\subseteq\mathbf{R}(I)$. We have $\mathbf{R}(I)\neq x\mathbf{R}(I)+\mathbf{R}(I)\subseteq I$, since I is an ideal of A with $\mathbf{R}(I)\subseteq I$. Obviously $x\mathbf{R}(I)+\mathbf{R}(I)$ is a right ideal of I and since

$$I(x\mathbf{R}(I)) = (Ix)\cdot\mathbf{R}(I) \subseteq I\cdot\mathbf{R}(I) \subseteq \mathbf{R}(I)$$

$x\mathbf{R}(I)+\mathbf{R}(I)$ is evidently a two-sided ideal of I.

Now define a mapping φ of $\mathbf{R}(I)$ onto $(x\mathbf{R}(I)+\mathbf{R}(I))/\mathbf{R}(I)$ by the equation

$$y\varphi = xy+\mathbf{R}(I) \quad (\text{for every } y\in\mathbf{R}(I)). \tag{$*$}$$

We shall show that φ is a ring homomorphism. Obviously, φ is an additive group homomorphism with $(\mathbf{R}(I))\varphi=x\mathbf{R}(I)+\mathbf{R}(I)/\mathbf{R}(I)$.

Furthermore, we have

$$xy_1y_2\in I\mathbf{R}(I) \subseteq \mathbf{R}(I).$$

Since $y_1y_2\varphi=xy_1y_2+\mathbf{R}(I)$ and $xy_1\in I$. Consequently $(y_1y_2)\varphi=0+\mathbf{R}(I)$.

On the other hand, we also have

$$(y_1\varphi)(y_2\varphi)\in\mathbf{R}(I)$$

since

$$(y_1\varphi)(y_2\varphi) = (xy_1+\mathbf{R}(I))(xy_2+\mathbf{R}(I)) = xy_1xy_2+\mathbf{R}(I)$$

and $xy_1x\in I$. Therefore $(y_1\varphi)(y_2\varphi)=0+\mathbf{R}(I)$.

Accordingly $(y_1y_2)\varphi=(y_1\varphi)(y_2\varphi)$ and thus φ is a ring homomorphism of the $\mathbf{R}$-radical ring $\mathbf{R}(I)$ onto the ring $(x\mathbf{R}(I)+\mathbf{R}(I))/\mathbf{R}(I)$. But since $I/\mathbf{R}(I)$ is $\mathbf{R}$-semisimple by axiom (iii) and since $(x\mathbf{R}(I)+\mathbf{R}(I))/\mathbf{R}(I)$ is an $\mathbf{R}$-radical ring by (i) and the above homomorphism, we get a contradiction to $x\mathbf{R}(I)\not\subseteq\mathbf{R}(I)$. Therefore $A\mathbf{R}(I)\subseteq\mathbf{R}(I)$, and similarly $\mathbf{R}(I)A\subseteq\mathbf{R}(I)$. Thus, $\mathbf{R}(I)$ is indeed an ideal of A. □

THEOREM 1.8. Every ideal of an $\mathbf{R}$-semisimple ring A is itself an $\mathbf{R}$-semisimple ring.

PROOF. Let $\mathbf{R}(I)$ be the $\mathbf{R}$-radical of the ideal I. Since $\mathbf{R}(I)$ is both an $\mathbf{R}$-ring and an ideal of A by theorem 1.7, we have $\mathbf{R}(I)\subseteq\mathbf{R}(A)$ by axiom (ii). Since $\mathbf{R}(A)=0$ it follows that $\mathbf{R}(I)=0$. □

Now we mention some consequences of theorem 1.8.

THEOREM 1.9. If B is an ideal of A, $\mathbf{R}$ is a radical property, B and A/B semisimple rings, then A is also $\mathbf{R}$-semisimple.

PROOF. Let $\mathbf{R}(A)$ be the $\mathbf{R}$-radical of A. We obviously have $\mathbf{R}(A)\nsubseteq B$ by the $\mathbf{R}$-semisimplicity of B. Consider the isomorphism

$$(\mathbf{R}(A)+B)/B \cong \mathbf{R}(A)/(\mathbf{R}(A)\cap B).$$

The ring on the left-hand side is semisimple by theorem 1.8, as it is an ideal of the semisimple ring A/B. The ring on the right-hand side is an $\mathbf{R}$-ring by axiom (i) as it is a homomorphic image of an $\mathbf{R}$-ring. Therefore $\mathbf{R}(A)/(\mathbf{R}(A)\cap B)=(0)$ and consequently $\mathbf{R}(A)\subseteq B$. Hence $\mathbf{R}(A)=0$ by the $\mathbf{R}$-semisimplicity of B. □

THEOREM 1.10. If B, C and D are ideals of the ring A, $\mathbf{R}$ is a radical property, and if B/C is $\mathbf{R}$-semisimple, then $(B\cap D)/(C\cap D)$ is also $\mathbf{R}$-semisimple.

PROOF. We have $(B\cap D)/C\cap D)\cong((B\cap D)+C)/C$ and $((B\cap D)+C)/C$ is an ideal of the semisimple ring B/C. Thus, theorem 1.8 can be applied. □

THEOREM 1.11. If $\mathbf{R}$ is a radical property and B_α are ideals in the ring A such that each factor ring A/B_α is $\mathbf{R}$-semisimple and $B_0=\bigcap_\alpha B_\alpha$ then A/B_0 is also $\mathbf{R}$-semisimple.

PROOF. Let $\mathbf{R}(A/B_0)$ be the $\mathbf{R}$-radical of A/B_0. If $\mathbf{R}(A/B_0)\neq(0)$ then there exists an ideal B_α satisfying $\mathbf{R}(A/B_0)\nsubseteq B_\alpha/B_0$. We can choose $\mathbf{R}(A/B_0)$ in the form K/B_0, where K is an ideal of A. We have $(K+B_\alpha)/B_\alpha\cong K/(K\cap B_\alpha)$ and $K\cap B_\alpha\supseteq K\cap B_0=B_0$. Accordingly $K/(K\cap B_\alpha)$ is a homomorphic image of $K/B_0=\mathbf{R}(A/B_0)$ and thus it is an $\mathbf{R}$-ring by axiom (i). On the other hand, $(K+B_\alpha)/B_\alpha$, being an ideal of the semisimple ring A/B_α, is semisimple by theorem 1.8. Consequently, we have $(K+B_\alpha)/B_\alpha=(0)$, $B_\alpha=K+B_\alpha\supseteq K$, and thus $K\subseteq\bigcap_\alpha B_\alpha=B_0$. Hence A/B_0 is semisimple. □

In other words theorem 1.11 means, that every subdirect sum (cf. p. 73) of $\mathbf{R}$-semisimple rings is $\mathbf{R}$-semisimple.

For a dualization we refer the reader to Szász and Wiegandt [1].

THEOREM 1.12. The radical $\mathbf{R}(A)$ of A is the intersection D of all ideals I_α of A such that A/I_α is $\mathbf{R}$-semisimple.

PROOF. Let D denote this intersection. Then $D\subseteq\mathbf{R}(A)$ obviously by axiom (iii).

But since A/D is **R**-semisimple by theorem 1.11, we have the relation $\mathbf{R}(A) \subseteq D$ because $\mathbf{R}(A)/D \subseteq A/D$. Consequently $\mathbf{R}(A) = D$. □

In some cases it is not easy to establish whether for a ring property **R** axioms (i), (ii) and (iii) are satisfied or not. Therefore it is useful to give for a radical property another system of axioms equivalent to the original system (i)–(iii). We have the following:

THEOREM 1.13. The ring property **R** is a radical property if and only if

(i′) Every homomorphic image A' of an **R**-ring A is an **R**-ring.

(ii′) If every non-zero homomorphic image of a ring A contains a non-zero **R**-ideal, then A is an **R**-ring.

PROOF. Let us suppose that **R** is a radical property. Then (i), (ii) and (iii) hold, and we shall show, that (i′) and (ii′) are satisfied. If (ii′) does not follow from (i), (ii) and (iii), then there exists a ring A, which is not an **R**-radical ring, although every non-zero homomorphic image A' of A contains a non-zero **R**-ideal. But this is impossible since if A is not an **R**-radical ring, then $\mathbf{R}(A) \neq A$ and $A/\mathbf{R}(A)$ is semisimple by (iii), and thus (0) is the only **R**-ideal of $A/\mathbf{R}(A)$. Therefore (ii′) follows from (i), (ii) and (iii).

Conversely, let us now assume that for **R** axioms (i′) and (ii′) are satisfied and show the validity of axioms (i), (ii) and (iii). The zero-ring is an **R**-ring by the validity of (i), since (0) trivially satisfies the condition occurring in the first part of (ii′).

In order to prove (ii) let S be the sum of all **R**-ideals of A. It is clear that S is an ideal of A. In case $S = (0)$, S is an **R**-ideal. In case $S \neq 0$, let S/U be a non-zero homomorphic image of S. Then because $S \neq U$ there exists an **R**-ideal V of A, which is not contained in U. We have $(V+U)/U \cong V/(V \cap U)$ by the isomorphism theorem. The right-hand side is a non-zero homomorphic image of the **R**-ring V since $V \cap U \neq V (\subsetneqq U)$, and thus $V/(V \cap U)$ is an **R**-ring by (i′). But then $(V+U)/U$ is a non-zero **R**-ideal of S/U, and therefore S is an **R**-ideal by (ii′). Now we prove that A/S is **R**-semisimple. If A/S is not **R**-semisimple, let W/S be the **R**-radical of A/S with $W \neq S$. Then W is an ideal of A satisfying $W \supset S$. Let W/X be an arbitrary non-zero factor ring of W. In case $X \supseteq S$, W/X is a homomorphic image of W/S; hence W/X is an **R**-ring by (i′) and therefore it is an **R**-ideal of W/X. In case $X \not\supseteq S$ we get from $X \cap S \subset X$ that the right-hand side of the isomorphism $(X+S)/X \cong S/(S \cap X)$ is a non-zero homomorphic image of the **R**-ring S. Thus $(X+S)/X$ is a non-zero **R**-ideal of W/S. Hence W is an **R**-ring by (ii′). But in case $W \neq S$ this is a contradiction since S is the sum of all **R**-ideals of A. Therefore $W = S$ and thus (iii) is proved. □

Applying axioms (i′) and (ii′) to **R**-semisimple rings we obtain the following interesting criterion:

THEOREM 1.14. For a radical property **R** the non-zero ring A is an **R**-radical ring if and only if A cannot be homomorphically mapped onto a non-zero **R**-semisimple ring.

PROOF. If A is an **R**-radical ring, then every homomorphic image of A is an **R**-ring by (i′) and thus A cannot be homomorphically mapped onto a non-zero **R**-semisimple ring.

Conversely, if A is not a radical ring, then by (ii′) not every non-zero homomorphic image contains a non-zero **R**-ideal. Such a homomorphic image is **R**-semisimple by definition. The ring $A/\mathbf{R}(A)$ is for instance such a homomorphic image. □

THEOREM 1.15. Let A be an arbitrary ring and **R** be an arbitrary radical property. We make correspond to every ideal I of A the ideals $\bar{I}$ and $\hat{I}$ uniquely as follows: $\varphi_1: I \to \bar{I} = \mathbf{R}(I)$ and $\varphi_2: I \to \hat{I}$, where $\hat{I}/I = \mathbf{R}(A/I)$. Then the mappings φ_1 and φ_2 are monotone in the lattice of ideals of the ring A, i.e. $I_1 \subseteq I_2$ implies $I_1\varphi_1 \subseteq \subseteq I_2\varphi_1$ and $I_1\varphi_2 \subseteq I_2\varphi_2$.

PROOF. If $I_1 \subseteq I_2$, the radical $\mathbf{R}(I_1)$ of I_1 is an ideal of I_2 by theorem 1.7. But $R(I_1)$ is also an **R**-ideal. Therefore $\mathbf{R}(I_1) \subseteq \mathbf{R}(I_2)$ by the definition of $\mathbf{R}(I_2)$ and thus $I_1\varphi_1 \subseteq I_2\varphi_1$.

Now suppose $I_1 \subseteq I_2$. Consider the isomorphism

$$(\hat{I}_1 + \hat{I}_2)/\hat{I}_2 \cong \hat{I}_1/(\hat{I}_1 \cap \hat{I}_2).$$

The right-hand side is an **R**-radical ring since we have

$$\hat{I}_1/(\hat{I}_1 \cap \hat{I}_2) \cong (\hat{I}_1/I_1)/((\hat{I}_1 \cap \hat{I}_2)/I_1) = \mathbf{R}(A/I_1)/((\hat{I}_1 \cap \hat{I}_2)/I_1)$$

with $I_1 \subseteq I_2 \subseteq \hat{I}_2$. The left-hand side is an **R**-semisimple ring by theorem 1.8 since $(\hat{I}_1 + \hat{I}_2)/\hat{I}_2$ is an ideal of the **R**-semisimple ring $A/\hat{I}_2$. It follows that $\hat{I}_1 = \hat{I}_1 \cap \hat{I}_2 \subseteq \hat{I}_2$. Therefore $I_1\varphi_2 \subseteq I_2\varphi_2$. □

By the above considerations one can establish a certain duality between the radical rings and semisimple rings where an ideal corresponds to a homomorphic image and conversely. An exact category-theoretical discussion of this duality can be found in Szász and Wiegandt [2]. We also refer to Ryabukhin [2]. For a category-theoretical dualization of subdirect sums we refer to Szász and Wiegandt [1].

§ 2. RELATIVE INDEPENDENCE OF AXIOMS (i), (ii) AND (iii) FOR THE AMITSUR–KUROSH RADICAL PROPERTY OF A RING

For the subject-matter of this paragraph, we refer to Szász [22]. We show the consistency and independence of the system of axioms (i), (ii) and (iii) discussed in § 1.

THEOREM 2.1. The system of axioms (i)–(iii) is consistent.

PROOF. Let **R** be for instance the property of rings A such that $A^2=A$ holds. Then the **R**-rings are the *idempotent rings*, i.e. the rings A with $A^2=A$. Then obviously (i) holds for **R**, since any homomorphic image of an idempotent ring is idempotent. Moreover, let $\mathbf{R}(A)$ be the sum of all idempotent ideals I_α of the ring A. Then in the expansion of the square $(\sum_\alpha I_\alpha)^2$ all ideals I_α^2 occur and even further ideals contained in some of the I_β. But we have $(\sum I_\alpha)^2 \supseteqq \sum I_\alpha$ since $I_\alpha^2=I_\alpha$. Also $(\sum I_\alpha)^2 \subseteqq \sum I_\alpha$ by the trivial relation $\sum I_\alpha \supseteqq (\sum I_\alpha)^2$. This implies the idempotence $(\mathbf{R}(A))^2=\mathbf{R}(A)$ of $\mathbf{R}(A)$. Therefore $\mathbf{R}(A)$ is an **R**-ideal. It is in fact the maximal **R**-ideal of A. Thus (ii) holds for **R**. Now we shall show the validity of (iii) for **R**. Let us assume that there exists an **R**-ideal $S/\mathbf{R}(A)$ in $A/\mathbf{R}(A)$. Then we have $S^2=S$ because $S^2+\mathbf{R}(A)=S$, $S\supseteqq\mathbf{R}(A)$ and $S^2\supseteqq(\mathbf{R}(A))^2=\mathbf{R}(A)$. Hence the equations $S=\mathbf{R}(A)$ and $\mathbf{R}(A/\mathbf{R}(A))=(0)$ follow by the maximality of $\mathbf{R}(A)$. Thus (iii) also holds for **R** and this shows the consistency of the system of axioms (i), (ii) and (iii). □

A ring A is said to be of *degree of nilpotence k* if there exists a natural number k such that $A^k=(0)$ and $A^{k-1}\neq(0)$. (Here A^k is the set of all sums $\sum a_1 a_2 \dots a_k$ with $a_i\in A$, and BC is, for ideals B and C of A, the set of all sums $\sum bc$ with $b\in B$ and $c\in C$.)

Every nilpotent ring is an **R**-semisimple ring where property **R** means the above idempotence. Every subset and hence also every ideal of a nilpotent ring is nilpotent. A cannot contain any non-zero idempotent ideal since $I^2=I$ and $I^k=(0)$ yield $I=I^2=\dots=I^k=(0)$. Here we have shown the semisimplicity of nilpotent rings for the idempotent radical because the nilpotent rings are radical rings for many important concrete notions of radical and because historically the radical has been first defined precisely with the help of the notion of nilpotence.

An element x is called *nilpotent* if $x^n=0$, and therefore any nilpotent element x is an n-th root of 0 (root=radix, in Latin). In every commutative ring the set of all nilpotent elements is a radical in the sense of Amitsur and Kurosh.

Now we shall prove the relative independence of axioms (i), (ii) and (iii). In what follows we use further definitions which we shall give only when their introduction becomes necessary.

THEOREM 2.2 (Szász [22]). The system of axioms (i), (ii) and (iii) is relatively independent.

PROOF. For an axiom $\mathfrak{X}$, where $\mathfrak{X}=$(i), (ii) or (iii), let $\overline{\mathfrak{X}}$ denote the negation of $\mathfrak{X}$. Since the formulation of (iii) requires the validity of (ii), the system (i), (ii) and (iii) naturally cannot be absolutely independent, since the validity of (iii) assumes the validity of (ii). Therefore, for the proof of the relative independence of (i), (ii) and (iii) we must show the consistency of the following five systems I, II, III, IV and V:

I. $(\overline{\text{i}})$, $(\overline{\text{ii}})$, $(\overline{\text{iii}})$

II. $(\overline{\text{i}})$, (ii), $(\overline{\text{iii}})$

III. $(\overline{\text{i}})$, (ii), (iii)

IV. (i), $(\overline{\text{ii}})$, $(\overline{\text{iii}})$

V. (i), (ii), $(\overline{\text{iii}})$.

DISCUSSION OF SYSTEM I. Let **S** be the ring property of being free from divisors of zero. (A ring A is said to be free from divisors of zero if $ab=0$ ($a\in A$, $b\in A$) implies that $a=0$ or $b=0$. If $a\neq 0$, $b\neq 0$ and $ab=0$ then a is said to be a left divisor of zero and b a right divisor of zero.) Then for **S** obviously $(\overline{\text{i}})$ holds, since the ring I of rational integers has no divisors of zero while the factor ring $I/(4)$ does. But for **S** $(\overline{\text{ii}})$ also holds. Let A be the direct sum of two subrings B and C having no divisors of zero, i.e. $A=B+C$, $B\cap C=BC=CB=(0)$, and $B+C$ is the set of all sums $b+c$ with $b\in B$ and $c\in C$. Then B and C are obviously S-ideals, but not maximal **S**-ideals in A. A has no absolutely maximal **S**-ideal and thus $(\overline{\text{ii}})$ holds. Therefore for this **S** $(\overline{\text{iii}})$ holds too, and this shows the consistency of the system I.

DISCUSSION OF SYSTEM II. Let **S** be the ring property that the ring is the sum of its nilpotent accessible subrings M_α (for the definition see what follows) for which $p^2M_\alpha=0$ but $pM_\alpha\neq 0$ for at least one index α where p is a fixed prime number. A is said to be the sum of the subsets T_α if every element $a\in A$ can be written as a finite sum

$$a = t_{\alpha_1}+t_{\alpha_2}+\ldots+t_{\alpha_n} \quad (t_{\alpha_i}\in T_{\alpha_i}).$$

If in the ring A there exists an ascending chain of subrings

$$M = A_n \subset A_{n-1} \subset \ldots \subset A_0 = A \qquad (*)$$

such that each subring A_i is a two-sided ideal in A_{i-1} then, following Baer [2], M is called an *accessible subring* in A. If $\overline{M}$ denotes the ideal generated by M in A it can be shown (by induction on the index n) that there exists an exponent

k such that $(\overline{M})^k \subseteqq M \subseteqq \overline{M}$. This implies that if every ideal of the ring is idempotent then every accessible subring is an ideal of the ring. It can be proved with the methods of Baer [2], that every nilpotent accessible subring can be embedded into a nilpotent ideal.

Let us consider those members X of the chain

$$M = A_n \subset A_{n-1} \subset \ldots \subset A_0 = A$$

for which $X \supseteqq M$ and XMX is a nilpotent ideal of X. The set S^* of these X is a further chain (**), and it is not empty. For, we have $M \in S^*$ since $M^3 = M \cdot M \cdot M$ is a nilpotent ideal in M. Let H be the union of an ascending subchain of S^*. Every element of HMH is of the form $t = \sum h_i m_i k_i$, $m_i \in M$ and $h_i, k_i \in H$, $i = 1, 2, \ldots$ $\ldots, j$. Therefore there exists a member $X \in S^*$ such that $h_i, k_i \in X$ and consequently $t \in XMX \subseteqq X$. Thus $H \in S^*$ in (**) there exists a maximal member W by Zorn's lemma. If $W \neq A$ then there exists an accessible subring V of A such that W is an ideal in V. Then $M \subseteqq W \subset V$ and

$$(VMV)^3 = (VMV^2)M(V^2MV) \subseteqq WMW.$$

Here WMW is a nilpotent ring with $WMW = W$ by the definition of W. But then VMV is also nilpotent satisfying the condition $VMV = V$ which contradicts the maximality of W. Consequently, we have $W = A$ and thus AMA is a nilpotent ideal of A. If $\overline{M} = M + AM + MA + AMA$ is the ideal of A generated by M then $(\overline{M})^3 \subseteqq AMA$. Accordingly $(\overline{M})^3$ and hence $\overline{M}$ is a nilpotent ideal of A. Moreover $p^2M = (0)$ implies $p^2\overline{M} = (0)$.

The sum $\mathbf{S}(A)$ of all $\mathbf{S}$-ideals of A is likewise an $\mathbf{S}$-ideal of A. If I is an $\mathbf{S}$-ideal then I is the sum of its nilpotent accessible subrings, and every nilpotent accessible subring of I is also a nilpotent accessible subring of $\mathbf{S}(A)$. Thus $\mathbf{S}(A)$ is an $\mathbf{S}$-ring. Now $\mathbf{S}(A)$ coincides with the sum $\overline{\mathbf{S}}(A)$ of all nilpotent ideals I satisfying $p^2I = 0$, since we have $\mathbf{S}(A) \supseteqq \overline{\mathbf{S}}(A)$ by the definition of accessible subrings and of property $\mathbf{S}$. On the other hand, the theorem of Baer, on the embeddability of nilpotent accessible subrings into a nilpotent ideal, yields $\mathbf{S}(A) \subseteqq \overline{\mathbf{S}}(A)$. Thus (ii) holds for property $\mathbf{S}$. Furthermore, let us consider the example of a zero ring $\{a\}$ with a cyclic additive group of order p^2. Then $\{a\}$ is an $\mathbf{S}$-ring, but its homomorphic image $\{a\}/p\{a\}$ is not an $\mathbf{S}'$-ring, for in $\{a\}/p\{a\}$ there exists no accessible subring M such that $pM \neq 0$. Therefore (i) holds for $\mathbf{S}$. (Here a ring is called a *zero-ring* if $ab = 0$ holds for every $a \in A$ and $b \in A$.) Finally, the zero-ring $\{b\}$ whose additive group is cyclic of order p^4 effectively shows that (iii) holds for $\mathbf{S}$. Thus, the system (i), (ii) and (iii) is also consistent.

Discussion of system III. Let a ring possess the property $\mathbf{S}$ if it is not isomorphic to a fixed simple ring A. Here *simple* means that the ring A satisfies $A^2 = A$ and that it contains no ideal different from (0) and A. Then the sum $B = A + A'$ satisfying $A' \cong A \cap A' = (0)$ is an S-ring but the homomorphic image $A \cong B/A'$ is not. Thus (ĩ) holds for S. Moreover, (ii) and (iii) are also valid for S.

DISCUSSION OF SYSTEM IV. Now let the S-rings be the nilpotent rings. If $A^k = (0)$ holds for a ring A then we have $B^k = (0)$ for every homomorphic image B of A, and thus S satisfies axiom (i). Now let A be the direct sum of subrings A_k satisfying $A_k^{k+1} = (0)$; $A_k^k \neq 0$. Then A contains S-ideals, but A does not contain any maximal S-ideal, since the sum of all S-ideals is the ring A itself which is not nilpotent as $A^k \supseteq A_k^k \neq (0)$. Therefore A is not an S-ring. Consequently (ĩi) and (ĩii) are valid for S.

DISCUSSION OF SYSTEM V. Let the S-rings be the rings which are the sums of their nilpotent accessible subrings. Then every homomorphic image of an S-ring is likewise an S-ring and thus (i) holds for S. Furthermore, the sum $S(A)$ of all S-ideals in every ring A coincides with sum $\overline{S}(A)$ of all nilpotent ideals of A as has been shown with the methods of Baer in the discussion of system II, and thus (ii) also holds for S. Now we give a ring A satisfying $S(A/S(A)) \neq (0)$, and showing thereby the validity of (ĩii) for S (cf. Szász [6]). This ring satisfies in addition the minimum condition on principal right ideals. Moreover every element x of A is nilpotent, i.e. $x^n = 0$ with $n = n(x)$ for every $x \in A$. The *minimum condition for principal right ideals* means that every descending chain of principal right ideals

$$(a_1)_r \supseteq (a_2)_r \supseteq (a_3)_r \supseteq \cdots$$

terminates after a finite number of steps where the principal right ideal $(a_i)_r$ is the intersection of all right ideals containing the element a_i of A. Let A be the ring generated by the elements a_0 and a_{ijk} satisfying the following defining relations:

$$a_0^2 = a_{i'i''k}^2 = a_{iik}^{2k} = a_0 a_{ijk} + \delta_{1i} a_{2jk} = \delta_{2j} a_{ijk} + a_{ijk} a_0 =$$
$$= a_{i_1 j_1 k_1} a_{i_2 j_2 k_2} + \delta_{j_1 i_2} \delta_{k_1 k_2} a_{i_1 j_2 k_1}^2 = 2a_0 + 2a_{ijk} = 0$$
$$(i, j, k = 1, 2, 3, \ldots)$$

where $i' \neq i''$; $1 \leq i \leq 2$; $1 \leq j \leq 2$ and δ_{ij} is the Kronecker-symbol (i.e. $\delta_{ij} = 0$ for $i \neq j$ and $\delta_{ij} = 1$ for $i = j$).

By a straightforward calculation it is easy to see that A is an associative ring. Moreover it is a *nil ring* (i.e. every element of A is nilpotent) and the minimum condition holds for principal right ideals since every infinite product $x_1 x_2 x_3 \ldots$ terminates after a finite number of steps in 0. The principal right ideal $(a_0)_r$ is not nilpotent since $a_{22k}^{2k-1} \neq 0$ and $a_{2jk} \in (a_0)_r$. Now, $\overline{S}(A)$ the sum of all nilpotent ideals, coincides with the sum $S(A)$ of all S-ideals, and $a_0 \notin \overline{S}(A)$ since if $a_0 \in S(A)$

then a_0 would lie in the sum of a finite number of nilpotent ideals of A and then $(a_0)_r$ would be nilpotent which is impossible. Moreover it can be checked by calculation that $S(A/S(A)) \neq (0)$ since we have $a_0 + S(A) \in S(A/S(A))$. Hence (iii) holds for this property S. This solves a problem verbally raised by O. Steinfeld in 1960.

Thus, we have shown the consistency of the systems I, II, III, IV and V, and this implies the relative independence of the system of axioms (i), (ii) and (iii). □

§ 3. THE UPPER RADICAL

For the subject-matter of this section we refer the reader to Kurosh [3].

We take in this section an arbitrary class **K** of rings as a base, defining a radical property **R** for which every ring in **K** is **R**-semisimple. For this purpose we need the following fundamental

THEOREM 3.1. The class **K** is the class of all **R**-semisimple rings for a radical property **R** if and only if **K** satisfies the following two conditions:

(j) Every ideal of a ring in **K** belongs to the class **K**.

(jj) If every non-zero ideal of a ring **A** can be homomorphically mapped onto a non-zero ring from **K** then A belongs to the class **K**.

REMARK: In Kurosh [3] a weaker condition is used instead of (j):

(j′) Every non-zero ideal I of a ring A in **K** can be homomorphically mapped onto a non-zero ring B in **K**.

According to Armendariz and Leavitt [1], (j) does not hold for all non-associative rings but (j′) holds for all rings.

It may be remarked that (j′) has a stronger connection with the axiom system (i), (ii) and (iii) for a radical property **R** than axiom (j); but (j) and (j′) are equivalent by theorems 1.8 and 3.1.

PROOF. Let **R** be a radical property, and let us assume that **K** is the class of all **R**-semisimple rings. Then every ideal I of $A \in \mathbf{K}$ is likewise **R**-semisimple by theorem 1.8. Thus $I \in \mathbf{K}$; and consequently, we have (j). In order to prove (jj) let A be not **R**-semisimple. Then A has a non-zero **R**-radical $\mathbf{R}(A)$, which cannot be homomorphically mapped, by theorem 1.14, onto a non-zero **R**-semisimple ring, consequently, onto a ring from **K**. Therefore (jj) also holds for K.

Conversely, let us suppose that a class **K** of rings is given satisfying conditions (j) and (jj). We define a radical property **R** for which **K** is precisely the class of all **R**-semisimple rings.

For, let A be an **R**-ring if A cannot be homomorphically mapped onto a non-zero ring from the class **K**.

In order to show that the ring property **R** is a radical property we verify axioms (i′) and (ii′) for **R**.

Condition (i′) obviously holds, since, if a homomorphic image B of a ring A can be homomorphically mapped onto a non-zero ring C in **K** then A can be homomorphically mapped onto C. In order to show (ii′), let A be a ring such that every non-zero homomorphic image of A contains non-zero **R**-ideal. If A is not an **R**-ring then A can be homomorphically mapped, by the definition of **R**, onto a non-zero ring A' in **K**. Also the homomorphic image A' has a non-zero **R**-ideal I by the assumed property of A. This leads to a contradiction, since I belongs to the class **K** by (j), but I being an **R**-ring, cannot be homomorphically mapped onto a non-zero ring of **K**. Therefore A is an **R**-ring and thus (ii′) is verified.

Finally, we show that every ring A of **K** is **R**-semisimple, and every **R**-semisimple ring is contained in **K**. Indeed, if A is a ring belonging to **K**, then every ideal J of A is also contained in **K** by (j). Since J can be identically mapped onto J itself, J is not an **R**-ring. Thus, **K** must be **R**-semisimple. On the other hand, if A is an **R**-semisimple ring then every non-zero ideal is **R**-semisimple by theorem 1.8. Thus, J can be homomorphically mapped onto a non-zero ring in **K**. Hence A is contained in **K** by (jj). Therefore **K** is precisely the class of all **R**-semisimple rings. □

REMARK 3.2. If **K** is a class of rings having no non-trivial ideals (different from (0) and A) then (i′) holds for this class **K**. But we can construct for every class **K** satisfying axiom (j) a class $\overline{\mathbf{K}}$ which is generally larger and satisfies both (j) and (jj).

THEOREM 3.3. If **K** is a class of rings satisfying axiom (j), and if $\overline{\mathbf{K}}$ is the class of rings with the property that each of their non-zero ideals can be homomorphically mapped onto a non-zero ring belonging to **K** then the class $\overline{\mathbf{K}}$ satisfies both axioms (j) and (jj).

PROOF. **K** is obviously contained in $\overline{\mathbf{K}}$ by axiom (j). Now, let A be a ring from $\overline{\mathbf{K}}$. Then every non-zero ideal can be homomorphically mapped onto a non-zero ring in **K** and thus also in $\overline{\mathbf{K}}$. Accordingly $\overline{\mathbf{K}}$ possesses the following property:

(j′) Every non-zero ideal of a ring from $\overline{\mathbf{K}}$ can be homomorphically mapped onto a non-zero ring in $\overline{\mathbf{K}}$.

Conversely, let A be a ring such that every non-zero ideal J of A can be homomorphically mapped onto a non-zero ring A' of $\overline{\mathbf{K}}$. Now A can be homomorphically mapped onto a non-zero ring in **K** by the definition of **K**. Consequently, A is contained in $\overline{\mathbf{K}}$.

It can be proved, just as in the proof of theorem 3.1, that there exists a radical property **R** for the class $\overline{\mathbf{K}}$ satisfying (j′) and (jj) such that $\overline{\mathbf{K}}$ consists precisely of all **R**-semisimple rings. But, then (j) also holds for the class $\overline{\mathbf{K}}$ (instead of the weaker axiom (j′)), by theorem 1.8. □

Now, let **K** be an arbitrary class of rings, $\overline{\mathbf{K}}$ be the extension of the class **K**, obtained by the adjunction of all accessible subrings of the rings belonging to **K**. Let $\overline{\overline{\mathbf{K}}}$ be the class of all rings with the property that their every non-zero ideal can be homomorphically mapped onto a non-zero ring from $\overline{\mathbf{K}}$. Then $\mathbf{K} \subseteq \overline{\overline{\mathbf{K}}}$, and $\overline{\overline{\mathbf{K}}}$ satisfies axioms (j) and (jj). Therefore, $\overline{\overline{\mathbf{K}}}$ defines by theorem 3.1 a radical property which all rings in $\overline{\overline{\mathbf{K}}}$ are **R**-semisimple for. This radical property **R** is called the *upper radical property* determined by the class **K**. The corresponding **R**-radical of a ring A is said to be the upper radical of A. Every ring in **K** is **R**-semisimple since $\mathbf{K} \subseteq \overline{\overline{\mathbf{K}}}$.

Let us explain the word "upper" in the expression above.

If $\mathbf{R}_1$ and $\mathbf{R}_2$ are two radical properties, we define $\mathbf{R}_1 \leqq \mathbf{R}_2$ if every $\mathbf{R}_1$-radical ring is also an $\mathbf{R}_2$-radical ring, $\mathbf{R}_1(A) \subseteq \mathbf{R}_2(A)$ in every ring A for the $\mathbf{R}_1$-radical and the $\mathbf{R}_2$-radical. This is equivalent to the fact that every $\mathbf{R}_2$-semisimple ring is also an $\mathbf{R}_1$-semisimple ring.

THEOREM 3.4. If **K** is an arbitrary class of rings, $\overline{\mathbf{K}}$ the class of all accessible subrings of rings from **K**, $\overline{\overline{\mathbf{K}}}$ the class of those rings whose every non-zero ideal can be homomorphically mapped onto a non-zero ring in **K**, and $\mathbf{R}'$ an arbitrary radical property for which all rings from **K** are $\mathbf{R}'$-semisimple then $\mathbf{R}' \leqq \mathbf{R}$, where **R** denotes the upper radical property determined by the class **K**. (Therefore, the **R**-radical rings are the rings which cannot be homomorphically mapped onto a non-zero ring in $\overline{\overline{\mathbf{K}}}$.)

PROOF. The class of all $\mathbf{R}'$-semisimple rings contains **K** and satisfies axioms (j) and (jj) by theorem 3.1. Furthermore it also contains $\overline{\overline{\mathbf{K}}}$. Now obviously $\mathbf{R}' \leqq \mathbf{R}$. □

§ 4. THE LOWER RADICAL

For the subject-matter of this section, we refer the reader to Amitsur [4, 5, 6] and Kurosh [3], Tangeman [1], Szász and Wiegandt [3].

Let **K** be an arbitrary class of rings. In this section we shall construct a radical property **R** for which all rings in **K** are **R**-radical rings and for which $\mathbf{R} \leqq \mathbf{R}'$, where $\mathbf{R}'$ is an arbitrary radical property such that all rings of **K** are $\mathbf{R}'$-radical rings. This construction uses transfinite methods, namely ordinal numbers. In the particular case when **K** consists of nilpotent rings this construction is due to

Baer [1]. Later on Amitsur [4] and Kurosh [3] gave a general construction of the lower radical property. Note that the term "upper radical property" is used by Amitsur. The construction of Kurosh has been sharpened by Suliński, Anderson and Divinsky [1] who have shown that in the radical construction in fact only ordinal numbers $\leqq \omega_0$ occur instead of arbitrary ordinal numbers, where ω_0 denotes the least infinite ordinal number, and, on the other hand, one gets on well even with smaller finite ordinal numbers for important and more special radical properties. Heinicke [1] has proved that in general ω_0 cannot be substituted by a smaller ordinal number.

Moreover, Ryabukhin [5] has shown that the construction of the lower radical need not terminate at the ω_0-th step for non-associative rings. Here we mention both the construction of Kurosh [3] and that of Suliński–Anderson–Divinsky.

A class **K** of rings is said to be *homomorphically closed,* if $A \in \mathbf{K}$ implies $A\varphi \in \mathbf{K}$ for every ring homomorphism φ. If **K** is an arbitrary class of rings then a ring A is said to have degree 1 over **K**, if $A=(0)$ or if A is a homomorphic image of a ring belonging to **K**. Every ring in **K** is obviously of degree 1 over **K**, and every homomorphic image of a ring of degree 1 over **K** is again a ring of degree 1 over **K**. We say that the ring A is of degree 2 over **K** if either every non-zero homomorphic image of A has a non-zero ideal of degree 1 over **K** or A itself is of degree 1 over **K**. If $\beta-1$ exists for an ordinal number β then the ring A is said to be of degree β over **K** if every non-zero homomorphic image of A contains a non-zero ideal which is a ring of degree $\beta-1$ over **K**. For a limit ordinal number β define a ring A to be of degree β over **K**, if A is a ring of degree α for some ordinal number $\alpha<\beta$.

Now we have the following:

PROPOSITION 4.1. Every homomorphic image of a ring of degree α over **K** is again a ring of degree α.

PROOF. The proposition is trivial for $\alpha=1$. Now suppose $\alpha\neq 1$ and that α is not a limit ordinal number. If the ring A is of degree α over **K** and A' is an arbitrary homomorphic image of A then every homomorphic image A'' of A' is again a homomorphic image of A and thus if $A''\neq(0)$, then A'' must contain a non-zero ideal of degree $\alpha-1$ over **K**. Accordingly, A' is of degree α over **K**. Now, let α be a limit ordinal number. Then A is of degree γ over **K** with $\gamma<\alpha$. Choose γ as small as possible. Then $\gamma\leqq\delta$ for every ordinal number $\delta<\alpha$ for which A is of degree δ over **K**. This choice of γ is always possible since every set of ordinal numbers is well-ordered for the order relation $\leqq$. But then every homomorphic image A' of A is a ring of degree δ by what has been said above and thus A' is of degree α over **K**. □

PROPOSITION 4.2. If $\alpha < \beta$, then every ring of degree α is also a ring of degree β over K.

PROOF. Clearly we can assume $\beta > 1$. If β is not a limit ordinal number and if A is of degree $\beta - 1$ over **K** then every homomorphic image of A is also of degree $\beta - 1$ over **K**. Thus A is a ring of degree β over **K**. Next let us assume that β is a limit ordinal number and A is a ring of degree α with $\alpha < \beta$. Then A is a ring of degree β by definition. □

DEFINITION 4.3. Let $\mathbf{K}^*$ be the class of rings which are of a certain degree over **K**. Let a ring A be an **R**-ring if and only if $A \in \mathbf{K}^*$. Then obviously $\mathbf{K} \leqq \mathbf{K}^*$. We now show that **R** is a radical property.

THEOREM 4.4. **R** is a radical property.

PROOF. We shall show that conditions (i′) and (ii′) of theorem 1.13 are satisfied. (i′) obviously holds by proposition 4.1. We now prove the validity of (ii′). Let A be a ring such that every non-zero homomorphic image A_α of A has a non-zero ideal J_α of degree β_α over **K**. Let γ be an ordinal number greater than or equal to β_α. Obviously, such a γ exists. Then every ideal J_α is of degree γ by proposition 4.2, and thus A is of degree $\gamma + 1$ over **K**. This implies that A is an **R**-ring, and thus (ii′) holds as well. Therefore **R** is a radical property. □

This radical property **R** is called the *lower radical property* determined by the class **K**; the corresponding **R**-radical of a ring A is said to be the *lower radical* of A. The word "lower" is explained by the following proposition:

THEOREM 4.5. If **K** is a class of rings, **R** is the lower radical property determined by the class **K** and $\mathbf{R}'$ an arbitrary radical property for which every ring belonging to **K** is an $\mathbf{R}'$-radical ring, then $\mathbf{R} \leqq \mathbf{R}'$.

PROOF. Both **R** and $\mathbf{R}'$ satisfy conditions (i′) and (ii′) of theorem 1.13. Every ring in **K** is an $\mathbf{R}'$-radical ring by (i′) and the assumption. We use induction and suppose that every ring of degree $\alpha < \beta$ is an $\mathbf{R}'$-ring. If β is a limit ordinal number and A is a ring of degree α where $\alpha < \beta$ then A is also an $\mathbf{R}'$-ring. However, if $\beta - 1$ exists and A is a ring of degree β then every non-zero homomorphic image of A contains a non-zero ideal I of degree $\beta - 1$, and hence I is an $\mathbf{R}'$-ring. Accordingly, A itself is an $\mathbf{R}'$-ring by theorem 1.13, and thus $\mathbf{R} \leqq \mathbf{R}'$ holds. □

We shall now discuss the construction given by Suliński, Anderson and Divinsky [1] for the lower radical property determined by a class **K**. This construction differs a little from the construction of Kurosh. Even the definition of a ring of degree 1 over **K**, given above, is different from that of Kurosh.

Let $\mathbf{K}=\mathbf{K}_0$ be an arbitrary class of rings and $\mathbf{K}_1$ be the class of all homomorphic images of rings from $\mathbf{K}$. Let us assume that the class $\mathbf{K}_\alpha$ is already defined for all ordinal numbers α with $1 \leqq \alpha < \beta$. Define $\mathbf{K}_\beta$ as the class of rings A for which every non-zero homomorphic image has a non-zero ideal I belonging to $\mathbf{K}_\alpha$ for some ordinal number $\alpha < \beta$.

Every class $\mathbf{K}_\beta$ is closed with respect to the construction of homomorphic images and $\mathbf{K}_\alpha \leqq \mathbf{K}_\beta$ for $\alpha \leqq \beta$. Furthermore, let $\mathbf{K}^{**}$ be $\cup \mathbf{K}_\alpha$. Then $\mathbf{K}^{**}$ determines a radical property $\mathbf{R}'$ which coincides with the lower radical property in the sense of Amitsur and Kurosh. These propositions can be similarly proved as above. The rings of $\mathbf{K}^{**}$ will be called $\mathbf{R}'$-radical rings. In the construction of $\mathbf{K}^{**}$ no distinction has been made between limit and non-limit ordinal numbers.

We now wish to prove $\mathbf{K}^{**}=\mathbf{K}_{\omega_0}$ by the methods of Suliński, Anderson and Divinsky. Here ω_0 denotes the least infinite ordinal number. In order to do this we need some preparation.

PROPOSITION 4.6. If A is a non-zero $\mathbf{R}'$-ring where $\mathbf{R}'$ denotes the lower radical property in the sense of Suliński, Anderson and Divinsky, determined by the class $\mathbf{K}$, then A contains a non-zero accessible subring B which belongs to $\mathbf{K}_1$.

PROOF. A belongs to a class $\mathbf{K}_\alpha$ for some α because $A \in \mathbf{K}^{**} = \cup \mathbf{K}_\alpha$. We choose α to be minimal. If $\alpha=1$ then $A \in \mathbf{K}_1$ and $B=A$ is the required accessible subring. Now let us assume that the proposition has already been proved for all rings $C \in \mathbf{K}_\gamma$ with $\gamma < \alpha$. Pick $A \in \mathbf{K}_\alpha$ and choose α to be minimal, i.e. $A \notin \mathbf{K}_\gamma$ for $\gamma < \alpha$. Every non-zero homomorphic image of A contains a non-zero ideal J in $\mathbf{K}_\gamma$ for some $\gamma < \alpha$ by the definition of $\mathbf{K}_\alpha$, and thus, in particular, A itself has an ideal $I \neq (0)$ which belongs to $\mathbf{K}_\gamma$ for an ordinal number $\gamma < \alpha$. Then I contains an accessible subring B in the class $\mathbf{K}_1$ by our assumption. But B is obviously an accessible subring of the ring A too, since, if the index of B in I equals j, then the index of B in A is $j+1$. □

PROPOSITION 4.7. If B is an accessible subring of the ring A, $\bar{B}$ is the two-sided ideal, generated by B in A and B belongs to $\mathbf{K}_1$, then $\bar{B}$ belongs to a class $\mathbf{K}_q$ for some finite q.

PROOF. Let $B=B_0 \subset B_1 \subset \ldots \subset B_n = A$, where B_i is an ideal of B_{i+1} and let n be chosen as small as possible. We prove our statement by induction on n. If $n=1$ then B is an ideal of A so that $\bar{B}=B$. Then $\bar{B}$ belongs to $\mathbf{K}_1$ by our assumption. Suppose now that the statement holds for all accessible subrings belonging to $\mathbf{K}_1$ with an index smaller than n.

Let B^* be the ideal generated by B in the ring B_{n-1}. We have $B^* \in \mathbf{K}_{q-1}$ for a finite number q by induction. Moreover, $B^* \leqq \bar{B} = A$, and B^* is an ideal of $\bar{B}$

since $\bar{B} \leqq B_{n-1}$. We obviously have

$$\bar{B} = B^* + B^* A + AB^* + AB^* A$$

since the ideal on the right-hand side contains $\bar{B}$ by $B \leqq B^*$, and it lies inside $\bar{B}$ because $B^* \leqq \bar{B}$.

We shall show that every non-zero homomorphic image of $\bar{B}$ has a non-zero ideal belonging to $\mathbf{K}_{q-1}$ where q is finite, and thus $\bar{B} \in \mathbf{K}_q$ will be proved.

Let $\bar{B}/W$ be an arbitrary non-zero homomorphic image of $\bar{B}$. If $B^* \nsubseteq W$, then $(B^* + W)/W$ is a non-zero ideal of $\bar{B}/W$ satisfying the isomorphism $(B^* + W)/W \cong B^*/(B^* \cap W)$. Then $B^*/(B^* \cap W) \in \mathbf{K}_{q-1}$ since $B^* \in \mathbf{K}_{q-1}$, and thus $(B^* + W)/W \in \mathbf{K}_{q-1}$.

Therefore, in what follows, we may assume $B^* \leqq W$.

If $B^* A \nsubseteq W$, then there exists an element $x \in A$ with $B^* x \nleqq W$. Then $(W + B^* x)/W$ is an ideal of $\bar{B}/W$ since

$$B^* x \cdot \bar{B} = B^* \cdot x\bar{B} \leqq B^* \cdot \bar{B} \leqq B^* \leqq W \quad \text{and} \quad \bar{B} \cdot B^* x \leqq B^* x.$$

We now show that the mapping

$$\varphi \colon b \to bx + W \quad (b \in B^*)$$

is a ring homomorphism of the ring $B^* \in \mathbf{K}_{q-1}$ onto the non-zero ideal $(B^* x + W)/W$ of $\bar{B}/W$. Then φ obviously maps B^* onto the whole ideal $(B^* x + W)/W$. Moreover, φ is an additive homomorphism. We have $(b_1 b_2)\varphi = 0 + W$ since

$$b_1 b_2 x \in B^* B^* A \leqq B^* \bar{B} \leqq B^* \leqq W$$

and also $b_1 \varphi \cdot b_2 \varphi = 0 + W$ because

$$b_1 x \cdot b_2 x \in B^* A \cdot B^* A \leqq B^* \cdot \bar{B} \leqq B^* \leqq W.$$

Accordingly, this yields $(b_1 b_2)\varphi = b_1 \varphi \cdot b_2 \varphi$ and thus φ is in fact a ring homomorphism. Therefore the ring $B^* \in \mathbf{K}_{q-1}$ can be homomorphically mapped onto the non-zero ideal $(B^* x + W)/W$ of $\bar{B}/W$ and thus a non-zero ideal of $\bar{B}/W$ belongs to $\mathbf{K}_{q-1}$. Similarly, in case $AB^* \leqq W$ a non-zero ideal of $\bar{B}/W$ belonging to $\mathbf{K}_{q-1}$ can be found.

Consequently, in what follows we may assume

$$B^* + AB^* + B^* A \leqq W.$$

Then $AB^* A \nsubseteq W$ since $W \neq \bar{B}$ and

$$\bar{B} = B^* + AB^* + B^* A + AB^* A.$$

Hence there exist elements x and y of A such that $xB^* y \nleqq W$. Then $(W + xB^* y)/W$ is an ideal of $\bar{B}/W$, since we have the relations

$$xB^* y \cdot \bar{B} \leqq AB^* \cdot A\bar{B} \leqq AB^* \bar{B} \leqq AB^* \leqq W,$$

and

$$\bar{B} \cdot xB^* y \leqq \bar{B} \cdot AB^* A \leqq \bar{B} \cdot B^* A \leqq B^* A \leqq W.$$

Now we define a mapping ψ of B^* onto the whole ideal $(W+xB^*y)/W$ of $\bar{B}/W$ by

$$\psi\colon b \to b\psi = xby+W.$$

Then ψ is obviously an additive homomorphism. We have $(b_1 b_2)\psi = 0+W$ since

$$xb_1 b_2 y \in AB^* B^* A \leqq \bar{B}B^* A \leqq B^* A \leqq W.$$

Moreover, the relation

$$xb_1 y \cdot xb_2 y \in AB^* AB^* A \leqq \bar{B}B^* A \leqq B^* A \leqq W$$

implies also $b_1\psi \cdot b_2\psi = 0+W$. Therefore ψ is a ring homomorphism, and thus the ring $B^* \in \mathbf{K}_{q-1}$ can be homomorphically mapped onto the non-zero ideal $(W+xB^*y)/W$ of $\bar{B}/W$ which belongs to $\mathbf{K}_{q-1}$.

This shows that $\bar{B}$ is contained in $\mathbf{K}_q$ for a finite q. □

Therefore we have the following

THEOREM 4.8. If $\mathbf{K}^{**} = \cup \mathbf{K}_\alpha$ is the class of all rings which are radical rings for the lower radical property determined by the class $\mathbf{K}$, then $\mathbf{K}^{**} = \mathbf{K}_{\omega_0}$ where ω_0 denotes the least infinite ordinal number.

PROOF. Obviously $\mathbf{K}_{\omega_0} \leqq \mathbf{K}^{**}$. Conversely, let A be a non-zero ring in the class $\mathbf{K}^{**}$. Then a non-zero accessible subring B of every homomorphic image A' of A belongs to $\mathbf{K}_1$ by proposition 4.6. Now A' has a non-zero ideal $\bar{B}$ with $\bar{B} \in \mathbf{K}_q$ for a finite q by proposition 4.7. Accordingly, A is contained in $\mathbf{K}_{\omega_0}$ by definition. This proves $\mathbf{K}^{**} = \mathbf{K}_{\omega_0}$. □

In an important special case we can say something more about $\mathbf{K}^{**}$. For this we need some preliminaries.

PROPOSITION 4.9 (Andrunakievich [11]). If B is an ideal in I, I is an ideal of the ring A, and $\bar{B}$ is the ideal of A generated by B in A, then $(\bar{B})^3 \leqq B$.

PROOF. We have $\bar{B} = B+AB+BA+ABA$ and thus $\bar{B} \leqq I$ implies

$$(\bar{B})^3 \subseteq I(B+BA+AB+ABA)I \leqq B+BAB = B. \square$$

A *radical class* determined by a radical property $\mathbf{R}$ is the *class of all* $\mathbf{R}$*-radical rings*. In particular, the lower (upper) radical class determined by a class $\mathbf{K}$ is the ring class correspondig to the lower (upper) radical property determined by $\mathbf{K}$. We have the following

THEOREM 4.10. Let $\mathbf{K}$ be a class of rings satisfying the following conditions:
(i) Every zero-ring A (i.e. every ring A such that $A^2=(0)$) belongs to $\mathbf{K}$.
(ii) Every ideal of a ring in $\mathbf{K}$ is again contained in $\mathbf{K}$.

Then $\mathbf{K}^{**}=\mathbf{K}_2$ for the lower radical class determined by $\mathbf{K}$.

REMARK. The $\mathbf{R}$-radical rings which are precisely the rings in $\mathbf{K}^{**}$, are called *supernilpotent* (cf. Chapter II).

PROOF. Let $A\in\mathbf{K}_3$. Then every non-zero homomorphic image A' of A contains a non-zero ideal I which belongs to $\mathbf{K}_2$. Moreover, I itself has a non-zero ideal B contained in $\mathbf{K}_1$. If $\bar{B}$ is the ideal, generated by B in A', then we have $B\leqq\bar{B}\leqq A'$, and B is an ideal of $\bar{B}$ since $\bar{B}\leqq I$ and B is an ideal of I. Then proposition 4.9 yields $(\bar{B})^3\leqq B$. Furthermore, $(\bar{B})^3\in\mathbf{K}_1$ since condition (i) of theorem 4.10 holds also for $\mathbf{K}_1$ if (i) holds for $\mathbf{K}$. If $(\bar{B})^3\neq(0)$ then A' contains a non-zero ideal $(\bar{B})^3\in\mathbf{K}_1$. If $(\bar{B})^3=(0)$ then $(\bar{B})^2$ is a zero-ring so that $(\bar{B})^2\in\mathbf{K}_1$ by condition (i) for $\mathbf{K}$. Finally, if $(\bar{B})^2=(0)$ then $\bar{B}$ already is a zero-ring, and this yields $\bar{B}\in\mathbf{K}_1$. Therefore, we have shown in all cases that A' has a non-zero ideal belonging to $\mathbf{K}_1$, and consequently we obtain $A\in\mathbf{K}_2$. □

Since the class $\mathbf{K}$ of all nilpotent rings satisfies conditions (i) and (ii) of theorem 4.10 and for this class we even have $\mathbf{K}=\mathbf{K}_1$, theorem 4.10 yields at once

PROPOSITION 4.11. If $\mathbf{K}$ is the class of all nilpotent rings, then the lower radical class $\mathbf{K}^{**}$, determined by $\mathbf{K}$, which corresponds to the lower nil radical of Baer [1], is precisely the class $\mathbf{K}_2$; i.e. it is the class of rings A such that every non-zero homomorphic image A' of A has a non-zero nilpotent ideal.

We shall show now the existence of a class $\mathbf{K}$ such that $\mathbf{K}^{**}=\mathbf{K}_3$ but $\mathbf{K}_2\neq\mathbf{K}_3$.

EXAMPLE 4.12 (Zassenhaus). Let $\mathbf{K}$ be the class which contains only the zero-ring defined upon the infinite cyclic group. Then the class $\mathbf{K}_1$ of all homomorphic images of rings from $\mathbf{K}$ contains all zero-rings with cyclic additive group. Moreover, the class $\mathbf{K}_2$ contains all nilpotent rings since every nilpotent ring $A\neq0$ has a non-zero ideal from $\mathbf{K}_1$; namely, if $A^n=0$ with $A^{n-1}\neq0$ then every cyclic additive group of A^{n-1} is an ideal of A and belongs to $\mathbf{K}_1$. Hence $\mathbf{K}^{**}=\mathbf{K}_3$ by proposition 4.11. We now give a ring B which belongs to $\mathbf{K}_3$ but not to $\mathbf{K}_2$.

Let B be the set of all formal finite sums $\sum r_\alpha x_\alpha$ where the index α is a rational number with $0<\alpha<1$, r_α is an element of a finite field and x_α are certain symbols for which the following multiplication is defined:

$$x_\alpha\cdot x_\beta=\begin{cases}x_{\alpha+\beta} & \text{if} \quad \alpha+\beta<1,\\ 0 & \text{if} \quad \alpha+\beta\geqq 1.\end{cases}$$

Then B obviously is an associative and commutative ring. Every element of B

is nilpotent. Indeed, if $n\alpha \geqq 1$ for a natural number n, then we have $x_\alpha^n = 0$. But B is not nilpotent since $x_{1/2} \cdot x_{1/4} \dots \cdot x_{1/2^n} \neq 0$ and $x_{1-1/2^n} \in B^n$ for every exponent n. Every non-zero homomorphic image of B contains a non-zero nilpotent ideal; but B itself does not contain an ideal which is a zero-ring defined upon a cyclic additive group. Accordingly we have $B \notin \mathbf{K}_2$ but $B \in \mathbf{K}_3$. Consequently $\mathbf{K}^{**} = \mathbf{K}_3 \neq \neq \mathbf{K}_2$.

It would be interesting to give a class $\mathbf{K}^{(n)}$ for $n = 4, 5, 6, \dots$ such that $\mathbf{K}^{**} = = \mathbf{K}_n^{(n)}$ and $\mathbf{K}^{**} \neq \mathbf{K}_{n-1}^{(n)}$. This problem is also due to Suliński, Anderson and Divinsky.

Now, following S. A. Tangeman's lower radical constructions in classes of rings (*Ring Theory,* Proc. Oklahoma Conference (1973) 173–183), we give below a survey of the most important recent results in the theory of lower radical constructions.

In *The hereditary property in the lower radical construction,* Canad. J. Math. *20* (1968) 474–476, E. P. Armendariz and W. G. Leavitt assert: Let $\mathbf{K} \subseteqq \mathbf{W}$, where $\mathbf{W}$ is an associative class, and assume $\mathbf{K}$ to be *hereditary* (i.e. every ideal of a ring in $\mathbf{K}$ is again in $\mathbf{K}$): Then $\mathbf{K}^{**} = \mathfrak{L}_\mathbf{W}\mathbf{K} = \mathbf{K}_3$.

Other results of this type may be found in

(i) Anderson–Divinsky–Suliński, *Lower radical properties for associative and alternative rings,* J. London Math. Soc. *41* (1966) 417–424;

(ii) S. E. Dickson, *A note on hypernilpotent radical properties for associative rings,* Canad. J. Math. *19* (1967) 447–448 and

(iii) R. L. Tangeman, *Strong heredity in radical classes,* Pacific J. Math. *42* (1972) 259–265.

In *A general theory of radicals II, Radicals in rings and bicategories,* Amer. J. Math. *76* (1954) 100–125, S. A. Amitsur provides the following interesting characterization of radical classes which is useful for giving another lower radical construction:

THEOREM 4.13. If $\mathbf{P} \subseteqq \mathbf{W}$ then $\mathbf{P}$ is a radical class in the class $\mathbf{W}$ if and only if the following conditions are satisfied:

(1) $\mathbf{P}$ is homomorphically closed.

(2) If I is an ideal of $A \in \mathbf{W}$, and if I, A/I are in $\mathbf{P}$, then $A \in \mathbf{P}$.

(3) A union of a chain of $\mathbf{P}$-ideals related to a ring $A \in \mathbf{W}$ is again a $\mathbf{P}$-ideal of A.

By using this result, in *Lower radicals in non-associative rings,* J. Austral. Math. Soc. *14* (1972) 419–423. D. Kreiling and R. L. Tangeman give the following construction for $\mathfrak{L}\mathbf{K} = \mathbf{K}^{**}$ from $\mathbf{K}$. Let $\mathbf{K} \subseteqq \mathbf{W}$ be arbitrary and define $\mathbf{K}_1 = \mathbf{H}(\mathbf{K})$ (all homomorphic images of rings in $\mathbf{K}$). Proceeding inductively, let us suppose

$\mathbf{K}_\alpha$ to be defined for all $\alpha<\beta$. If β is not a limit ordinal, define $\mathbf{K}_\beta=\{A|A\in\mathbf{W}, I, A/I\in\mathbf{K}_{\beta-1}$ for some ideal I of $A\}$. For limit ordinals β, put $\mathbf{K}_\beta=\{A|A\in\mathbf{W}, A$ is a union of a chain of ideals each of which belongs to one of the classes $\mathbf{K}_\alpha$, $\alpha<\beta\}$. Then $\mathfrak{L}_\mathbf{W}\mathbf{K}$ is the union of all the classes $\mathbf{K}_\beta$. This construction will be referred to as the extension-union construction.

Again, the construction is monotone in the sense that $\alpha<\beta$ implies $\mathbf{K}_\alpha\subseteq\mathbf{K}_\beta$, but conditions are unknown to guarantee that $\mathfrak{L}\mathbf{K}=\mathbf{K}_\beta$ holds for any particular ordinal number β. An example in the above paper of Kreiling and Tangeman shows that hypotheses of the type mentioned in the above publication of Armendariz and Leavitt will not be very useful in this direction.

In his paper *On the construction of lower radical properties,* Pacific J. Math. *28* (1969) 393–395, Y. L. Lee constructs $\mathfrak{L}\mathbf{K}$ from $\mathbf{K}$ in the case when $\mathbf{K}$ is contained in an associative class; but W. G. Leavitt and Y. L. Lee in *A radical coinciding with the lower radical in associative and alternative rings,* Pacific J. Math. *30* (1969) 459–461, show that this construction does not in general yield $\mathfrak{L}\mathbf{K}$. We use here the notation of the above publication of Leavitt and Lee. Let $\mathbf{W}$ be an arbitrary *universal class* (i.e. closed for homomorphic images and for ideals), $\mathbf{K}\subseteq\mathbf{W}$ and $\mathbf{K}_0=\mathbf{H}(\mathbf{K})$. Then we can define inductively (this is finite induction) $\mathbf{K}_n=\{A|A\in\mathbf{W}, A$ has a non-zero ideal $I\in\mathbf{K}_{n-1}\}$, $\mathbf{K}_\omega=\bigcup_n\mathbf{K}_n$, and $\mathbf{Y}(\mathbf{K})=\{A|A\in\mathbf{W}, A/I\in\mathbf{K}_\omega$ for all ideals I of $A\}$.

It has been shown that $\mathbf{Y}(\mathbf{K})$ is a radical class containing $\mathbf{K}$, and that $\mathbf{Y}(\mathbf{K})=\mathfrak{L}\mathbf{K}$ if the semisimple class $S\mathfrak{L}\mathbf{K}$ is hereditary. Since radicals in alternative classes are known to have hereditary semisimple classes (see T. Anderson, N. Divinsky and A. Suliński, *Hereditary radicals in associative and alternative rings,* Canad. J. Math. *17* (1965) 594–603), it follows that in such classes $\mathbf{Y}(\mathbf{K})=\mathfrak{L}\mathbf{K}$. It has also been shown by Leavitt and Lee that $\mathbf{Y}(\mathbf{K})$ defines the same radical class as that constructed by J. F. Watters in *Lower radicals in associative rings,* Canad. J. Math. *21* (1969) 466–476.

In *Strong radical properties of alternative and associative rings* written by N. Divinsky, J. Krempa and A. Suliński, J. Algebra *17* (1971) 369–388, a radical is defined to be *strong* if $\mathbf{P}(A)$ contains all the one-sided $\mathbf{P}$-ideals of A, and a construction is given for the lower strong radical.

In T. L. Jenkins and D. Kreiling, *On the construction of the Kurosh lower radical class for associative rings,* J. Austral. Math. Soc. *13* (1972) 362–364; R. F. Rossa, *Radical properties involving one-sided ideals,* Pacific J. Math. *49* (1973) 467–471, and R. L. Tangeman, *Strong heredity in radical classes,* Pacific J. Math. *42* (1972) 259–265, further constructions are given for radical classes, and conditions are also stated under which these constructions yield the lower radical class.

The following result is valid in an arbitrary universal class (see A. E. Hoffman and W. G. Leavitt, *Properties inherited by the lower radical,* Portugaliae Math. *27* (1968) 63–66): If **K** is hereditary, then so is $\mathfrak{L}\mathbf{K}$.

This is proved by showing in the Anderson–Divinsky–Suliński construction that each of the classes $\mathbf{K}_\beta$ is hereditary if **K** is. The theorem is reproved by D. Kreiling and R. L. Tangeman, in *Lower radicals in non-associative rings,* J. Austral. Math. Soc. *14* (1972) 419–423, using the extension-union construction. In *More properties inherited by the lower radical,* Proc. Amer. Math. Soc. *33* (1972) 247–249, R. F. Rossa notices that the latter proof with slight modifications establishes the following result: Let **W** be a universal class which is closed under subrings and let $\mathbf{K}\subseteq\mathbf{W}$. If **K** is hereditary for subrings (left ideals, right ideals), then so is $\mathfrak{L}_\mathbf{W}\mathbf{K}$.

As corollaries to results of this type, there are constructions for minimal radical classes with given properties containing a given class **K**. The idea is first to close **K** under the desired property and then pass to the lower radical. An example from Rossa's above publication will suffice: suppose **W** to be a universal class which is closed under right ideals and $\mathbf{K}\subseteq\mathbf{W}$. We wish to construct a radical class **P** which is hereditary for right ideals and minimal among the classes in **W** containing **K**. We simply define $\mathbf{G}_1=\mathbf{K}$ and $\mathbf{G}_n=\{R|R\in\mathbf{W},\ R$ is a right ideal of some ring in $\mathbf{G}_{n-1}\}$ for $n\geqq 2$. Then the desired radical **P** is just $\mathfrak{L}\mathbf{G}(\mathbf{K})$ with $\mathbf{G}(\mathbf{K})=\bigcup_n \mathbf{G}_n$.

Another desirable property for a radical **P** is that $\mathbf{P}(I)=I\cap\mathbf{P}(A)$ should hold whenever I is an ideal of A. This is known to be true for hereditary radicals in associative or alternative classes. Amitsur has shown that in arbitrary universal classes, this condition is equivalent to both the classes **P** and $S\mathbf{P}$ (the class of all **P**-semisimple rings) being hereditary. In *Strongly hereditary radicals*, Proc. Amer. Math. Soc. *21* (1969) 703–705, W. G. Leavitt has defined a radical **P** to be *strongly hereditary* if $\mathbf{P}(I)=I\cap\mathbf{P}(A)$ whenever I is an ideal of A and **P** also satisfies the following condition:

(a) if $J\in\mathbf{P}$ is an ideal of an ideal of some $A\in\mathbf{W}$, then $J'\in\mathbf{P}$, where J' is the ideal of A generated by J.

THEOREM 4.14. A hereditary radical class **P** in a universal class **W** is strongly hereditary if and only if it has the property

(a) If $I\in\mathbf{P}$ is an ideal of an ideal of some $A\in\mathbf{W}$, then $I'\in\mathbf{P}$ where I' is the ideal of A generated by I.

Leavitt then proceeds to construct the unique minimal strongly hereditary radical class containing a given class **K**. An alternative construction of this class comes from the following result of R. L. Tangeman (*Strong heredity in radical classes*, Pacific J. Math. *42* (1972) 259–265): if **K** is homomorphically closed and satisfies property (a), then $\mathbf{K}^{**}=\mathfrak{L}\mathbf{K}$ satisfies property (a). This proof uses the

Anderson–Divinsky–Suliński construction. An unpublished proof using the extension-union construction can be found in R. F. Rossa's doctoral dissertation (University of Oklahoma, 1971). One-sided versions of most of these results are discussed in R. F. Rossa, *Radical properties involving one-sided ideals*, Pacific J. Math. *49* (1973) 467–471.

In their work *On intersections and unions of radical classes*, J. Austral. Math. Soc. *13* (1972) 354–356. Y. L. Lee and R. E. Propes have shown that if **K** and **M** are homomorphically closed hereditary classes of associative rings then $\mathfrak{L}(\mathbf{K}\cap\mathbf{M})=\mathfrak{L}\mathbf{K}\cap\mathfrak{L}\mathbf{M}$. This proof uses the Anderson–Sivinsky–Suliński construction.

Another interesting result is due to A. E. Hoffman and W. G. Leavitt (*Properties inherited by the lower radical,* Portugaliae Math. *27* (1968) 63–66). It says that if **K** is a homomorphically closed class and **M** is a hereditary class such that $\mathbf{K}\cap\mathbf{M}=0$ then $\mathbf{M}\cap\mathfrak{L}\mathbf{K}=0$. Here 0 is the class containing the zero-ring (0) only.

We finish this paragraph with the discussion of hereditary radicals following Szász and Wiegandt [3] and using both lower and upper radicals. We recall that a radical is said to be hereditary if every ideal of a radical ring is again a radical ring. First we show that there may be relatively many radical classes between two radical classes. In order to show this let **F** be a class of fields (but not necessarily a set in the sense of the Zermelo–Fraenkel set theory) and let $U\mathbf{F}$ denote the upper radical, determined by **F**. Further let us consider a subset $\mathbf{F}(\mathfrak{m})\subseteq\mathbf{F}$ of cardinality $\mathfrak{m}$.

THEOREM 4.15. Let $\mathbf{P}$ ($\subseteq U\mathbf{F}$) be a radical class. If $\mathbf{Q}$ is the lower radical $L(\mathbf{P}\cup\mathbf{F}(\mathfrak{m}))$ belonging to $\mathbf{P}\cup\mathbf{F}(\mathfrak{m})$ then the interval $[\mathbf{P},\mathbf{Q}]$ (i.e. the class of all radicals **R** with $\mathbf{P}\subseteq\mathbf{R}\subseteq\mathbf{Q}$) has exactly $2^{\mathfrak{m}}$ radical classes. Moreover, all radicals of $[\mathbf{P},\mathbf{Q}]$ are hereditary if and only if **P** is hereditary.

PROOF. Let **R** be a radical of $[\mathbf{P},\mathbf{Q}]$. Let us take into consideration the subset $\mathbf{F}(\mathfrak{m})$ consisting of all fields of $\mathbf{F}(\mathfrak{n})$ which are ideals in some ring of R. (Here $\mathfrak{n}$ denotes the cardinality of $\mathbf{F}(\mathfrak{m})$). We remark that if $B\in\mathbf{F}(\mathfrak{n})$ and B is an ideal of a ring A then A always splits into a direct sum $A=B\oplus C$ because B is a ring with unity element. Hence $A\in R$ implies $B\in R$. Thus $\mathbf{P}\cup\mathbf{F}(\mathfrak{n})=R$. Therefore $L(\mathbf{P}\cup\mathbf{F}(\mathfrak{n}))\subseteq R$.

If A is an **R**-ring not belonging to **P** then the ring $A_1=A/\mathbf{P}(A)$ is not zero and it has zero **P**-radical. However $R(A_1)=A_1$. Since **Q** is a lower radical, it is the union $\mathbf{Q}=\mathbf{N}_\omega$, where $N_1=\mathbf{P}\cup\mathbf{F}(\mathfrak{n})$, and $\mathbf{N}_k$ is defined just as in the Anderson–Divinsky–Suliński construction. Hence $\mathbf{Q}(A_1)=A_1$ and by proposition 4.6 there exists a non-zero accessible subring A_n of A_1 such that A_n belongs to $N_1=\mathbf{P}\cup\mathbf{F}(\mathfrak{N})$. If $A_n\in\mathbf{P}$ then the **P**-radical $\mathbf{P}(A_{n-1})$ is an ideal of A_{n-2} by theorem 1.7 and $0\neq A_n\subseteq\mathbf{P}(A_{n-1})$. Hence $A_{n-3},\dots,A_1$ also have non-zero **P**-radicals which contradicts the choice of A_1. Thus, A_1 has an accessible subring $A_n \in \mathbf{F}(\mathfrak{m})$. Hence A_n is a field, and by proposition 4.9 we have $A_n = A_n^3 \leqq A_n^{*3} \leqq A_n$ for the ideal A_n^* generated by A_n in A_1. Consequently, A_n is an ideal of A_1 which implies $A_n \in \mathbf{F}(\mathfrak{n})$. Hence clearly $\mathbf{F} \leqq L(\mathbf{P} \cup \mathbf{F}(\mathfrak{n}))$.

Since $\mathbf{F}(\mathfrak{m})$ has exactly $2^{\mathfrak{m}}$ different subsets $\mathbf{F}(\mathfrak{n})$, $\mathfrak{n} \leqq \mathfrak{m}$, the first statement of theorem 4.15 is proved. The second statement follows immediately from the fact that a lower radical $\mathfrak{L}\mathbf{N}$ is hereditary if $\mathbf{N}$ is hereditary (see E. P. Armendariz–W.G. Leavitt, *The hereditary property in the lower radical construction,* Canad. J. Math. *20* (1968) 474–476)). □

As an immediate consequence theorem 4.15 yields

COROLLARY 4.16. If $\mathbf{P} \leqq U\mathbf{F}$ and the subclass $\mathbf{G}$ of $\mathbf{F}$ is not a set, then the radicals of $[\mathbf{P}, \mathbf{Q}]$ form a class and not a set for $\mathbf{Q} = \mathfrak{L}(\mathbf{P} \cup \mathbf{G})$. If $\mathbf{P}$ is hereditary, then the radicals of $[\mathbf{P}, \mathbf{Q}]$ are also hereditary.

Here we are going to prove the existence of a unique maximal hereditary radical contained in a radical class $\mathbf{R}$. We shall then characterize it.

THEOREM 4.17. If $\mathbf{R}$ is an arbitrary radical class then there exists a unique maximal hereditary radical $\mathbf{H_R} \leqq \mathbf{R}$. Moreover, $\mathbf{H_R}$ is the union of all hereditary radicals $\mathbf{H} \leqq \mathbf{R}$.

PROOF. If $\mathbf{H_R}$ is the union of all hereditary radicals contained in $\mathbf{R}$ then $\mathbf{H_R}$ is a homomorphically closed hereditary class. Thus the lower radical $L\mathbf{H_R}$ is hereditary and it satisfies $\mathbf{H_R} \leqq L\mathbf{H_R} \leqq \mathbf{H_R}$. Hence $\mathbf{H_R} = L\mathbf{H_R}$. □

THEOREM 4.18. If $\mathbf{R}$ is an arbitrary radical class, then the maximal hereditary radical $\mathbf{H_R}$ consists of all rings each non-zero accessible subring of which has no non-zero homomorphic image in the semisimple class $S\mathbf{R}$.

PROOF. It is well known that $\mathbf{H_R}$ is exactly the upper radical $US\mathbf{R}$ of the semisimple class $S\mathbf{R}$. Let $\mathbf{A}$ denote the class of all rings each non-zero accessible subring of which has no non-zero homomorphic image in the class $S\mathbf{R}$. We have to show $\mathbf{A} = \mathbf{H_R}$. Obviously, the class $\mathbf{A}$ is homomorphically closed and hereditary, so the lower radical $L\mathbf{A}$ is again hereditary. Furthermore, $\mathbf{A} \leqq L\mathbf{A} \leqq US\mathbf{R} = \mathbf{R}$. Hence $\mathbf{A} \leqq L\mathbf{A} \leqq \mathbf{H_R}$. On the other hand, let us take into consideration a ring $A \in \mathbf{H_R}$. A cannot be homomorphically mapped onto any non-zero ring of $S\mathbf{R} \leqq S\mathbf{H_R}$. Furthermore, since $\mathbf{H_R}$ is hereditary, every accessible subring of A belongs to $\mathbf{H_R}$ and cannot be homomorphically mapped onto a non-zero ring of $S\mathbf{R}$. This implies $A \in \mathbf{A}$. Therefore, $\mathbf{H_R} \leqq A$. □

In theorem 4.15 we have seen examples for classes $[\mathbf{P}, \mathbf{Q}]$ of radicals consisting of hereditary radicals. Generally, a class [**P**, **Q**] of radicals may consist of non-hereditary radicals, except **P** and **Q**.

EXAMPLE 4.19. Let $\mathbf{P}=\{0\}$ be the zero-radical class and **Q** be the lower radical determined by the zero-ring $Z(p)$ over a cyclic group of prime order p. If $\mathbf{H}\neq\mathbf{P}$ is a hereditary radical of [**P**, **Q**] then every ring $A\in\mathbf{H}$ belongs to **Q**, and thus A has an accessible subring isomorphic to $Z(p)$. Since **H** is hereditary, $Z(p)\subseteq\mathbf{H}$, and $\mathbf{Q}=LZ(p)\subseteq\mathbf{H}$ follows. On the other hand it is easy to see that the lower radical $LZ(p^\infty)$ determined by the zero-ring $Z(p^\infty)$ over the quasicyclic group $C(p^\infty)$ (cf. § 6) consists exactly of those rings which are discrete direct sums of rings isomorphic to $Z(p^\infty)$. Since $LZ(p^\infty)\neq\mathbf{Q}$, $LZ(p^\infty)$ is not hereditary. Hence [**P**, **Q**] does not contain hereditary radicals different from **P** and **Q** but it contains the non-hereditary radical $LZ(p^\infty)$.

If $\mathbf{P}\subseteq\mathbf{Q}$ are arbitrary radicals, we can define two operations in [**P**, **Q**]. The intersection $\bigcap_{i\in I}\mathbf{R}_i$ of a family $\{\mathbf{R}_i\}_{i\in I}$ of radicals from [**P**, **Q**] is again a radical and it belongs to [**P**, **Q**]. If $\mathbf{R}_1, \mathbf{R}_2\in[\mathbf{P}, \mathbf{Q}]$ then the sum $\mathbf{R}_1+\mathbf{R}_2$ can be defined as the lower radical $\mathfrak{L}(\mathbf{R}_1\cup\mathbf{R}_2)$. (Another representation of $\mathbf{R}_1+\mathbf{R}_2$ is given by Leavitt; cf. § 5). If **Q** is hereditary, then we define the *hereditary closure* $\overline{\mathbf{R}}$ of some $\mathbf{R}\in[\mathbf{P}, \mathbf{Q}]$ to be the intersection of all hereditary radicals of [**P**, **Q**] containing **R**. Obviously, $\overline{\mathbf{R}}$ is the lower radical determined by the rings of **R** and their ideals. It is also obvious that the hereditary closure operation *in* [**P**, **Q**] is topological in the sense that $\overline{\mathbf{R}}_1+\overline{\mathbf{R}}_2=\overline{\mathbf{R}_1+\mathbf{R}_2}$. These considerations combine with theorem 4.17 to give

COROLLARY 4.20. If the radical **R** is non-hereditary then there are hereditary radicals **P** and **Q** such that neither [**P**, **R**], nor [**R**, **Q**] contains hereditary radicals except **P** and **Q**, respectively. Namely, one can choose $\mathbf{P}=\mathbf{H}_{\mathbf{R}}$ and $\mathbf{Q}=\overline{\mathbf{R}}$.

As pointed out by Gardner, the whole class [**P**, **Q**] may contain other hereditary radicals, too. For instance, let **R** be the radical class of rings on groups of the form $A_p\oplus A_q$ where p, q are different primes, A_p is a direct sum of copies of $C(p^\infty)$ and A_q is a direct sum of copies of $C(q^\infty)$. Then the maximal hereditary radical class that **R** contains is the class **0** consisting of the ring 0 only, while the minimal hereditary radical $\overline{\mathbf{R}}$ containing the radical **R** is the class $\mathbf{B}(p, q)$ of prime radical rings on direct sums of p-groups and q-groups. If $\mathbf{B}(p)$ denotes the class of prime radical p-rings then $\mathbf{B}(p)$ is hereditary and $\mathbf{B}(p)\in[0, \mathbf{B}(p, q)]$.

Hereditary radicals appear usually in lower radical constructions. Armendariz [3] and Ryabukhin [8] have characterized the semisimple classes with hereditary upper radicals. Here we continue with further investigations about the hereditariness of upper radicals.

THEOREM 4.21. Let **M** be a hereditary class. The upper radical $U\mathbf{M}$ is hereditary if and only if $U\mathbf{M}$ consists exactly of all rings, all non-zero accessible subrings of which have non-zero homomorphic images in the class $\mathbf{M}$.

PROOF. By theorem 4.18, the radical $U\mathbf{M}$ is hereditary if and only if $U\mathbf{M}$ consists of all rings each accessible subring of which has no non-zero homomorphic image in the semisimple class $SU\mathbf{M}=\overline{\mathbf{M}}$. Hence, if $A\in U\mathbf{M}$, then obviously no accessible subring of A can be homomorphically mapped onto a ring of $\mathbf{M}\subseteq\overline{\mathbf{M}}$. If $A\notin U\mathbf{M}$, then there exists an accessible subring B of A having a non-zero homomorphic image B_1 in $\overline{\mathbf{M}}$. Moreover, B_1 can be homomorphically mapped onto a non-zero ring of $\mathbf{M}$. □

In the following two theorems we present sufficient conditions for the hereditariness of upper radical classes. In what follows, $\mathbf{M}$ will always denote a hereditary class of rings. Let us consider the following conditions:

(i) $\mathbf{M}$ is homomorphically closed.
(ii) If L is an ideal of an ideal I of a ring A such that $I/L\in\mathbf{M}$ then L is an ideal of A.
(iii) If I is a non-zero ideal of a ring A and $I\in\mathbf{M}$ then there exists a proper ideal N of A such that $I+N=A$.
(iv) If I is an ideal of A with $I\in\mathbf{M}$ and I_0 denotes the annihilator of I in A then $A/I_0\in\mathbf{M}$.

Let us recall that the *two-sided annihilator* of I is defined to be $I_0=\{x|x\in A, xI=Ix=(0)\}$.

As to the conditions (i)–(iv), we can observe that for instance every class of simple rings with unity element (see. §§ 6, 32 and 39) satisfies all these conditions.

THEOREM 4.22. If the hereditary class $\mathbf{M}$ fulfils conditions (i), (ii) and (iii) then the upper radical class $U\mathbf{M}$ is hereditary.

PROOF. Let us suppose that $U\mathbf{M}$ is not hereditary, and let $A\in U\mathbf{M}$ be a ring with a non-zero ideal $I\notin U\mathbf{M}$. Now I can be homomorphically mapped onto a non-zero ring of $\mathbf{M}$, i.e., there exists an ideal L of I such that $(0)\neq I/L\in\mathbf{M}$. By condition (ii) L is an ideal of A. Hence I/L is an $\mathbf{M}$-ideal of A/L and thus by (iii) there exists an ideal N/L of A/L such that $A/L=I/L+N/L$ and $N/L\neq A/L$. This implies that $I+N=A$ and $L\subseteq I, N$. Thus, it follows that

$$(0)\neq A/N=(I+N)/N\cong I(I\cap N)$$

and $I/(I\cap N)$ is a homomorphic image of $I/L\in\mathbf{M}$. Hence we have $A/N\cong I/(I\cap N)\in$ $\in\mathbf{M}$ by (i), contradicting the assumption $A\in U\mathbf{M}$. □

THEOREM 4.23. If the hereditary class $\mathbf{M}$ fulfils conditions (ii) and (iv), then the upper radical class $U\mathbf{M}$ is hereditary.

PROOF. If $U\mathbf{M}$ is not hereditary then there exists a ring $A \in U\mathbf{M}$ with an ideal $0 \neq I \notin U\mathbf{M}$. By theorem 1.7 the $U\mathbf{M}$-radical $U\mathbf{M}(I)$ of I is an ideal of A, and $A' = A/U\mathbf{M}(I)$ belongs to $U\mathbf{M}$ but it has an ideal $(0) \neq I' = I/U\mathbf{M}(I)$ belonging to $SU\mathbf{M} = \overline{\mathbf{M}}$. Since any non-zero ideal of I' can be homomorphically mapped onto a non-zero ring of $\mathbf{M}$, there exists an ideal L' of I' such that $I'/L' \in \mathbf{M}$. But I'/L' is a homomorphic image of I, so that there exists an ideal L of I with $I/L \cong I'/L' \in \mathbf{M}$. Hence, condition (ii) implies that L is an ideal of A. Let us take into consideration $A'' = A/L$ and $I'' = I/L$. For the annihilator I_0'' of I'' in A'' condition (iv) implies that $(0) \neq A/I \cong A''/L'' \in \mathbf{M}$, contradicting the assumption $A \in U\mathbf{M}$. □

Concerning theorem 4.23 we can prove that in some cases the rings of the semisimple class $\overline{\mathbf{M}}$ are subdirect sums of $\mathbf{M}$-rings. In order to pursue this aim we need the concept of ideal quotient (see also § 11). As usual, the *two-sided ideal quotient* $L:I$ of L and the ideal $I \subseteq A$ is defined to be the set

$$L:I = \{x \mid x \in A,\ xI \subseteq L \text{ and } Ix \subseteq L\}.$$

Obviously, if L is also an ideal of A, then $L:I$ is an ideal of A. Furthermore, let us take into consideration the following condition:

(v) If L and I are ideals of the ring A such that $I/L \in \mathbf{M}$ then $(L:I) \cap I = L$.

Let us remark that if $\mathbf{M}$ is a class of *simple rings with unity element* then condition (v) is also fulfilled.

THEOREM 4.24. If the hereditary class $\mathbf{M}$ satisfies conditions (ii), (iv) and (v), then every ring of the semisimple class $\overline{\mathbf{M}} = SU\mathbf{M}$ is a subdirect sum of $\mathbf{M}$-rings (cf. § 12).

PROOF. Consider a ring $A \in \overline{\mathbf{M}}$. Since A has a non-zero homomorphic image in $\mathbf{M}$, there exists an ideal M of A with $(0) \neq A/M \in \mathbf{M}$. Let $\{M_\alpha\}$ be the set of all ideals of A such that $A/M_\alpha \in \mathbf{M}$. Taking $I = \bigcap_\alpha M_\alpha$ as a base, it is sufficient to verify that $I = (0)$. Suppose $I \neq (0)$. Since $\overline{\mathbf{M}}$ is a hereditary class, $A \in \overline{\mathbf{M}}$ implies $I \in \overline{\mathbf{M}}$. Consequently, there exists an ideal L of I with $(0) \neq I/L \in \mathbf{M}$. Hence, by condition (ii), L is an ideal of A. Moreover, the annihilator of I/L in A/L is $(L:I)/L$. Hence condition (iv) implies

$$A/(L:I) \cong (A/L)/((L:I)/L) \in \mathbf{M}.$$

Thus $(L:I) \in \{M_\alpha\}$ and so $I \subseteq (L:I)$. By (v) we obtain $I \leqq (L:I) \cap I = L \leqq I$, which contradicts the choice of L. □

We point out that it has been shown by Szász and Wiegandt [3] that none of the conditions (i), (ii), (iii), (iv) and (v) holds necessarily.

§ 5. SETS OF RADICAL CLASSES

In this section we discuss what can be said about the intersection and the union of radical classes. In this investigation we rely upon the results and notation of the interesting paper [1] of Leavitt.

As we stated it in § 4 a class $\mathbf{K}$ of rings is said to be homomorphically closed, if $A \in \mathbf{K}$ implies $A\varphi \in \mathbf{K}$ for every ring homomorphism φ. Further, a class $\mathbf{K}$ is called hereditary, if $A \in \mathbf{K}$ implies $I \in \mathbf{K}$ for every ideal I of A.

A class which is homomorphically closed and hereditary is said to be a *universal class*.

For an arbitrary class $\mathbf{K}$ of rings and for a subclass $\mathbf{P}$ of $\mathbf{K}$ we define

$$S\mathbf{P} = \{A \mid A \in \mathbf{K},\ I \notin \mathbf{P} \text{ for every ideal } I \text{ of } A\}, \tag{1}$$

$$U\mathbf{P} = \{A \mid A \in \mathbf{K},\ A\varphi \notin \mathbf{P} \text{ for every non-zero homomorphism } \varphi \text{ of } A\}. \tag{2}$$

REMARK 5.1. For the sake of simplicity, in what follows we choose the same notation for a radical property and the corresponding radical class. If $\mathbf{P}$ is a radical class, then $S\mathbf{P}$ is, by the definitions given in § 1, a class of $\mathbf{P}$-semisimple rings, and also $US\mathbf{P}=\mathbf{P}$ by theorem 1.14. If, on the other hand, $\mathbf{P}$ is a complete class of semisimple rings then $U\mathbf{P}$ is a radical class and $SU\mathbf{P}=\mathbf{P}$ by theorem 3.1.

Let $\mathbf{P}$ be a radical class and $\mathbf{P}(A)$ the $\mathbf{P}$-radical of the ring A. If H is an ideal of A such that $A/H \in S\mathbf{P}$, then we have $\mathbf{P}(A) \subseteqq H$ because $\mathbf{P}(A) \not\subseteqq A$ would imply that $(\mathbf{P}(A)+H)/H \cong \mathbf{P}(A)/(\mathbf{P}(A) \cap H)$ is a non-zero $\mathbf{P}$-ideal of A/H which contradicts our assumption.

REMARK 5.2. If $\mathbf{K}$ is an arbitrary class of rings, denote the lower radical class determined by $\mathbf{K}$ in the sense of the preceding section by $L\mathbf{K}$. This coincides with $\cup \mathbf{K}_\alpha$ where $\mathbf{K}_\alpha$ is the class of rings of degree α over $\mathbf{K}$. Then we have from the definitions (1) and (2) the relation $\mathbf{K}_2 = US\mathbf{K}_1$ and thus

$$L\mathbf{K} = \mathbf{K} \cup US\mathbf{K} \cup USUS\mathbf{K} \cup USUSUS\mathbf{K} \cup \ldots .$$

Now, let $\{\mathbf{P}_j\}$ $(j \in I)$ be a set of radical subclasses of the class $\mathbf{K}$. Then we have the following

THEOREM 5.3. If $\mathbf{D} = \bigcap_{j \in I} \mathbf{P}_j$ then $\mathbf{D}$ is a radical class and

$$\mathbf{D}(A) \subseteqq \bigcap_{j \in I} \mathbf{P}_j(A) \tag{3}$$

for every ring A. If every class $\mathbf{P}_j$ is hereditary then $\mathbf{D}$ is also hereditary and in (3) the equality prevails.

PROOF. Since every class $\mathbf{P}_j$ is homomorphically closed, so is $\mathbf{D}$. If every non-zero homomorphic image $A\varphi$ of A has a non-zero $\mathbf{D}$-ideal I, then I is a $\mathbf{P}_j$-ideal for every $\mathbf{P}_j$ and thus A is by theorem 1.13 a $\mathbf{P}_j$-ring for every $\mathbf{P}_j$. However then A is a $\mathbf{D}$-ring and thus $\mathbf{D}$ is a radical class, again by theorem 1.13. Moreover, if every class $\mathbf{P}_j$ is hereditary then so is $\mathbf{D}$. Now $\mathbf{H}=\bigcap \mathbf{P}_j(A)$ is an ideal in every $\mathbf{P}_j(A)$ which yields $\mathbf{H}\in\mathbf{D}$ and $\mathbf{H}\subseteqq\mathbf{D}(A)$ since $\mathbf{D}$ is hereditary. On the other hand we have $\mathbf{D}(A)\subseteqq\mathbf{H}$ and thus $\mathbf{H}=\mathbf{D}(A)$. □

This ideal $\mathbf{D}(A)$ is called the *intersection radical* of the set $\{\mathbf{P}_j\}$.

We now construct in a ring A a system of subrings V_γ for every ordinal number in order to give a more explicit description of $\mathbf{D}(A)$.

We take $V_1=A$. Let us assume that the subrings V_α are already defined for all ordinal numbers $\alpha<\beta$ where β is a fixed ordinal number. Then we take

$$V_\beta=\begin{cases}\bigcap\limits_{\alpha<\beta} V_\alpha & \text{if } \beta \text{ is a limit ordinal number;}\\ \mathbf{P}_j(V_{\beta-1})\neq V_{\beta-1} & \text{if } \beta-1 \text{ and such } \mathbf{P}_j \text{ exists;}\\ V_{\beta-1} & \text{otherwise.}\end{cases} \tag{4}$$

Then there exists an ordinal number γ such that $\mathbf{P}_j(V_\gamma)=V_\gamma$ for every $j\in I$.

We remark that definition (4) gives a great freedom to the ordinals β of the first type. Nevertheless, the following interesting theorem holds:

THEOREM 5.4. If the subrings V_β of the ring A are defined by (4) then there exists an ordinal number γ satisfying $\mathbf{D}(V_\gamma)=V_\gamma$, and we also have $\mathbf{D}(A)=V_\gamma$.

PROOF. There exists an ordinal number with $\mathbf{P}_j(V_\gamma)=V_\gamma$ for every $j\in I$. Then we obviously have $V_\gamma=\mathbf{D}(V_\gamma)$ by $V_\gamma\in\mathbf{D}$.

Moreover, we have $\mathbf{D}(A)\subseteqq V_1$, since $V_1=A$ by definition. We use induction. Assume that $\mathbf{D}(A)\subseteqq V_\alpha$ for every ordinal number $\alpha<\beta$ where β is fixed. If β is a limit ordinal then we have $\mathbf{D}(A)\subseteqq V_\gamma$ by (4). But if $\beta-1$ exists then $\mathbf{D}(A)\subseteqq V_{\beta-1}$ by assumption and $\mathbf{D}(A)$ is a $\mathbf{P}_j$-ideal of $V_{\beta-1}$ since $\mathbf{D}\subseteqq\mathbf{P}_j$. Hence it follows that $\mathbf{D}(A)\subseteqq\mathbf{P}_j(V_{\beta-1})=V_\beta$. Thus, we have $\mathbf{D}(A)\subseteqq V_\beta$ for every β. Accordingly, $\mathbf{D}(A)\subseteqq V_\gamma$ for the above ordinal number γ. But we have already seen that V_γ is a $\mathbf{D}$-ring. Since by theorem 1.7 the radical of an ideal of A is again an ideal of the ring A by induction we infer that V_γ is an ideal of A and as it is a $\mathbf{D}$-ring, it follows that $V_\gamma\subseteqq\mathbf{D}(A)$. Together with the relation $\mathbf{D}(A)\subseteqq V_\gamma$ this gives $\mathbf{D}(A)=V_\gamma$. □

Now we discuss the union radical of a ring. Let $\{\mathbf{P}_j\}$ $(j\in I)$ be again a set of radical classes. We define $\mathbf{H}=\bigcup\mathbf{P}_j$. In general $\mathbf{H}$ is not a radical class of rings. However the rings of degree 2 over $\mathbf{H}$ do form a radical class.

We now prove

THEOREM 5.5. $S\mathbf{H}=\bigcap\limits_{j\in I} S\mathbf{P}_j$ and $\mathbf{H}\subseteqq US\mathbf{H}$ hold for the union $\mathbf{H}=\bigcup\mathbf{P}_j$ $(j\in I)$ of the radical classes $\mathbf{P}_j$ $(j\in I)$.

PROOF. The first relation is clear, since a ring A has no non-zero ideal in $\mathbf{H}$ if it has no non-zero ideal in any P_j $(j\in I)$. Since $\mathbf{H}$ is homomorphically closed, $\mathbf{H}=\mathbf{H}_1$ and we can apply the final proposition of remark 5.2.

Let $\mathbf{H}(A)$ be the sum of all ideals of the ring A which belong to $U S\mathbf{H}$. Then theorem 5.5 implies that $\mathbf{H}(A)\supseteqq\sum\mathbf{P}_j(A)$. □

Now we have the following

THEOREM 5.6. $S\mathbf{H}$ is a class of semisimple rings and $\mathbf{H}(A)=\mathbf{H}^*(A)$.

PROOF. Let $A\in S\mathbf{P}_j$ for every $j\in I$. Since $S\mathbf{P}_j$ is a semisimple class of the radical class $\mathbf{P}_j$, $S\mathbf{P}_j$ is hereditary by theorem 1.8. Consequently every ideal of $A\in\bigcap S\mathbf{P}_j$ belongs again to $\bigcap S\mathbf{P}_j$. Thus $S\mathbf{H}$ is also hereditary by theorem 5.5. On the other hand, assume $A\notin S\mathbf{P}_j$ for an index $j\in I$. Then A has a non-zero ideal B no non-zero homomorphic image $B\varphi$ of which belongs to $S\mathbf{P}_j$. But then $B\varphi\notin\bigcap S\mathbf{P}_j= =S\mathbf{H}$. Therefore $S\mathbf{H}$ is, by theorem 3.1, a semisimple class.

Thus $US\mathbf{H}$ is an upper radical class. Let $\mathbf{J}(A)$ be the radical of A with respect to $US\mathbf{H}$. Since $\mathbf{H}_2=US\mathbf{H}\subseteqq\mathbf{H}^*$ and $\mathbf{H}^*$ is the lower radical property, the minimality of $\mathbf{H}^*$ yields $\mathbf{H}^*=US\mathbf{H}$. □

The radical $\mathbf{J}(A)$ is called the *union radical* of A for the set $\{\mathbf{P}_j\}$.

Now we shall give $\mathbf{J}(A)$ in a little more explicit way. Our procedure is similar to the transfinite construction of the intersection radical of a ring.

We define the ideals W_β in a ring as follows:

We put $W_1=0$. Let us assume that the ideals W_α are already defined for all ordinal numbers $\alpha<\beta$ where β is an arbitrary fixed ordinal number.

Then put $W_\beta=\bigcup W_\alpha$ for a limit ordinal number. If β is not a limit ordinal number then we take

$$W_\beta/W_{\beta-1}=\mathbf{P}_j(A/W_{\beta-1}) \tag{5}$$

with a radical property $\mathbf{P}_j$ for which $\mathbf{P}_j(A/\mathbf{W}_{\beta-1})\neq(0)$ (if such a $\mathbf{P}_j$ exists).

Since the ideals of A form a set there exists an ordinal number γ such that

$$\mathbf{P}_j(A/W\gamma)=(0) \quad \text{holds for every} \quad j\in I. \tag{6}$$

Also here we have a freedom to choose the radical properties $\mathbf{P}_j$ in the case of a non-limit ordinal number and hence $\mathbf{W}_\gamma$ does not seem to be unique. Nevertheless the following holds:

THEOREM 5.7. If W_γ is the ideal defined above in a ring A, and it fulfils condition (6) $\mathbf{P}_j(A/W_\gamma)=(0)$ for every $j\in I$, then $W_\gamma=\mathbf{J}(A)$.

PROOF. Obviously $W_1\in US\mathbf{H}$. Let us assume that $W_\alpha\in US\mathbf{H}$ for every ordinal number $\alpha<\beta$. If β is a limit ordinal, theorem 1.3 yields $W_\beta\in US\mathbf{H}=\mathbf{H}^*$. But if

$\beta-1$ exists then

$$W_\beta/W_{\beta-1} = \mathbf{P}_j(A/W_{\beta-1}) \quad \text{for some} \quad j \in I.$$

Let W_β/K be an arbitrary non-zero homomorphic image of W_β. Then there exists an ordinal number α with $W_\alpha \subseteq K$ and $W_{\alpha+1} \nsubseteq K$. Since $W_{\alpha+1} \cap K \supseteq W_\alpha$, the right-hand side of the isomorphism

$$(W_{\alpha+1}+K)/K \cong W_{\alpha+1}/(W_{\alpha+1} \cap K)$$

is a homomorphic image of the $\mathbf{P}_j$-radical ring $W_{\alpha+1}/W$ and thus $(W_{\alpha+1}+K)/K$ is a non-zero $\mathbf{P}_j$-ideal of W_β/K. Since K is an arbitrary ideal of W_β, we have $W_\beta \in U S\mathbf{H}$. An induction yields $W_\beta \in U S\mathbf{H}$ for every β, and thus $W_\gamma \in U S\mathbf{H}$. But this relation implies $W\gamma \subseteq \mathbf{J}(A)$.

Moreover, we also have $A/W_\gamma \in \cap S\mathbf{P}_j$ whence it follows that $\mathbf{J}(A) \subseteq W_\gamma$ by the end of remark 5.1. Therefore $W_\gamma = \mathbf{J}(A)$. □

§ 6. DECOMPOSITIONS OF THE CLASS OF ALL RINGS HAVING NO NON-TRIVIAL IDEALS

For the subject-matter of this section we refer the reader to Kurosh [3].

Let $\mathbf{E}$ denote the class of rings which contain only the trivial ideals. Then $\mathbf{E}$ is the union of the subclass $\mathbf{E}_1$ of all simple rings and the subclass $\mathbf{E}_2$ of all zero-rings of prime order. Isomorphic rings of $\mathbf{E}$ will be identified.

In this section we explain the connection between the decompositions of the class $\mathbf{E}$ and the corresponding lower and upper radical properties determined by them.

A ring in $\mathbf{E}$, i.e. a ring having no non-trivial ideals, is either an $\mathbf{R}$-radical ring, or else it is $\mathbf{R}$-semisimple for every radical property $\mathbf{R}$ since the radical $\mathbf{R}(A)$ is an ideal of the ring A. Accordingly, every radical property $\mathbf{R}$ induces a decomposition of the class $\mathbf{E}$ into two disjoint subclasses $\mathbf{E}'$ and $\mathbf{E}''$, i.e. $\mathbf{E}' \cup \mathbf{E}'' = \mathbf{E}$ and $\mathbf{E}' \cap \mathbf{E}'' = \emptyset$, where $\mathbf{E}'$ denotes the system of all $\mathbf{R}$-radical rings in $\mathbf{E}$, and $\mathbf{E}''$ the system of all $\mathbf{R}$-semisimple rings in $\mathbf{E}$. $\mathbf{E}'$ is called the *lower class*, and $\mathbf{E}''$ the *upper class*.

But the converse statement also holds. If an arbitrary decomposition $\mathbf{E}' \cup \mathbf{E}'' = \mathbf{E}$ with $\mathbf{E}' \cap \mathbf{E}'' = \emptyset$ of the class $\mathbf{E}$ is given, then there exists a radical property $\mathbf{R}$ such that $\mathbf{E}'$ is the lower class, i.e. the class of all $\mathbf{R}$-radical rings, and $\mathbf{E}''$ is the upper class, i.e. the class of all $\mathbf{R}$-semisimple rings.

Thus, we have the following

THEOREM 6.1. If $\mathbf{E}' \cup \mathbf{E}'' = \mathbf{E}$ with $\mathbf{E}' \cap \mathbf{E}'' = \emptyset$ is an arbitrary decomposition of the class $\mathbf{E}$ of all rings having no non-trivial ideals, then a radical property $\mathbf{R}$

exists for which $\mathbf{E}'$ is a class of certain $\mathbf{R}$-radical rings and $\mathbf{E}''$ is a class of certain $\mathbf{R}$-semisimple rings. (Therefore, in general, $\mathbf{E}' \neq L\mathbf{E}'$ and $\mathbf{E}'' \neq S\mathbf{E}''$.)

PROOF. Let $\mathbf{R}'$ be the lower radical property, determined by the class $\mathbf{E}'$. Then every ring in $\mathbf{E}'$ is an $\mathbf{R}'$-radical ring. Now, if A is an arbitrary $\mathbf{R}'$-radical ring in $\mathbf{E}$, then A is of some degree α over $\mathbf{E}'$. Let β be the least ordinal number such that the ring A is of degree β over the class $\mathbf{E}'$. Then β is not a limit ordinal number. If $\beta > 1$ then every non-zero homomorphic image of A has a non-zero ideal of degree $\beta - 1$ over $\mathbf{E}'$. Then A itself is of degree $\beta - 1$ over $\mathbf{E}'$ since $A \in \mathbf{E}$, which contradicts the minimality in case $\beta > 1$. Therefore we have $\beta = 1$, whence $A \in \mathbf{E}$ implies $A \in \mathbf{E}'$. Consequently, every ring in $\mathbf{E}''$ is an $\mathbf{R}'$-semisimple ring.

Since the rings in $\mathbf{E}''$ do not contain non-trivial ideals, condition (j) of theorem 3.1 is fulfilled and thus by theorem 3.3, $\mathbf{E}''$ can be embedded into a larger class $\mathbf{K}''$ such that $\mathbf{K}''$ satisfies both (j) and (jj). Then $\mathbf{K}''$ determines a radical property $\mathbf{R}''$ for which every ring in $\mathbf{K}''$ and hence also in $\mathbf{E}''$ is $\mathbf{R}''$-semisimple. Now, if A is an $\mathbf{R}''$-semisimple ring in $\mathbf{E}$ then A can be homomorphically mapped onto a non-zero ring from the class $\mathbf{E}''$. However, since $A \in \mathbf{E}$, this homomorphism is an isomorphism, and this yields $A \in \mathbf{E}''$. Hence every ring in $\mathbf{E}'$ is an $\mathbf{R}''$-radical ring. □

Obviously $\mathbf{R}' \leqq \mathbf{R}''$ by theorems 3.4 and 4.5. Now we may ask whether $\mathbf{R}' = \mathbf{R}''$ or $\mathbf{R}' < \mathbf{R}''$ for every decomposition or for some of them. If the subclass $\mathbf{E}'$ is empty, then every ring in $\mathbf{E}$ is $\mathbf{R}'$-semisimple, and only (0) is an $\mathbf{R}'$-radical ring. But there exists a non-zero ring A which is a radical ring for the upper radical property determined by $\mathbf{E}'' = \mathbf{E}$. Such a ring $A \neq (0)$ cannot be homomorphically mapped onto a ring in $\mathbf{E}$, and this fact shows that $\mathbf{R}' < \mathbf{R}''$ for this trivial decomposition of $\mathbf{E}$.

We now give an example of a ring having no maximal ideals which cannot therefore be homomorphically mapped onto a ring belonging to the class $\mathbf{E}$.

EXAMPLE 6.2. The zero-ring $Z(p^\infty)$ is such a ring A. We have $A^2 = (0)$. The additive group of this ring is the Prüferian group $C(p^\infty)$. Now $C(p^\infty)$ is isomorphic to the factor group P/I where P is the additive group of all rational numbers with a denominator which is a power of a fixed prime number p, and I denotes the additive group of all integers. Obviously $I \subseteqq P$. However $C(p^\infty)$ is isomorphic to the multiplicative group of the union of all p-th, p^2-th, p^3-th, ..., p^n-th, ... complex roots of unity.

This group $C(p^\infty)$ has interesting properties:

(1) Every ring over the additive group $C(p^\infty)$ is a zero-ring.

The group $C(p^\infty)$ can be generated by the elements $a_1, a_2, \ldots, a_n, \ldots$ and with the defining relations $pa_1 = 0, \ldots, pa_{n+1} = a_n, \ldots$. Now if $i \geqq j$ we have for the product the equation

$$a_i a_j = p^j a_{i+j} a_j = a_{i+j}(p^j a_j) = a_{i+j} \cdot 0 = 0,$$

and thus every product xy equals 0 since we have $x=ka_i$ and $y=la_j$ for some natural numbers, k, l, i, j.

(2) $C(p^\infty)$ is a group which is itself infinite, but contains only finite proper subgroups, namely the cyclic groups of order p^k $(k=1, 2, 3, \ldots)$. In the class of all *Abelian* groups every infinite group without infinite proper subgroups is isomorphic to a group $C(p^\infty)$ (see Szélpál [1]). A similar fact also holds for some larger classes of non-commutative groups, e.g. for the *RN-groups, FC-groups, locally finite S-groups, groups having no maximal normal subgroups.* For the proof we refer the reader to the paper [13] of Szász and for definitions to the book [1] of Kurosh.

(3) Every non-zero homomorphic image of $C(p^\infty)$ is isomorphic to $C(p^\infty)$. Indeed, the kernel of a homomorphism φ of $C(p^\infty)$ with $C(p^\infty)\varphi\neq 0$ is a cyclic subgroup $\{a_n\}$ of $C(p^\infty)=\{a_1, a_2, \ldots, a_n, \ldots\}$ where $pa_1=0$ and $pa_{n+1}=a_n$ for $n=$ $=1, 2, 3, \ldots$. Then we have $C(p^\infty)\varphi=\{a_{n+1}, \ldots\}$ with $pa_{n+1}\in\{a_n\}$, $pa_{m+n+1}=$ $=a_{m+n}$, and the mapping $a_i\varphi=a_{i+n}$ is obviously an isomorphism of $C(p^\infty)$ onto the homomorphic image $C(p^\infty)\varphi$. This mapping is also a ring isomorphism, if we regard $C(p^\infty)$ as a zero-ring $Z(p^\infty)$.

We have the following important

EXAMPLE 6.3. Let Z^∞ be the zero-ring over the infinite cyclic additive group C^∞. Z^∞ can be generated by an element a. If I is a non-zero ideal of Z^∞ and m the least natural number for which $ma\in I$ then I is generated by ma. But then the mapping

$$\varphi\colon\ ka \to kma$$

realizes a ring isomorphism of Z^∞ onto its ideal I. The map φ is an isomorphism since it is a homomorphism and C^∞ is torsion-free.

Now we can give more exact answer to the problem raised above as to whether $\mathbf{R}'=\mathbf{R}''$ or $\mathbf{R}'<\mathbf{R}''$. We have the following

THEOREM 6.4. $\mathbf{R}'<\mathbf{R}''$ *for every decomposition* $\mathbf{E}'\cup\mathbf{E}''=\mathbf{E}$ *(with* $\mathbf{E}'\cap\mathbf{E}''=\emptyset$*) of the class* $\mathbf{E}$ *of all rings having no non-trivial ideals where* $\mathbf{R}'$ *denotes the lower radical property determined by* $\mathbf{E}'$ *and* $\mathbf{R}''$ *denotes the upper radical property determined by* $\mathbf{E}''$.

PROOF. Let us assume first that $\mathbf{E}''$ does not contain a zero-ring of prime order. In this case we shall prove that the zero-ring Z^∞ is both an $\mathbf{R}'$-semisimple ring and an $\mathbf{R}''$-radical ring.

If Z^∞ is not an $\mathbf{R}'$-semisimple ring then Z^∞ has a non-zero $\mathbf{R}'$-ideal B. But, since the ideal B is isomorphic to the ring Z^∞, Z^∞ itself is an $\mathbf{R}'$-radical ring. Consequently, Z^∞ is of some degree over $\mathbf{E}'$. Let α be a minimal ordinal number such that Z^∞ is of degree α over the class $\mathbf{E}'$. Then α is not a limit ordinal number. If $\alpha>1$ then every non-zero homomorphic image of Z^∞, and thus Z^∞ itself, contains an ideal C of degree $\alpha-1$. But, since Z^∞ is isomorphic to the ideal C, Z^∞ itself is of degree $\alpha-1$ over $\mathbf{E}''$, which contradicts the minimality of α if $\alpha>1$. Therefore we have $\alpha=1$ but this is impossible, since Z^∞ contains non-trivial ideals. Hence Z^∞ is an $\mathbf{R}'$-semisimple ring.

Now we shall show that Z^∞ is an $\mathbf{R}''$-radical ring. If Z^∞ is not an $\mathbf{R}''$-radical ring, then Z^∞ can be homomorphically mapped onto a ring in $\overline{\mathbf{E}''}$ and thus also in $\mathbf{E}''$, since $\overline{\mathbf{E}''}$ consists of all rings in which every non-zero ideal can be homomorphically mapped onto a ring in $\mathbf{E}''$. But such a homomorphic image of Z^∞ is again a zero-ring in $\mathbf{E}''$ which contradicts our assumption that $\mathbf{E}''$ does not contain a zero-ring of prime order. Therefore Z^∞ is an $\mathbf{R}''$-radical ring.

Now suppose that $\mathbf{E}''$ contains a zero-ring of prime order. We shall show in this case that $Z(p^\infty)$ is both an $\mathbf{R}'$-semisimple ring and an $\mathbf{R}''$-radical ring. We first prove the $\mathbf{R}'$-semisimplicity of $Z(p^\infty)$. Let $\{a_n\}$ be an arbitrary non-zero proper ideal of $Z(p^\infty)$. Then $\{a_n\}/\{a_{n+1}\}$ is a zero-ring of prime order p, and thus $\{a_n\}$ can be homomorphically mapped onto a non-zero ring in $\mathbf{E}''$, and hence onto an $\mathbf{R}'$-semisimple ring. Accordingly, by theorem 1.14 $\{a_n\}$ is not an $\mathbf{R}'$-ring. Consequently, $Z(p^\infty)$ cannot contain an $\mathbf{R}'$-ideal except possibly itself. If $Z(p^\infty)$ is an $\mathbf{R}'$-radical ring, then $Z(p^\infty)$ is of some degree over $\mathbf{E}'$. Let α be the minimal ordinal number such that $Z(p^\infty)$ is of degree α over $\mathbf{E}'$. Then α is not a limit ordinal number. If $\alpha>1$, then $Z(p^\infty)$ contains a non-zero ideal I of degree $\alpha-1$ over $\mathbf{E}'$. But by what has been said above, $I=\{a_n\}$ is not an $\mathbf{R}'$-ring and this yields $I=Z(p^\infty)$, which contradicts the minimality of α if $\alpha>1$. Therefore we have $\alpha=1$, which is impossible, since $Z(p^\infty)$ has non-trivial ideals. Thus, we have proved that $Z(p^\infty)$ is an $\mathbf{R}'$-semisimple ring.

Now, if $Z(p^\infty)$ is not an $\mathbf{R}''$-radical ring then $Z(p^\infty)$ can be homomorphically mapped onto a non-zero ring in $\overline{\mathbf{E}''}$, where $\overline{\mathbf{E}''}$ is the class of all rings such that all their non-zero ideals can be homomorphically mapped onto rings in $\mathbf{E}''$. But then $Z(p^\infty)$ can be homomorphically mapped onto a ring in $\mathbf{E}''$, and since every non-zero homomorphic image of $Z(p^\infty)$ is isomorphic to $Z(p^\infty)$, we have the contradiction $Z(p^\infty)\in\mathbf{E}'$. Thus, we have shown that $Z(p^\infty)$ is an $\mathbf{R}''$-radical ring. Therefore $\mathbf{R}'<\mathbf{R}''$ in all cases. □

§ 7. THE RADICAL OF AN ALGEBRA OVER AN OPERATOR DOMAIN AND THE RADICAL OF A FULL MATRIX RING

For the subject-matter of this section we refer the reader to Divinsky and Suliński [1], and Amitsur [5].

Let A be a ring and Φ an arbitrary set. We say that Φ is a *set of operators*, or in short an operator domain for A, if a product $\varphi a\in A$ is defined for every element $\varphi \in \Phi$ and every element $a \in A$ such that the conditions

$$\varphi(a+b) = \varphi a + \varphi b,$$
$$\varphi(ab) = (\varphi a) b = a(\varphi b)$$

are fulfilled for all $\varphi \in \Phi$ and $a, b \in A$. In particular, if Φ is a commutative ring, then A is called an *algebra over* Φ. But in general, Φ need not be an algebraic structure and thus not a ring either.

In what follows, we suppose that $\varphi I \subseteqq I$ for every $\varphi \in \Phi$ and for every ideal of the ring A with the operator domain Φ.

The following result is due to Divinsky and Suliński [1].

THEOREM 7.1. If A is a ring with the operator domain Φ and $\mathbf{R}$ is a radical property then the $\mathbf{R}$-radical of A considered as a ring with operator domain exists and this $\mathbf{R}$-radical coincides with the $\mathbf{R}$-radical of the ring considered without operator domain.

PROOF. Let I be an ideal of the ring A considered without operator domain. Let us assume that I is an $\mathbf{R}$-ring. Put $I_\varphi = I + \varphi I$ for every operator $\varphi \in \Phi$. We have $a\varphi i = \varphi a i$ for every $a \in A$ and $i \in I$, so the element $a\varphi i$ is contained in φI. Thus obviously I_φ is a left ideal of A. Therefore $AI_\varphi = I_\varphi$. Similarly $(\varphi I)A = \varphi(IA) \subseteqq \varphi I$ can also be verified. Therefore I_φ is an ideal of A.

Now we show that I_φ is an $\mathbf{R}$-ideal of A. If h is the natural homomorphism of I_φ onto the factor ring I_φ/I, then we define for every $i \in I$ a mapping of I onto I_φ/I by the relation

$$g\colon i \to g(i) = h(\varphi i).$$

Obviously, g is unique and because

$$g(i_1+i_2) = h(\varphi i_1) + h(\varphi i_2) = g(i_1) + g(i_2)$$

g is an additive homomorphism.

Furthermore $i_1 \varphi^2 i_2 \in I$ yields

$$g(i_1) g(i_2) = h(\varphi i_1) \cdot h(\varphi i_2) = h(\varphi i_1 \cdot \varphi i_2) = h(i_1 \varphi^2 i_2).$$

On the other hand

$$g(i_1 i_2) = h\varphi(i_1 i_2) = h((\varphi i_1) i_2)$$

since $(\varphi i_1) i_2 \in I$. Consequently we have $g(i_1 i_2) = g(i_1) g(i_2)$ in the factor ring I_φ/I and thus g is a ring homomorphism of I onto I_φ/I. But, since I is an $\mathbf{R}$-ring, I_φ/I is also an $\mathbf{R}$-ring by the definition of a radical property in § 1. But then I_φ is also an $\mathbf{R}$-ring by theorem 1.2, since both I and I_φ/I are $\mathbf{R}$-rings.

Now, if $\mathbf{R}(A)$ is the $\mathbf{R}$-radical of the ring A, then $I_\varphi \subseteqq \mathbf{R}(A)$ for every $\mathbf{R}$-ideal I and every $\varphi \in \Phi$, by definition of the radical. In particular, also $(\mathbf{R}(A))_\varphi = \mathbf{R}(A)$, whence $\mathbf{R}(A) \subseteq (\mathbf{R}(A))_\varphi$ implies the equality $\mathbf{R}(A) = (\mathbf{R}(A))_\varphi$. Therefore we have $\varphi\mathbf{R}(A) \subseteq \mathbf{R}(A)$ for every $\varphi \in \Phi$. □

Now we shall prove a theorem for the radical of a full matrix ring A_n over a ring A.

THEOREM 7.2 (Amitsur [5]). Let A be a ring and $\mathbf{R}$ a radical property which fulfils the following conditions:

(i) If B is an $\mathbf{R}$-ring then so is B_n.
(ii) If C is an $\mathbf{R}$-semisimple ring then so is C_n.

Then $\mathbf{R}(A_n) = (\mathbf{R}(A))_n$.

PROOF. $(\mathbf{R}(A))_n$ is an $\mathbf{R}$-ring in A_n by condition (i), and since $\mathbf{R}(A)$ is an ideal of A, by the multiplication of matrices $(\mathbf{R}(A))_n$ is an $\mathbf{R}$-ideal of A_n. Therefore we have $(\mathbf{R}(A))_n \subseteq \mathbf{R}(A_n)$ by the definition of the radical. On the other hand, we obviously have $(A/\mathbf{R}(A))_n \cong A_n/(\mathbf{R}(A))_n$. Since $A/\mathbf{R}(A)$ is $\mathbf{R}$-semisimple, condition (ii) yields that $A_n/(\mathbf{R}(A))_n$ is also $\mathbf{R}$-semisimple. But by theorem 1.12 $\mathbf{R}(A_n)$ is the intersection of all ideals K of A_n for which A_n/K is $\mathbf{R}$-semisimple. Therefore we have $\mathbf{R}(A_n) \subseteq (\mathbf{R}(A))_n$. Thus $\mathbf{R}(A_n) = (\mathbf{R}(A))_n$. □

A consequence of theorem 7.2. is

PROPOSITION 7.3. If $\mathbf{R}$ is a radical property satisfying conditions (i) and (ii) of the above theorem then the full matrix ring A_n is an $\mathbf{R}$-ring if and only if A is an $\mathbf{R}$-ring.

PROOF. If A is not an $\mathbf{R}$-ring, then we have $A \neq \mathbf{R}(A)$ and $\mathbf{R}(A_n) \neq A_n$ since $\mathbf{R}(A_n) = (\mathbf{R}(A))_n$. Thus A_n is not an $\mathbf{R}$-ring. But if A is an $\mathbf{R}$-ring then by condition (i) A_n is an $\mathbf{R}$-ring. □

§ 8. A PROPERTY OF HEREDITARY RADICALS AND OF STRONG SEMISIMPLICITIES

A radical property $\mathbf{R}$ is said to be *hereditary* if every ideal of an $\mathbf{R}$-ring is itself an $\mathbf{R}$-ring.

Now we prove the following

THEOREM 8.1 (Amitsur [4]). Let A be a ring, L_1 and L_2 ideals of A and $\mathbf{R}$ a hereditary radical property. Let $\hat{I}_1$ and $\hat{I}_2$ be the (possibly greatest) ideals of A, uniquely determined by $\hat{I}_i/I_i = \mathbf{R}(A/I_i)$. Then $(\widehat{I_1 \cap I_2}) = \hat{I}_1 \cap \hat{I}_2$.

Proof. First let us assume $I_1 \subseteq I_2$. Then $\hat{I}_1 \subseteq \hat{I}_2$ holds, since the mapping $\varphi: I \to \hat{I}$ is monotone by theorem 1.15. Therefore, since $I_i \supseteq I_1 \cap I_2$ we have $\hat{I}_i \supseteq \widehat{(I_1 \cap I_2)}$, and thus $\hat{I}_1 \cap \hat{I}_2 \supseteq \widehat{I_1 \cap I_2}$. On the other hand, both

$$(\hat{I}_1 \cap \hat{I}_2)/(\hat{I}_1 \cap I_2) \quad \text{and} \quad (\hat{I}_1 \cap I_2)/(I_1 \cap I_2)$$

are **R**-radical rings by the hereditary property of **R** and because

$$(\hat{I}_1 \cap \hat{I}_2)/((\hat{I}_1 \cap \hat{I}_2) \cap I_i) \cong ((\hat{I}_1 \cap \hat{I}_2) + I_i)/I_i \leqq \hat{I}_i/I_i.$$

Moreover, both the factor rings $(\hat{I}_1 \cap \hat{I}_2)/(I_1 \cap \hat{I}_2)$ and $(I_1 \cap \hat{I}_2)/(I_1 \cap I_2)$ are **R**-radical rings by the hereditary property of **R** and the relation

$$(\hat{I}_1 \cap I_2)/((\hat{I}_1 \cap I_2) \cap I_1) \cong ((\hat{I}_1 \cap I_2) + I_1)/I_1 \subseteq \hat{I}_1/I_1.$$

Thus, by theorem 1.2, $(\hat{I}_1 \cap \hat{I}_2)/(I_1 \cap I_2)$ is also an **R**-radical ring. The definition of the radical and that of $\hat{I}$ yield now $\hat{I}_1 \cap \hat{I}_2 \subseteq \widehat{(I_1 \cap I_2)}$ which together with $\widehat{(I_1 \cap I_2)} \subseteq \subseteq \hat{I}_1 \cap \hat{I}_1$ implies the equality $\hat{I}_1 \cap \hat{I}_2 = \widehat{(I_1 \cap I_2)}$. □

In other words, for hereditary radical properties **R**, the mapping $I \to \hat{I}$ is an intersection endomorphism in the lattice of ideals of the ring A.

We wish now to dualize theorem 8.1. For this we need some preliminaries.

If **R** is a radical property $A/\mathbf{R}(A)$ is **R**-semisimple for every ring A. A homomorphic image B of $A/\mathbf{R}(A)$ will be not **R**-semisimple in general. For example, if $\mathfrak{J}$ is the ring of the rational integers and **R** denotes the lower radical property determined by the class of all nilpotent rings, then the homomorphic image $\mathfrak{J}/4\mathfrak{J}$ is not **R**-semisimple, since $\mathfrak{J}/4\mathfrak{J}$ contains the nilpotent ideal $2\mathfrak{J}/4\mathfrak{J}$, although $\mathfrak{J}$ obviously is **R**-semisimple.

For a radical property **R**, an **R**-semisimplicity is called *strong* ***R**-semisimplicity*, if every homomorphic image of every **R**-semisimple ring is again **R**-semisimple. This notion plays an important role in Chapter II. If **R** is the upper radical property, determined by the class of all Boolean rings, then every **R**-semisimple ring is also strongly **R**-semisimple. (As it is well known, a ring A is called a *Boolean ring* if $a^2 = a$ for every element $a \in A$. These rings are commutative.)

We recall that the lattice of all ideals of a ring is *modular*. This condition is as follows:

If I_1 and I_2 are ideals of A such that $I_1 \geqq I_2$, then

$$I_1 \cap (I_2 + I_3) = I_2 + (I_1 \cap I_3)$$

for every ideal I_3 of A.

If **R** is an arbitrary Amitsur–Kurosh radical property and I an arbitrary ideal of the ring A, then the radical $\bar{I}=\mathbf{R}(I)$ of I is again an ideal of the ring A by theorem 1.7. Therefore

$$\varphi\colon\ I \to \bar{I} = \mathbf{R}(I)$$

is a mapping of the lattice of all ideals of A into itself satisfying $I\varphi \subseteqq I$ and $I\varphi^2 = I\varphi$. For the radical properties **R** for which every **R**-semisimple ring is strongly **R**-semisimple, we have the following

THEOREM 8.2 (Szász [24]). If **R** is a radical property for which every **R**-semisimple ring is strongly **R**-semisimple, A is an arbitrary ring, I an ideal of A and $\varphi I = \bar{I} = \mathbf{R}(I)$, then the mapping

$$\varphi\colon\ I \to \mathbf{R}(I)$$

is a union endomorphism of the lattice of all ideals of A; i.e. we have $\overline{I_1+I_2} = \bar{I}_1+\bar{I}_2$ for arbitrary ideals I_1 and I_2 of A.

PROOF. By theorem 1.15 $I_1 \subseteqq I_2$ always implies $\bar{I}_1 \subseteqq \bar{I}_2$ for arbitrary ideals I_1 and I_2 of A.

Since $I_1+I_2 \supseteqq I_j$ and $I_j \supseteqq I_1 \cap I_2$ we obtain $\overline{I_1+I_2} \supseteqq \bar{I}_1+\bar{I}_2$ and $\bar{I}_1 \cap \bar{I}_2 \supseteqq \overline{I_1 \cap I_2}$ for an arbitrary radical property **R**. The strong **R**-semisimplicity of the **R**-semisimple rings has not yet been used. This latter condition will be used only for the proof of the converse relation $\overline{I_1+I_2} \subseteqq \bar{I}_1+\bar{I}_2$.

The factor ring $I_1/(I_1 \cap (\bar{I}_1+\bar{I}_2))$ is a homomorphic image of the **R**-semisimple ring $I_1/\bar{I}_1$ because $I_1 \cap (\bar{I}_1+I_2) = \bar{I}_1+(I_1 \cap I_2)$. Thus $I_1/(I_1 \cap)\bar{I}_1+I_2))$ is also **R**-semisimple in consequence of the strong **R**-semisimplicity.

But the isomorphism theorem yields

$$(I_1+I_2)/(\bar{I}_1+I_2) \cong I_1/(I_1 \cap (\bar{I}_1+I_2)),$$

and therefore $(I_1+I_2)/(\bar{I}_1+I_2)$ is **R**-semisimple.

On the other hand, the factor ring $I_2/(I_2 \cap (\bar{I}_1+\bar{I}_2))$ is a homomorphic image of the **R**-semisimple ring $I_2/\bar{I}_2$ since $I_2 \cap (\bar{I}_1+\bar{I}_2) = \bar{I}_2+(\bar{I}_1 \cap I_2)$. Also $I_2/(\bar{I}_2+(\bar{I}_1 \cap I_2)$ is **R**-semisimple because of the strong **R**-semisimplicity. By the isomorphism theorem, we have

$$(\bar{I}_1+I_2)/(\bar{I}_1+\bar{I}_2) = ((\bar{I}_1+\bar{I}_2)+I_2)/(\bar{I}_1+\bar{I}_2) \cong I_2/(I_2 \cap (\bar{I}_1+\bar{I}_2))$$

and thus $(\bar{I}_1+I_2)/(\bar{I}_1+\bar{I}_2)$ is **R**-semisimple. By theorem 1.9 the **R**-semisimplicity of $(\bar{I}_1+I_2)/(\bar{I}_1+\bar{I}_2)$ and $(I_1+I_2)/(\bar{I}_1+I_2)$ imply that $(I_1+I_2)/(\bar{I}_1+\bar{I}_2)$ is also **R**-semisimple. But $(\overline{I_1+I_2})/(\bar{I}_1+\bar{I}_2)$ is an **R**-ideal. Hence $\overline{I_1+I_2} = \bar{I}_1+\bar{I}_2$, and thus $\overline{I_1+I_2} = \bar{I}_1+\bar{I}_2$. □

§ 9. RADICALS AND OPERATOR MODULES

For the subject-matter of this section we refer the reader to Andrunakievich and Ryabukhin [3].

We shall show in this section that the Amitsur–Kurosh theory of radicals can also be formulated in the language of operator modules and representations. In this discussion the "general modules" which are generalizations of prime modules and irreducible modules, play an important role.

Let A be an arbitrary (associative) ring. By an A-module we shall always mean a *right A-module*. The subset $(0:M)_A = \{x;\, x \in A,\, Mx = (0)\}$ of the ring A is called the annihilator of the A-module M. If $MA = (0)$ we call M a trivial A-module and if $(0:M)_A = 0$ we call it a faithful A-module. If $\mathbf{M}_A$ is an arbitrary class of A-modules, we define the kernel of the class $\mathbf{M}_A$ to be the intersection

$$\ker \mathbf{M}_A = \bigcap \{(0:M)_A;\ M \in \mathbf{M}_A\}.$$

If $\mathbf{M}_A$ is empty, we put $\ker \mathbf{M}_A = A$. If $\ker \mathbf{M}_A = (0)$ the class $\mathbf{M}_A$ is called *faithful.* We mention the well-known

PROPOSITION 9.1. If A is a ring, B is an ideal of A and M is an A/B-module, then M is an A-module under the operation $ma = m(a+B)$ and then $B \subseteq (0:M)_A$. Conversely, if M is an A-module and B is an ideal of A such that $B \subseteq (0:M)_A$, then M is an A/B-module with the operation $m(a+B) = ma$. Here a subset N of M is an A/B-submodule of M if and only if N is an A-submodule of M. Moreover, $(0:M)_{A/B} = ((0:M)_A)/B$.

We associate to every ring A an (eventually empty) class $\mathbf{M}_A$ of non-trivial A-modules and denote $\mathbf{M} = \bigcup \mathbf{M}_A$ where A runs over all associative rings.

A class $\mathbf{M}$ is called a *general class of modules,* if the following conditions are fulfilled:

(1) $M \in \mathbf{M}_{A/B}$ implies $M \in \mathbf{M}_A$.

(2) $M \in \mathbf{M}_A$ and $B \leqq (0:M)_A$ imply $M \in \mathbf{M}_{A/B}$ for the ideal B of A.

(3) If $\mathbf{M}_A$ is a faithful class then $\mathbf{M}_B$ is non-empty for every non-zero ideal B of A.

(4) If $\mathbf{M}_B$ is non-empty for every non-zero ideal B of A, then $\mathbf{M}_A$ is faithful.

If M is a general class of modules, then by the **M**-*radical* $\mathbf{M}(A)$ of the ring A we mean the kernel of the class $\mathbf{M}_A$, i.e.

$$\mathbf{M}(A) = \bigcap \{(0:M)_A;\ M \in \mathbf{M}_A\}.$$

If $\mathbf{M}_A$ is empty, A is called an **M**-radical ring. If $\mathbf{M}_A$ is faithful we call A **M**-*semi-simple*.

Clearly, A is an $\mathbf{M}$-radical ring if and only if $A=\mathbf{M}(A)$. For, if $\mathbf{M}_A$ is empty, then we have $\mathbf{M}(A)=\ker \mathbf{M}_A=A$. But if $\mathbf{M}_A$ is non-empty, then we have $\mathbf{M}(A)\leqq \leqq(0:M)_A \subsetneqq A$ since $(0:M)_A\neq A$ for every $M\in\mathbf{M}_A$.

The definition of $\mathbf{M}(A)$ and condition (1) in the definition of a general class of modules yield

PROPOSITION 9.2. Every homomorphic image of an $\mathbf{M}$-radical ring is again an $\mathbf{M}$-radical ring.

Furthermore we obtain the following

PROPOSITION 9.3. If B is an ideal of the ring A with $B\subseteqq\mathbf{M}(A)$ then

$$\mathbf{M}(A/B) = \mathbf{M}(A)/B.$$

PROOF. If $\mathbf{M}_A$ is empty, then proposition 9.2 implies

$$\mathbf{M}(A/B) = A/B = \mathbf{M}(A)/B.$$

If $\mathbf{M}_A$ is non-empty, then since $B\subseteqq\mathbf{M}(A)$ we have by the application of conditions (1) and (2) and proposition 9.1

$$\mathbf{M}(A/B) = \cap\{(0:M)_{A/B};\ M\in\mathbf{M}_{A/B}\} = \cap\{(0:M)_{A/B};\ M\in\mathbf{M}_A\} = $$
$$= (\cap\{(0:M)_A;\ M\in\mathbf{M}_A\})/B = \mathbf{M}(A)/B.\square$$

The previous results imply

PROPOSITION 9.4. The factor ring $A/\mathbf{M}(A)$ of an arbitrary ring with respect to its $\mathbf{M}$-radical is $\mathbf{M}$-semisimple. A ring A is an $\mathbf{M}$-radical ring if and only if it cannot be homomorphically mapped onto a non-zero $\mathbf{M}$-semisimple ring. □

PROPOSITION 9.5. The $\mathbf{M}$-radical $\mathbf{M}(A)$ of an arbitrary ring A coincides with the intersection of all ideals T_α of A such that A/T_α is $\mathbf{M}$-semisimple.

PROOF. Put $D=\cap T_\alpha$, where A/T_α is $\mathbf{M}$-semisimple. Then $\mathbf{M}(A)\supseteqq D$ since $\mathbf{M}(A/\mathbf{M}(A))=(0)$. We take $A_\alpha=A/T_\alpha$. Then we have

$$\cap\{(0:M)_{A_\alpha};\ M\in\mathbf{M}_{A_\alpha}\} = (0);\ (0:M)_{A_\alpha} = (0:M)_A/T_\alpha.$$

Hence $\cap\{(0:M)_A/T_\alpha;\ M\in\mathbf{M}_{A\alpha}\}=(0)$ and thus $\cap\{(0:M)_A;\ M\in A_\alpha\}=T_\alpha$. Therefore condition (1) yields $T_\alpha\geqq\cap\{(0:M)_A;\ M\in\mathbf{M}_A\}$. Consequently $T_\alpha\geqq D\geqq\mathbf{M}(A)$. □

PROPOSITION 9.6. Every subdirect sum A of $\mathbf{M}$-semisimple rings A_α is $\mathbf{M}$-semisimple (see § 12).

PROOF. There exist ideals T_α in A such that $\cap T_\alpha=0$ and $A_\alpha\cong A/T_\alpha$. The result essentially follows from proposition 9.5. □

If $\mathbf{M}$ is a general class of modules, we denote by $N(\mathbf{M})$ the class of all rings which have a faithful A-module in $\mathbf{M}_A$.

PROPOSITION 9.7. Every $\mathbf{M}$-semisimple ring is a subdirect sum of rings from the class $N(\mathbf{M})$.

PROOF. If A_α has a faithful A_α-module M from $\mathbf{M}$, then $(0:M)_{A_\alpha}=(0)$, and thus $\mathbf{M}(A_\alpha)=(0)$. But then, by proposition 9.6, every subdirect sum of the rings A_α is $\mathbf{M}$-semisimple. Conversely, if $\mathbf{M}(A)=(0)$, then A is a subdirect sum $A/(0:M)_A=\bar{A}$, where by proposition 9.1 and condition (1), the statements $M\in\mathbf{M}_{\bar{A}}$ and

$$(0:M)_{\bar{A}} = (0:M)_A/(0:M)_A = (0).\square$$

PROPOSITION 9.8. A ring A is $\mathbf{M}$-semisimple if and only if there exists a faithful A-module M such that it is the discrete direct sum of A-modules from $\mathbf{M}_A$.

PROOF. If A_α has a faithful A_α-module M in $\mathbf{M}$, then $(0:M)_A=\bigcap(0:M_\alpha)_A=(0)$, and thus A is $\mathbf{M}$-semisimple. Conversely, if $\mathbf{M}(A)=\bigcap\{(0:M)_A;\ M\in\mathbf{M}_A\}=(0)$, then ideals of the form $B=(0:M)_A$ with $M\in\mathbf{M}_A$ can be provided with indices $\gamma\in\Gamma$. Then $M=\sum_{\gamma\in\Gamma}\oplus M_\gamma$ is the desired module since $M_\gamma\in\mathbf{M}_A$ and $(0:M)_A=$ $=\bigcap_{\gamma\in\Gamma}(0:M_\gamma)_A.\square$

PROPOSITION 9.9. Every $\mathbf{M}$-radical ideal B of a ring A is contained in the $\mathbf{M}$-radical $\mathbf{M}(A)$ of A.

PROOF. We suppose that $B\subsetneqq\mathbf{M}(A)$. Then there exists an A-module $M\in\mathbf{M}_A$ with $B\subsetneqq(0:M)_A$ and thus

$$B/(0:M)_B = B/((0:M)_A\cap B) \cong (B+(0:M)_A)/(0:M)_A$$

is a non-zero $\mathbf{M}$-radical ideal of the $\mathbf{M}$-semisimple ring $\bar{A}=A/(0:M)_A$. This contradiction proves $B\subseteqq\mathbf{M}(A).\square$

PROPOSITION 9.10. The sum T of arbitrary $\mathbf{M}$-radical ideals I_α is an $\mathbf{M}$-radical ideal. The extension A of an $\mathbf{M}$-radical ring B by an $\mathbf{M}$-radical ring A/B is an $\mathbf{M}$-radical ring. $\mathbf{M}(A)$ is the sum of all $\mathbf{M}$-radical ideals of A.

PROOF. I_α is a radical ideal in the ring T also and thus by proposition 9.9 $\mathbf{M}(T)\supseteqq I_\alpha$. Hence $\mathbf{M}(T)\supseteqq T\supseteqq\mathbf{M}(T)$.

Suppose that $\mathbf{M}_A$ is not empty and $M\in\mathbf{M}_A$. Then, by proposition 9.9, $B\leqq\mathbf{M}(A)$ and thus $MB=0$. By condition (2) we have $M\in\mathbf{M}_{A/B}$ which is a contradiction.

The sum T of all $\mathbf{M}$-radical ideals I_α of A is an $\mathbf{M}$-radical ideal by the above and A/T has no $\mathbf{M}$-radical ideal. Hence $\mathbf{M}(A)=T.\square$

THEOREM 9.11. Every **M**-radical is a radical property in the sense of Amitsur and Kurosh. Conversely, for every radical property **R** there exists a general class of operator modules such that $\mathbf{R}(A)$ coincides with the **M**-radical. A class **N** of rings coincides with the class $N(\mathbf{M})$ for a general class **M** if and only if the following conditions are fulfilled:

(1) Every non-zero ideal of an arbitrary ring from **N** can be homomorphically mapped onto a non-zero ring from **N**.
(2) Every **R**-semisimple ring is a subdirect sum of rings from **N**, where **R** denotes the upper radical property determined by the class **N**.

PROOF. If $\mathbf{R}(A)$ is an **M**-radical, then **R** is a radical property in the sense of Amitsur and Kurosh by propositions 9.2, 9.4 and 9.10.

Conversely, let **R** be an arbitrary radical property and **N** be the class of all **R**-semisimple rings. Let $\mathbf{M}_A$ denote the class of all A-modules **M** for which $\bar{A}= =A/(0:M)_A \in \mathbf{N}$. Then we have $\bar{A}/(0:M)_{\bar{A}} \cong A/(0:M)_A$. Moreover, for $\mathbf{M}_A$ conditions (1) and (2) of the definition of a general class of modules are fulfilled. We remark that conditions (3) and (4) have not been used in the proofs of propositions 9.2–9.8.

We shall verify that a ring is **R**-semisimple if and only if it has a faithful module in **M**. For, if $(0:M)_A=(0)$ for a module $M \in \mathbf{M}_A$, then A is **R**-semisimple by the definition of $\mathbf{M}_A$. Conversely, if A is **R**-semisimple, embed A into a ring M with unity element. Then M is an A-module, moreover it is a faithful A-module with $M \in \mathbf{M}_A$.

Assume that the class $\mathbf{M}_A$ is faithful, i.e. $\cap\{(0:M)_A;\, M \in \mathbf{M}_A\}=(0)$. A is a subdirect sum of the **R**-semisimple rings $\bar{A}=A/(0:M)_A$, whence A is itself **R**-semisimple. By the preceding, there exists a faithful A-module $M \in M_A$. If B is a non-zero ideal of A, then $MB \neq (0)$. Thus B can be homomorphically mapped onto the ring

$$\bar{B} = B/(B \cap (0:M)_A) \cong (B+(0:M)_A)/(0:M)_A$$

which is a non-zero ideal of the **R**-semisimple ring $\bar{A}=A/(0:M)_A$. Therefore $\bar{B}$ can be homomorphically mapped onto a non-zero **R**-semisimple ring which has a faithful module in M. Hence $\mathbf{M}_{\bar{B}} \neq \emptyset$ and thus $\mathbf{M}_B \neq \emptyset$. Consequently condition (3) of the definition of a general class is also fulfilled. Now, conversely, assume $\mathbf{M}_B \neq \emptyset$ for every non-zero ideal B of the ring A. Then B can be homomorphically mapped onto an **R**-semisimple ring and thus B is not a radical ring. Hence the ring A is **R**-semisimple, and the class $\mathbf{M}_A$ is faithful, since there exists a faithful A-module in $\mathbf{M}_A$. Therefore condition (4) also holds.

Then **M** is a general class of the modules $M=N(\mathbf{M})$, Moreover, the ring A is an **M**-radical ring if and only if A is an **R**-radical ring.

Now, let **K** be a class of rings such that every non-zero ideal of a ring in **K** can be homomorphically mapped onto a non-zero ring in **K**. Let **R** be the upper radical property determined by the class **K**, for which, by assumption, every **R**-semisimple ring is a subdirect sum of rings from **K**. Then there exists a class **M** of modules such that $\mathbf{K}=N(\mathbf{M})$. The converse statement holds by condition (3) and proposition 9.7. □

§ 10. THE DEFINITION OF A MARANDA-MICHLER QUASIRADICAL

In this section we shall discuss the "inner" definition of a quasiradical. This definition rests upon the notion of the preradical introduced by Maranda [1] in the theory of modules. It will be shown that every hereditary radical property is a complete hereditary quasiradical and conversely.

By contrast, we termed the definition of an Amitsur–Kurosh radical property introduced in § 1 an "outer" property since this definition uses a class **K** of **R**-rings which fulfil axioms (i), (ii) and (iii), and the **R**-radical $\mathbf{R}(A)$ of an arbitrary ring A is constructed by means of this class. Here the radical $\mathbf{R}(A)$ of every ring A is always an **R**-ring.

A function f defined on a class **K** of rings is called a *preradical,* if f associates to every ring A a uniquely determined ideal fA of A such that

$$(fA)\mu \subseteq f(A\mu)$$

holds for every ring homomorphism μ. A preradical f on **K** is a *quasiradical* on **K** if

$$f(A/f(A)) = (0)$$

for every ring A.

If f is a preradical on **K**, $A\in\mathbf{K}$ is called an *f-ring* if $A=fA$. An ideal I of a ring A is said to be an f-ideal if I is an f-ring. A preradical on **K** is called *hereditary* (see Andrunakievich [11, 12]) if for every ring $A\in\mathbf{K}$ every ideal of A contained in fA is an f-ideal. If, for every ring $A\in\mathbf{K}$, every f-ideal is contained in fA, then the preradical f is called *complete.*

The complete hereditary preradicals and quasiradicals have been discussed by Michler [2] in detail. But the quasiradicals were called "*radicals*" by Michler. In what follows, we shall study the most important properties of the complete hereditary preradicals according to the interesting paper of Michler. We shall also give a connection between the definition of a radical property and of a quasiradical of a ring.

We shall discuss this latter topic, and let **R** be a radical property fulfilling axioms (i), (ii) and (iii), given in § 1. Moreover, let $\mathbf{R}(A)$ be the **R**-radical of a ring A. Then $\mathbf{R}(A)$ is an ideal of A, and the mapping

$$f_R\colon A \to \mathbf{R}(A)$$

is a preradical on the class of all rings, since we have $(\mathbf{R}(A))\mu = \mathbf{R}(A\mu)$ for every ring homomorphism μ of A. As $\mathbf{R}(A/\mathbf{R}(A)) = 0$, $\mathbf{R}$ is a quasiradical of A. But conversely, if f is a quasiradical, then, in general fA need not be an f-ideal of A. This is illustrated by the following example.

Let fA denote the intersection of all maximal right ideals of the ring A if A has maximal right ideals and take $fA = A$ if A does not contain a proper maximal right ideal. Then $f\colon A \to fA$ is a preradical. We have $(fA)\mu \leqq f(A\mu)$ since for a maximal right ideal R of A and for a homomorphism μ of A $R\mu$ is either a maximal right ideal of $A\mu$ or else $R\mu$ coincides with $A\mu$. Conversely, the largest set mapped onto a maximal right ideal of $A\mu$ is a maximal right ideal of A and thus $(fA)\mu \subseteqq f(A\mu)$. Moreover, we also have $f(A/f(A)) = 0$, since fA is contained in every maximal right ideal R of A. Therefore fA is a quasiradical. But in general fA is not an f-ideal of A, as it is shown by the example of the zero-ring $Z(p^2)$ with a cyclic additive group $\{a\}$ of order p^2 where p is a prime number. We have namely

$$f\{a\} = \{pa\}, \quad \text{and} \quad ff\{a\} = f\{pa\} = (0).$$

Therefore, in general, fA is not an f-ideal of A.

Moreover, in general, fA need not contain all f-ideals of A.

For example, let E be a simple ring with minimum condition on principal right ideals which does not satisfy the maximum condition on subrings. Let K_1 be a finite field. Let the class $\mathbf{K}$ consist of $A_0 = E \oplus K_1$ (the ring direct sum of E and K_1) and all ideals and homomorphic images of A_0. Then we define a preradical on $\mathbf{K}$ as follows:

$$fA = \begin{cases} A & \text{if } A \in \mathbf{K} \text{ is a ring with maximum condition on subrings,} \\ (0) & \text{otherwise.} \end{cases}$$

Then f is a quasiradical on $\mathbf{K}$, which is "inductive" in the sense that the sum of f-ideals is again an f-ideal. But f is not a radical property in the sense of Amitsur and Kurosh, since we have $fK_1 = K_1 \neq (0)$, however $fA_0 = 0$ and K_1 is an ideal of A_0.

Therefore there exists a quasiradical which is not a radical property in the sense of Amitsur and Kurosh and thus the notions of radical property and quasiradical are different. But we shall see (proposition 10.5) that every hereditary and complete quasiradical is a hereditary radical property.

Next we characterize the hereditary and complete quasiradicals by means of some additive properties of the function f.

Let $\mathbf{K}$ be a class of rings with the property that if a ring belongs to $\mathbf{K}$ then all its ideals and homomorphic images also belong to $\mathbf{K}$.

Then a preradical f on **K** is called *additive* if the sum of two f-ideals which are contained in a **K**-ideal Q of a ring $A \in \mathbf{K}$, is again an f-ideal of A.

The preradical f on **K** is called *inductive* if the sum of all f-ideals which are contained in a **K**-ideal of a ring A is again an f-ideal of A.

We now give an example of an additive preradical which is not inductive.

Let the function f be defined as follows:

$$fA = \begin{cases} A & \text{if } \ A \text{ is a nilpotent ring,} \\ (0) & \text{otherwise.} \end{cases}$$

Then f is obviously a preradical and f is nilpotent, since the sum of two nilpotent ideals I_1 and I_2 is nilpotent because of the isomorphism $(I_1+I_2)/I_1 \cong \cong I_2/(I_1 \cap I_2)$ and theorem 1.2. Namely, both I_1 and $(I_1+I_2)/I_1$ are nilpotent, and the latter is isomorphic to the homomorphic image $I_2/(I_1 \cap I_2)$ of I_2. On the other hand, the sum of an arbitrary number of nilpotent ideals is in general not nilpotent in rings having no maximal nilpotent ideals (see in the proof of theorem 2.2, the ring considered in the discussion of system IV).

PROPOSITION 10.1. If f is a preradical on **K** such that all ideals of A are contained in $f(A)$ whenever A is contained in $f(A)$ then the following statements are equivalent:

(1) f is inductive on **K**.

(2) f is additive on **K**, and the union of every ascending chain of f-ideals of every ring $A \in \mathbf{K}$ is again an f-ideal of A.

PROOF. (1) obviously implies (2) by definition. Next let us assume (2). Then since $f(0)=(0)$, the set $\mathfrak{M}$ of all f-ideals of A is not empty so by (2) Zorn's lemma can be applied. Therefore there exists a maximal f-ideal M in $\mathfrak{M}$. If M does not contain an f-ideal K of A, then by (2) $K+M \neq M$ and $K+M$ is an f-ideal of A, which contradicts the maximality of M. Accordingly, the sum of all f-ideals contained in a **K**-ideal of A, is an f-ideal. □

PROPOSITION 10.2. For a preradical f on **K** the following conditions are equivalent:

(1) f is complete and hereditary.

(2) (a) f is inductive on **K**;

(b) for every ring $A \in \mathbf{K}$ we have $fA = \{x;\ x \in A, (x) = f(x)\}$, where (x) is the ideal generated by x.

(3) (a) f is additive on **K**;

(b) the union of every ascending chain of f-ideals of every ring $A \in \mathbf{K}$ is again an f-ideal of A;

(c) a principal ideal (x) of a ring $A \in \mathbf{K}$ is an f-ideal if and only if x is contained in fA.

PROOF. Obviously (2b) and (3c) mutually imply themself, and proposition 10.1 states the equivalence of (2a) to (3a) and (3b). Accordingly (2) and (3) are mutually equivalent.

Now, if (1) holds, and $\mathfrak{M}$ is an arbitrary set of f-ideals X of the ring A, then every X belongs to fA, since f is complete. The hereditary property of f and the inclusion $\sum X \leqq fA$ imply that $\sum X$ is an f-ideal. Therefore f is inductive on $\mathbf{K}$. Moreover, (2b) follows at once from (1).

Conversely, let us assume (2). If Y is an ideal of A contained in fA, then every principal ideal (y) of A with $y \in Y$ is by (2b) an f-ideal of A. Hence $Y = \sum_{y \in Y} (y)$ implies $fY = Y$ by (2a). Therefore, f is hereditary. If X is an f-ideal of A, then f being hereditary, the relation $f(x) = (x)$ holds for all $x \in X$. Another application of (2b) gives $X \subseteq fA$. □

PROPOSITION 10.3. For a preradical f on $\mathbf{K}$ the following conditions are equivalent:

(1) f is a quasiradical on $\mathbf{K}$.
(2) $(fA)\mu = f(A\mu)$ for every ring homomorphism μ satisfying the condition $\ker \mu \subseteq fA$.

PROOF. If (1) holds, and if $A \in \mathbf{K}$ is a ring, μ is a homomorphism of A onto $A\mu$ with $\ker \mu \subseteq fA$, then $(fA)\mu \subseteq f(A\mu)$ follows from the definition of the preradical. Likewise by this definition we have

$$(f(A/\ker \mu))/(fA/\ker \mu) \subseteq f((A/\ker \mu)/(fA/\ker \mu)).$$

We then obtain

$$(f(A/\ker \mu))/(fA/\ker \mu) = f(A/fA) = 0$$

because

$$(A/\ker \mu)/(fA/\ker \mu) \cong A/fA$$

and f is a quasiradical by (1). Thus (1) implies (2). The converse statement is trivial. □

PROPOSITION 10.4. A quasiradical on $\mathbf{K}$ is complete if and only if A/fA has no non-zero f-ideals for every ring A.

PROOF. If $A \in \mathbf{K}$ and if f is complete, then every f-ideal X of A/fA is contained in $f(A/fA)$. But, since f is a quasiradical, $f(A/fA) = 0$. Therefore $\overline{X} = (0)$ and thus $X \subseteq fA$.

Conversely, if the condition of the proposition holds, then the completeness of f can be shown as follows. If I is an f-ideal of A, we have

$$(I+fA)/fA \cong fI/(fI \cap fA) = I/(I \cap fA).$$

Since $I \cap fA$ is an ideal of A contained in I and $I=fI$, proposition 8.3 yields

$$f(I/(I \cap fA)) = fI/(I \cap fA).$$

Hence $I/(I \cap fA) = f(I/I \cap fA)$. Then the isomorphism

$$(I+fA)/fA \cong I/(I \cap fA)$$

guarantees that $(I+fA)/fA$ is an f-ideal of a/fA. Therefore we have $I \subseteq fA$. □

The following yields an inner characterization of the hereditary radical.

THEOREM 10.5. For a preradical on **K** the following conditions are equivalent.

(1) f is hereditary and A/fA has no non-zero f-ideals for every ring $A \in \mathbf{K}$.
(2) f is a hereditary and complete quasiradical.
(3) For every ring $A \in \mathbf{K}$ the following hold
  (a) if $B \supseteq C$ are ideals of A such that B/C is an f-ideal of A/C and C is an f-ideal of A, then B is also an f-ideal of A;
  (b) the union of every ascending chain of f-ideals of A is again an f-ideal;
  (c) every principal ideal of A contained in fA is an f-ideal;
  (d) in A/fA (0) is the unique f-ideal.
(4) f is inductive on **K** and for every ring $A \in \mathbf{K}$ the following hold
  (a) if μ is a homomorphism of A onto $A\mu$ with $\ker \mu \subseteq fA$, then $(fA)\mu = f(A\mu)$;
  (b) $fA = \{x;\, x \in A,\ (x)$ is an f-ideal of $A\}$.
(5) f is a hereditary radical property in A.

PROOF. Since together with A A/fA also belongs to **K** and f is a hereditary preradical on **K**, it follows that $f(A/fA)$ is an f-ideal in A/fA, whence (1) implies $f(A/fA) = (0)$. Therefore f is a quasiradical. Proposition 10.4 yields the completeness of f. On the other hand, since (1) follows at once from (2), (1) and (2) are equivalent.

Now, if (2) holds then so does condition (3c). Also statement (1) implies (3d). Since f is complete and hereditary it follows by the statement (2a) of proposition 10.2 that f is inductive on **K**. Hence proposition 10.1 yields (3b). Next let $B \geqq C$ be ideals of A such that $fC = C$ and $f(B/C) = B/C$. Then $C \subseteq fA$ and $B/C \subseteq f(A/C)$, since f is complete. Now proposition 10.3 implies

$$B/C \subseteq f(A/C) = fA/B.$$

Therefore $B \subseteq fA$. The hereditary property of f gives $fB=B$, which shows that (2) implies (3).

If (3) is fulfilled, and if B and C are two f-ideals of a ring $A \in \mathbf{K}$, then, f being a preradical, we have

$$(B+C)/C \cong B/(B \cap C) = fB/(fB \cap C) \subseteq f(B/(B \cap C)) = B/(B \cap C).$$

Since $B/(B \cap C) = f(B/(B \cap C))$ we have

$$f((B+C)/C) = (B+C)/C,$$

whence by (3a) $B+C$ is an f-ideal of A. Therefore f is additive on $\mathbf{K}$. This and (3b) imply by proposition 10.1 that f is inductive on $\mathbf{K}$. Now fA is an f-ideal because of (3c) and the relation $fA = \sum_{x \in fA} (x)$. Together with A every homomorphic image $A\mu$ of A also belongs to $\mathbf{K}$. Consequently $f(A/fA)$ is also an f-ideal of the ring A/fA, whence (3d) implies $f(A/fA)=(0)$. Thus f is a quasiradical, which is complete by proposition 10.4, whence condition (4b) also holds by (3c). Proposition 10.3 yields (4a), and thus (3) implies statement (4), whence we have condition (2) by proposition 10.2. But (5) is obviously equivalent to (1), (2), (3) or (4). □

PROPOSITION 10.6. If f a quasiradical with the domain $\mathbf{K}$, then for every ring $A \in \mathbf{K}$, fA is the intersection of all ideals M of A such that $f(A/M)=(0)$ (see theorem 1.12).

PROOF. If $\mathfrak{M}$ is the set of all ideals M of A such that $f(A/M)=(0)$ holds, then we have $fA \in \mathfrak{M}$, since f is a quasiradical. This yields $\bigcap_{M \in \mathfrak{M}} M \subseteq fA$. Next, if $fA \subsetneqq \bigcap_{M \in \mathfrak{M}} M$, then there exists an $M \subset \mathfrak{M}$ such that $fA \subsetneqq M$, and we have

$$(0) \neq (fA+M)/M \subseteq f(A/M) = (0),$$

since f is a preradical, and this contradiction proves proposition 10.6. □

As a consequence we have

PROPOSITION 10.7. If f is a hereditary and complete quasiradical on $\mathbf{K}$, then for every ring A, fA is the sum of all f-ideals of A and the intersection of all ideals M of A with $f(A/M)=(0)$, and thus fA is an f-ideal.

PROOF. As a consequence of the completeness and the hereditary property of f, we have by theorem 10.5 (4b) that fA is the sum of all f-ideals of A, and fA is an f-ideal. □

If f is a preradical, then the mapping $\mathbf{s}_f$, which orders to every ring $A \in \mathbf{K}$ the sum of all f-ideals, is a preradical on $\mathbf{K}$. Namely, every f-ideal of a ring A is mapped by every ring homomorphism μ onto an f-ideal of $A\mu$, since $I\mu = (fI)\mu \leqq f(I\mu) \subseteq$

$\subseteqq I\mu$ implies $I\mu = f(I\mu)$. The preradical $\mathbf{s}_f$ is in general not a quasiradical, as the following function f shows:

$$fA = \begin{cases} A & \text{if } A \text{ is nilpotent,} \\ (0) & \text{otherwise.} \end{cases}$$

But with the help of $\mathbf{s}_f$ a uniquely determined quasiradical f^* satisfying $\mathbf{s}_f A \subseteqq \subseteqq f^* A$ for every ring $A \in \mathbf{K}$ can be constructed by f and K. We define an ascending chain of ideals $M\beta$ of A as follows:

(1) $M_0 = (0)$,
(2) if $\beta = \alpha + 1$, then M_β is the unique ideal of A such that $\mathbf{s}_f(A/M_\alpha) = M_\beta/M_\alpha$,
(3) if β is a limit ordinal number, then $M_\beta = \bigcup_{\alpha<\beta} M_\alpha$.

Since A is a set, there exists a (least) ordinal number γ such that $M_{\gamma+1} = M_\gamma$, and then we put $f^*A = M_\gamma$.

For a detailed discussion of f^* we refer the reader to the paper [2] of Michler, mentioned above.

REMARK 10.8. As we have seen in § 10, every hereditary complete quasiradical is also a radical property. Let A be a ring, f a hereditary quasiradical on the class of all (associative) rings, and $\mathfrak{P}(A/f)$ the (eventually empty) set of all prime ideals P of A such that $f(A/P) = (0)$. For a subset of $\mathfrak{P}(A/f)$ we put

$$\Delta\mathfrak{A} = \begin{cases} \bigcap_{P\in\mathfrak{A}} P & \text{if } \mathfrak{A} \text{ is not empty,} \\ A & \text{otherwise.} \end{cases}$$

For the *closure* $\overline{\mathfrak{A}}$ of $\mathfrak{A}$ we take

$$\overline{\mathfrak{A}} = \{P;\ P\in\mathfrak{P}(A/f), P \supseteqq \Delta\mathfrak{A}\}.$$

The set $\mathfrak{A}$ is said to be *closed*, if $\overline{\mathfrak{A}} = \mathfrak{A}$. We have the following properties for $\overline{\mathfrak{A}}$:

(1) $\overline{\emptyset} = \emptyset$, where $\emptyset$ is the empty set.
(2) $\overline{\mathfrak{A}} \supseteqq \mathfrak{A}$.
(3) $\overline{\overline{\mathfrak{A}}} = \overline{\mathfrak{A}}$.
(4) $\overline{\mathfrak{A} \cup \mathfrak{L}} = \overline{\mathfrak{A}} \cup \overline{\mathfrak{L}}$.

Thus $\mathfrak{P}(A/f)$ is a topological space, namely the *Stone space* which always is a T_0-space. $\mathfrak{P}(A/f)$ has been called the f-structure space of the ring A by Michler [3]. This space is a generalization of the structure space in the sense of Jacobson [5]. A sufficient condition is given by Michler [3] for $\mathfrak{P}(A/f)$ to be a T_1-space. If M is an accessible subring of A and

$$\mathfrak{P}(M) = \{P;\ P\in\mathfrak{P}(A/f),\ P \supseteqq M\}$$

then the mapping

$$\mu: P \rightarrow P \cap M$$

is a homeomorphism of the subset $\mathfrak{P}(M)$ of $\mathfrak{P}(A/f)$ onto the f-structure space $\mathfrak{P}(M/f)$ of the ring M.

PROBLEM 1.† Discuss as a generalization of the radical rings the class of **T**-rings satisfying the following conditions:

(a) Every homomorphic image of a **T**-ring is again a **T**-ring.

(b) If every non-zero homomorphic image B of a ring A has a non-zero **T**-subring which is an accessible subring, then A itself is a **T**-ring.

PROBLEM 2. Substitute in problem 1 the requirement "accessible subring" by *"quasiideal"* (see Steinfeld [2]) or by "(m, n)-ideal" (see Lajos [1], Lajos and Szász [6]) for fixed m and n.

PROBLEM 3.† Discuss the radical properties for which the radical $\mathbf{R}(A)$ of an arbitrary ring A always contains every **R**-right ideal and every **R**-left ideal.

PROBLEM 4. Which radical properties **R** satisfy the condition that the sum of two arbitrary **R**-right ideals in an arbitrary ring is again an **R**-right ideal?

PROBLEM 5. Which radical properties **R** satisfy the condition that the union of an arbitrary chain of **R**-right ideals an **R**-right ideal?

PROBLEM 6. Which radical properties have the property that $\mathbf{R}(A)$ is invariant in every ring with respect to all endomorphisms of the additive group $(A, +)$?

PROBLEM 7. Discuss the question analogous to problem 6, now with respect to the multiplicative semigroup $(A, \cdot)$ of the ring, and to the *adjoint semigroup* $(A, \circ)$ of the ring, where $x \circ y = x + y - xy$.

PROBLEM 8. For which classes **K** of rings is the following condition satisfied: If a ring A is an **R**-radical ring for the lower radical property, determined by the class **K**, then A is a *transfree image* (Szász and Wiegandt [1]) of the free product of certain rings in **K**.

PROBLEM 9. Investigate the rings, in which the property to be an ideal is transitive. That is, an ideal of an ideal must always be an ideal of the ring. (This is important for the transfree images.)

PROBLEM 10. Discuss the rings whose lattice of ideals is *cocompactly generated,* i.e. in which every ideal is an intersection of cocompact ideals K of A. (An ideal K is *cocompact*, if $\bigcap_{\gamma \in \Gamma} L_\gamma \leqq K$ for the ideals L_γ of A implies the existence a finite number of ideals $L_{\gamma'}$, $(\gamma' \in \Gamma' \subset \Gamma)$ with $\bigcap_{\gamma' \in \Gamma'} L_{\gamma'} \leqq K$. This is important again for the transfree images. If every ideal is cocompact, then the minimum condition holds for ideals.)

PROBLEM 11. (Anderson, Divinsky and Suliński). Does there exist, for every natural number n, a class $\mathbf{K}_n$ of rings with the following property: If $\mathbf{R}_n$ is the lower radical property determined by the class $\mathbf{K}_n$, then every $\mathbf{R}_n$-radical ring has exactly degree n over $\mathbf{K}_n$?

PROBLEM 12. How can a necessary and sufficient condition be formulated for a radical property $\mathbf{R}$ in order that $\mathbf{R}(I_1)+\mathbf{R}(I_2)=\mathbf{R}(I_1+I_2)$ should always be valid for arbitrary ideals I_1 and I_2 of an arbitrary ring?

PROBLEM 13. For a radical property $\mathbf{R}$ and a ring A let $\hat{I}$ denote the ideal of A, uniquely determined by $\hat{I}/I=\mathbf{R}(A/I)$. How can a necessary and sufficient condition be formulated for a radical property in order that $\widehat{(I_1\cap I_2)}=\hat{I}_1\cap\hat{I}_2$ should always be valid for arbitrary ideals I_1 and I_2 of an arbitrary ring?

PROBLEM 14. Discuss (in connection with problems 12 and 13) the questions:

(a) For which radical properties $\mathbf{R}(I_1\cap I_2)=\mathbf{R}(I_1)\cap\mathbf{R}(I_2)$ (also) holds for arbitrary ideals I_1 and I_2 of an arbitrary ring?

(b) For which radical properties $\mathbf{R}$ $\widehat{(I_1+I_2)}=\hat{I}_1+\hat{I}_2$ (also) holds for arbitrary ideals I_1 and I_2 of an arbitrary ring?

PROBLEM 15. Let $\mathbf{K}$ be a class of rings such that every non-zero ideal of an arbitrary ring from $\mathbf{K}$ can be homomorphically mapped onto a non-zero ring from $\mathbf{K}$, and let $\mathbf{R}$ be the upper radical property determined by the class $\mathbf{K}$. How can a necessary and sufficient condition be formulated in order that every $\mathbf{R}$-semisimple ring A should be a subdirect sum of rings $A_\alpha\in\mathbf{K}$ such that A also contains the discrete direct sum $\sum A_\alpha$?

CHAPTER II

THEORY OF THE SUPERNILPOTENT AND SPECIAL RADICALS

In this chapter the definition and the most important properties of the supernilpotent and special radicals will be discussed. Obviously, there exist radicals which are not supernilpotent. Every special radical is supernilpotent, and by a beautiful non-trivial example of Ryabukhin there exist supernilpotent but non-special radicals. It must be mentioned that the nil radicals, the Jacobson radical and the Brown–McCoy radical, to be discussed later, are special radicals. We give criteria for a radical to be supernilpotent, and special, respectively. The special classes of subdirectly irreducible rings with idempotent heart will be discussed in detail. Supplementing radicals, a dual pair of radicals, the antisimple radical and the subidempotent radical will also be treated in this chapter. We discuss in detail the connection between supernilpotent radicals and direct decompositions of a ring with minimum condition on principal right ideals and between special radicals and special modules, respectively. Finally some open problems will be mentioned. We remark here that, unlike in Chapter I, in what follows we do not distinguish between the notions of radical property and radical.

§ 11. SUPERNILPOTENT AND SPECIAL RADICALS

For the subject-matter of this section and the rest of this chapter, we principally refer the reader to Andrunakievich [8, 11, 12, 13] and Ryabukhin [3].

Let **R** be a radical property in the sense of Amitsur and Kurosh. Then **R** is called *supernilpotent* if the following conditions are fulfilled:

(1) **R** is *hereditary,* i.e. every ideal of an **R**-ring is again an **R**-ring.
(2) Every nilpotent ring is an **R**-ring.

If **R** is the lower radical property determined by the class of all zero-rings, then, as we shall show later, both (1) and (2) are satisfied, and therefore this **R** yields an example for a supernilpotent radical.

On the other hand, if **R** denotes the property of being *idempotent,* i.e. $A^2=A$ for the ring A, then **R** is a radical property for which neither (1) nor (2) is satisfied. For example, the ring I of the rational integers is idempotent, but no proper ideal of I is idempotent. Moreover, nilpotent rings are semisimple for this idempotent radical property, since no non-zero ideal can be at the same time nilpotent and idempotent, and nilpotence is hereditary.

PROPOSITION 11.1. A radical property $\mathbf{R}$ is hereditary if and only if for every ring A and every ideal B of A the equality $\mathbf{R}(B)=B\cap\mathbf{R}(A)$ holds.

PROOF. If $\mathbf{R}$ is hereditary, then $D=B\cap R(A)$ is an ideal of A contained in $\mathbf{R}(A)$. It is thus an ideal of $\mathbf{R}(A)$. Hence by the hereditary property of $\mathbf{R}$ it follows that D is an $\mathbf{R}$-ring. Since D is an $\mathbf{R}$-ideal of B, we have $D\leqq\mathbf{R}(B)$. However by theorem 1.7, $\mathbf{R}(B)$ is an ideal of the ring A contained in B. This yields $\mathbf{R}(B)\leqq \leqq B\cap\mathbf{R}(A)=D$, and so $D=\mathbf{R}(B)=B\cap\mathbf{R}(A)$.

Conversely, if the equality $\mathbf{R}(B)=B\cap\mathbf{R}(A)$ holds for a radical property $\mathbf{R}$ and for every ideal B of every ring A, then taking $\mathbf{R}(A)=A$ we obviously have $\mathbf{R}(B)=B\cap\mathbf{R}(A)=B\cap A=B$, implying that $\mathbf{R}$ is hereditary. □

If a class $\mathbf{M}$ of rings is hereditary (i.e. every ideal of every ring from $\mathbf{M}$ again belongs to $\mathbf{M}$), then by Armendariz and Leavitt [2] the lower radical class $L\mathbf{M}$ is also hereditary.

Next we give some characterizations of supernilpotent radicals. The first characterization is due to Andrunakievich [13], the second to Ryabukhin [3]. We need some preliminaries before discussing these characterizations.

PROPOSITION 11.2. If $\mathbf{R}(A)$ is a radical containing every nilpotent ideal of the ring $\mathbf{R}$ then $B^n\leqq\mathbf{R}(A)$ implies $B\leqq\mathbf{R}(A)$ where B is an ideal and n a natural number.

PROOF. By the natural homomorphism $x\to\bar{x}+\mathbf{R}(A)$ we have $\bar{B}=(B+ +\mathbf{R}(A))/\mathbf{R}(A)$, and thus $\bar{B}^n=(0)$ since $B^n\leqq\mathbf{R}(A)$. But $A/\mathbf{R}(A)$ is $\mathbf{R}$-semisimple, and so our condition on $\mathbf{R}(A)$ implies $\bar{B}=(0)$ and thus $B\leqq\mathbf{R}(A)$. □

For an ideal B of A we define

$$B^{-1}\cdot\mathbf{R}(B)=\{x;\ x\in A,\ Bx\leqq\mathbf{R}(B)\}$$
$$\mathbf{R}(B)\cdot B^{-1}=\{y;\ y\in A,\ yB\leqq\mathbf{R}(B)\}$$
$$\mathbf{R}(B):B\ =B^{-1}\mathbf{R}(B)\cap\mathbf{R}(B)B^{-1}.$$

Thus

$$\mathbf{R}(B):B\ =\{z;\ z\in A,\ Bz+zB\leqq\mathbf{R}(B)\}.$$

Hence we have the following

PROPOSITION 11.3. If $\mathbf{S}(A)$ is a supernilpotent radical in the ring A, we have

$$B^{-1}\mathbf{S}(B)=\mathbf{S}(B):B=\mathbf{S}(B)\cdot B^{-1}$$

for every ideal B of A.

PROOF. Theorem 1.7 and condition $\mathbf{S}(B)=B\cap\mathbf{S}(A)$ imply that $\mathbf{S}(B)$ is an ideal of A, and thus $B^{-1}\mathbf{S}(B)$ and $\mathbf{S}(B)B^{-1}$ are ideals of A. Moreover,

$$(B^{-1}\mathbf{S}(B)B)^2 = B^{-1}\mathbf{S}(B)B\cdot B^{-1}\mathbf{S}(B)B \leqq BB^{-1}\mathbf{S}(B)B \leqq \mathbf{S}(B)$$

and thus, by proposition 11.2, $B^{-1}\mathbf{S}(B)B\leqq\mathbf{S}(B)$. Hence $B^{-1}\mathbf{S}(B)\leqq\mathbf{S}(B)B^{-1}$. Similarly $\mathbf{S}(B)B^{-1}\leqq B^{-1}\mathbf{S}(B)$. Thus $\mathbf{S}(B):B=B^{-1}\mathbf{S}(B)=\mathbf{S}(B)B^{-1}$. □

Now the following fundamental theorem can be proved:

THEOREM 11.4 (Andrunakievich [13]). An Amitsur–Kurosh radical $\mathbf{S}$ is supernilpotent if and only if the relation

$$\mathbf{S}(A) = \varphi^{-1}(\mathbf{S}(A/B))\cap(\mathbf{S}(B):B) \qquad (*)$$

holds for every ring A and for every ideal B of A, where φ denotes the natural homomorphism $x\to x+B$ of A onto A/B.

PROOF. Suppose $(*)$ holds for every ring A and for every ideal B of A. We shall show that $\mathbf{S}$ is supernilpotent. If $B=A$ we deduce from $(*)$

$$\mathbf{S}(A) = \mathbf{S}(A):A, \qquad (**)$$

since $\varphi^{-1}(0)=A$. If A is the zero-ring, $A^2=(0)$ so we have the relation $\mathbf{S}(A):A=A$. Thus $\mathbf{S}(A)=A$, by $(**)$. Therefore, every zero-ring is an $\mathbf{S}$-radical ring. If A is a nilpotent ring, then the $\mathbf{S}$-semisimple factor ring $\bar{A}=A/\mathbf{S}(A)$ is again nilpotent. Let $\bar{A}^n=(0)$, $\bar{A}^{n-1}\neq(0)$. Then $\bar{A}^{n-1}$ is a zero-ring. Therefore it is an $\mathbf{S}$-ring, and also an ideal of $\bar{A}$. If $\bar{A}\neq 0$ the above contradicts the semisimplicity of $\bar{A}$. Therefore $\bar{A}=(0)$ and thus $A=\mathbf{S}(A)$ is an $\mathbf{S}$-ring. Therefore every nilpotent ring is an $\mathbf{S}$-radical ring. Next we shall show that $\mathbf{S}$ is hereditary. Put $\mathbf{S}(A)=A$ and let B be an ideal of A. Then we have $\mathbf{S}(A/B)=A/B$. $(*)$ gives the relation

$$A = A\cap[\mathbf{S}(B):B] = \mathbf{S}(B):B.$$

Because $AB+BA\leqq\mathbf{S}(B)$ we have $B^2\leqq\mathbf{S}(B)$. Therefore by proposition 11.2 we have $B\leqq\mathbf{S}(B)$. Thus $B=\mathbf{S}(B)$. Consequently $\mathbf{S}$ is hereditary and thus supernilpotent.

Conversely, let $\mathbf{S}$ be now a supernilpotent radical. We shall show, that $(*)$ holds for $\mathbf{S}$. Let us denote by A_0 the right-hand side of $(*)$. This is a subring of A. First we shall show that $\mathbf{S}(A)\leqq A_0$. Note that $\varphi(\mathbf{S}(A))\leqq\mathbf{S}(A/B)$ yields $\mathbf{S}(A)\leqq$ $\leqq\varphi^{-1}\mathbf{S}(A/B)$. By proposition 11.3 we have $\mathbf{S}(A)\leqq\mathbf{S}(B):B$. Thus

$$\mathbf{S}(A) \subseteqq A_0 = \varphi^{-1}(\mathbf{S}(A/B))\cap(\mathbf{S}(B):B).$$

Next shall show that $A_0\cap B=\mathbf{S}(B)$. We have

$$A_0\cap B \supseteqq \mathbf{S}(A)\cap B = \mathbf{S}(B),$$

and, on the other hand, the definition of A_0 yields

$$B(A_0 \cap B) \subseteq B((\mathbf{S}(B):B) \cap B) = \mathbf{S}(B).$$

So $(A_0 \cap B)^2 \subseteq \mathbf{S}(B)$, whence $A_0 \cap B \subseteq \mathbf{S}(B)$ by proposition 11.2. But since $A_0 \cap B \supseteq \mathbf{S}(B)$ in fact $\mathbf{S}(B) = A_0 \cap B$. Since A_0 is an ideal of A, $\mathbf{S}(A) \subseteq A_0$ and $\mathbf{S}(A_0) = A_0 \cap \mathbf{S}(A)$ obviously imply $\mathbf{S}(A) = \mathbf{S}(A_0)$. Because $\mathbf{S}(B) = A_0 \cap B$ and $\mathbf{S}(A) = \mathbf{S}(A_0)$ we have

$$A_0/\mathbf{S}(A_0) \cong (A_0/(A_0 \cap B))/(\mathbf{S}(A)/\mathbf{S}(B)). \qquad (***)$$

The left-hand side of $(***)$ is an $\mathbf{S}$-semisimple ring. Since $A_0/(A_0 \cap B) \cong (A_0 + B)/B = \varphi(A_0) \subseteq \mathbf{S}(A/B)$ the right-hand side of $(***)$ is by the hereditary property of $\mathbf{S}$, a homomorphic image of an $\mathbf{S}$-ring and therefore an $\mathbf{S}$-radical ring. But this is possible only if $A_0 = \mathbf{S}(A_0)$. Since $\mathbf{S}(A) = \mathbf{S}(A_0)$ we have $A_0 = \mathbf{S}(A)$. Therefore the ring on the right-hand side of $(*)$ coincides with $\mathbf{S}(A)$, which shows the validity of $(*)$ for supernilpotent radicals, and this completes the proof of theorem 11.4. □

In Chapter IV the Jacobson radical $\mathbf{J}$ will be discussed in detail. As is well known, this radical is supernilpotent, and in every ring A the radical $\mathbf{J}(A)$ coincides with the intersection of some prime ideals of A. It should be remarked that the relation $(*)$ has been already proved by Szendrei [1] for the Jacobson radical, as for a well-defined concrete radical. Moreover, Steinfeld [1] has shown a generalization of the relation $(*)$ in some lattice ordered semigroups for those special radical elements which are the intersections of prime absorbents. (These latter correspond to the prime ideals in rings.)

We shall now give another characterization of supernilpotent radicals. For this we define a weakly special class of rings (cf. Ryabukhin [3]).

A class $\mathbf{K}$ of rings is called *weakly special* if the following conditions are fulfilled for $\mathbf{K}$:

(1) Rings contained in $\mathbf{K}$ have no non-zero nilpotent ideals.
(2) Every ideal of a ring in $\mathbf{K}$ belongs to $\mathbf{K}$.
(3) If an ideal B of a ring A with $B^* = (0)$ is contained in $\mathbf{K}$, where $B^* = (0)B^{-1} \cap B^{-1}(0)$, then A also belongs to $\mathbf{K}$.

We have the following

THEOREM 11.5 (Ryabukhin [3]). A radical $\mathbf{S}$ is supernilpotent if and only if $\mathbf{S}$ coincides with the upper radical determined by a weakly special class $\mathbf{K}$. Then in every ring A, $\mathbf{S}(A)$ is the intersection of all those ideals T_α of A for which A/T_α is contained in $\mathbf{K}$, and thus every $\mathbf{S}$-semisimple ring is a subdirect sum of rings from $\mathbf{K}$.

PROOF. Let **S** be a supernilpotent radical and **K** the class of all **S**-semisimple rings. Then **S** obviously coincides with the upper radical determined by **K**. We shall show that **K** is a weakly special class.

Since **K** consists of **S**-semisimple rings and **S** is a supernilpotent radical, (1) holds for K.

Furthermore by theorem 1.8 **K** satisfies condition (2) too.

Let $\mathbf{S}(A)$ be the radical of a ring A containing the **S**-semisimple ideal B with $B^*=(0)$. Then we have $\mathbf{S}(A)\cap B=\mathbf{S}(B)=(0)$ and thus $\mathbf{S}(A)\cdot B=B\cdot\mathbf{S}(A)=(0)$. Now $\mathbf{S}(A)\leqq B^*=(0)$ yields $\mathbf{S}(A)=(0)$, $A\in\mathbf{K}$, and hence (3) also holds for **K**.

Therefore the class **K** of all **S**-semisimple rings is weakly special for every supernilpotent radical **S**.

Conversely, let a weakly special class **K** be given. We define **S** as the upper radical property determined by the class **K**. Then **S** is the class of all rings which cannot be homomorphically mapped onto a non-zero ring from **K**. Then **S** is, in fact, a radical property by condition (2) for **K**.

We shall now show that **S** is supernilpotent. First we prove that **S** is hereditary.

Let A be a ring, B an ideal of A and C an ideal of B such that $B/C\in\mathbf{K}$. We have

$$(AC)^2 \leqq (ACA)\,C \subseteqq BC \leqq C.$$

By condition (1) for the weakly special class **K** this implies $AC\leqq C$. Similarly $CA\leqq C$. Consequently C is an ideal of A and thus

$$T = C{:}B = \{x;\ x\in A,\ xB+Bx \subseteqq C\}$$

is an ideal of A. Then $C\leqq T$ and in the factor ring $\bar{A}=A/C$, T/C coincides with the annihilator $\bar{B}^*$ of $\bar{B}=B/C$. Therefore we have

$$A/T \cong \bar{A}/\bar{B}^*.$$

Then $\bar{A}/\bar{B}^*$ contains the ideal

$$(\bar{B}+\bar{B}^*)/B^* \cong \bar{B}\in\mathbf{K}$$

with $\bar{B}\cap\bar{B}^*=(0)$, and thus $\bar{B}^*=(0)$. Condition (3) for **K** yields $\bar{A}\in\mathbf{K}$. Consequently, A is not an **S**-ring, if an ideal B of A is not an **S**-ring. But this means precisely that **S** is hereditary.

Since, by condition (1) for **K**, a nilpotent ring cannot be homomorphically mapped onto a non-zero ring from **K**, every nilpotent ring is an **S**-radical ring. Thus **S** is supernilpotent.

Now let $\{T_\alpha\}$ be the set of all ideals T_α of a ring A. for which $A_\alpha=A/T_\alpha\in\mathbf{K}$. Then we have $\mathbf{S}(A)\leqq B=\bigcap T_\alpha$. Let us suppose $\mathbf{S}(A)\neq B$. In this case we have $\mathbf{S}(B)=B\cap\mathbf{S}(A)=\mathbf{S}(A)\neq B$. Thus B is not an **S**-ring. Consequently there exists an ideal C of B with $(0)\neq B/C\in\mathbf{K}$. By an argument similar to the preceding C must be an ideal of A. We put $T=C{:}B=\{x;\, x\in A,\, xB+Bx\leqq C\}$. Then T is an ideal of A with $A/T \in \mathbf{K}$. Consequently $T \in \{T_\alpha\}$. Since $B/C \in \mathbf{K}$ we have $T \cap B = (C:B) \cap B = C$, and thus

$$B = \cap T_\alpha = B \cap T = B \cap (C:B) = C \subset B \quad \text{a contradiction.}$$

Hence, $B = \mathbf{S}(A)$. □

We shall now define the special radicals. For this we need the notions of prime ideal and prime ring.

An ideal P of a ring A is called a *prime ideal* of A if $BC \leqq P$ implies $B \leqq P$ or $C \leqq P$ for every pair of ideals B, C of A. A ring A is said to be a *prime ring* if the zero-ideal (0) is a prime ideal in A. Obviously, a ring A is just a prime ring if and only if the product of two arbitrary non-zero ideals B and C is non-zero. Therefore, in a prime ring only the zero-ideal (0) is nilpotent. A ring without divisors of zero is a prime ring, but the converse statement in general does not hold.

A class $\mathbf{K}$ of rings is called a *special class*, if $\mathbf{K}$ fulfils the following conditions:

(1) every ring A in $\mathbf{K}$ is a prime ring,
(2) every ideal of a ring in $\mathbf{K}$ is again a ring in $\mathbf{K}$,
(3) if the ideal I of a ring A is contained in $\mathbf{K}$, and if $I^* = \{x;\, x \in A,\, xI = Ix = (0)\}$ is the annihilator of I in A, then A/I^* is a ring in $\mathbf{K}$.

We have the following

PROPOSITION 11.6. Every special class $\mathbf{K}$ of rings is weakly special.

PROOF. Since a prime ring has no non-zero nilpotent ideals, condition (1) of a weakly special class is satisfied for $\mathbf{K}$. Condition (2) of a weakly special class holds trivially for $\mathbf{K}$.

Now, if I is an ideal of A contained in $\mathbf{K}$, and I^* is its annihilator in A, then we have $A/I^* \in \mathbf{K}$ by condition (3) for a special class $\mathbf{K}$. Therefore $A \in \mathbf{K}$ if $I^* = (0)$. Thus condition (3) of a weakly special class follows. □

Every upper radical determined by a special class is called a *special radical*. As an important consequence of proposition 11.6, we have

THEOREM 11.7. Every special radical is supernilpotent.

PROOF. By proposition 11.6 every special class is weakly special, and by theorem 11.5 a radical is supernilpotent if and only if it is an upper radical determined by a weakly special class. □

THEOREM 11.8. Every special radical $\mathbf{S}(A)$ of a ring A is the intersection of all prime ideals P_α of A such that A/P_α belongs to the special class $\mathbf{K}$ determining $\mathbf{S}$.

PROOF follows from the second part of theorem 11.5. □

We have already seen that every special radical is supernilpotent. Next we investigate further the connection between supernilpotent and special radicals.

THEOREM 11.9. Every supernilpotent radical **S** can be extended to a special radical **S**′, i.e. $\mathbf{S} \leqq \mathbf{S}'$, where **S**′ is the upper radical determined by the class **K** of all **S**-semisimple prime rings, and **S**′ is the least special radical with $\mathbf{S}' \geqq \mathbf{S}$.

PROOF. First we shall show that the class **K** of all **S**-semisimple prime rings is a special class for every supernilpotent radical **S**.

By definition every ring in **K** is a prime ring. Moreover, every ideal of a ring in **K** is **S**-semisimple by theorem 1.8.

Now in general, every ideal B of a prime ring A is again a prime ring. For, if C and D are non-zero ideals of the ring B with $CD=(0)$, then a contradiction can be derived as follows. $BC \neq (0)$, since B is an ideal of A and the right annihilator of B in A is an ideal of the prime ring A. Similarly we have $XB \neq (0)$ for every non-zero subset X of A. Thus $BCB \neq (0)$ since $BC \neq 0$. But $BCB \leqq C$, since C is an ideal of B. Consequently we have $BCB \cdot D \leqq C \cdot D = (0)$. Since BCB is a non-zero ideal of A, and A is a prime ring, $(BCB) \cdot D = (0)$ is impossible for $D \neq 0$. Thus B is also a prime ring.

Hence every ideal of every ring in **K** belongs again to **K**. For, by theorem 1.8 every ideal of an **S**-semisimple ring is **S**-semisimple.

Now, let B be an **S**-semisimple prime ring, B an ideal of the ring A and B^* the annihilator of B in A.

We shall show A/B^* is a prime ring. For this it is sufficient to check that B^* is a prime ideal in A. If C and D are ideals of A with $CD \leqq B^*$, then we have $BC \cdot DB \leqq B \cdot B^*B = (0)$. But since BC and DB are ideals in the prime ring B, we have $BC=(0)$ or $DB=(0)$. If $BC=(0)$ then $(CB)^2 = C(BC)B = 0$. Therefore $CB=(0)$ and $C \leqq B^*$, since the prime ring B has no non-zero nilpotent ideal. If $DB=(0)$, we similarly have $D \leqq B^*$. Thus A/B^* is a prime ring.

It still remains to prove that A/B^* is an **S**-semisimple ring. $(B+B^*)/B^*$ is an ideal of the factor ring A/B^*. Now

$$(B+B^*)/B^* \cong B/(B \cap B^*) \cong B$$

since, B being a prime ring, $(B \cap B^*)^2 = (0)$ implies $B \cap B^* = (0)$. Hence, $(B+B^*)/B^*$ is a non-zero **S**-semisimple ideal of A/B^*. But, since **S** is hereditary,

$$\mathbf{S}((B+B^*)/B^*) = \mathbf{S}(A/B^*) \cap (B+B^*)/B^* = (\bar{0}).$$

This implies $\mathbf{S}(A/B^*) \cdot ((B+B^*)/B^*) = (\bar{0})$. Therefore $\mathbf{S}(A/B^*) = (0)$, since by the preceding A/B^* is a prime ring and $(B+B^*)/B^* \neq (\bar{0})$. Thus, we have proved that **K** is a special class of rings. **K** is a subclass of the class **K**′ of all **S**-semisimple rings, and $\mathbf{S}$ is just the upper radical determined by the class $\mathbf{K}'$. Since the special radical $\mathbf{S}'$ is the upper radical determined by $\mathbf{K}$, and $\mathbf{K}' \geqq \mathbf{K}$, we have $\mathbf{S} \leqq \mathbf{S}'$.

Finally, if $\mathbf{T}$ is a special radical with $\mathbf{T} \geqq \mathbf{S}$, then $\mathbf{T}$ is an upper radical determined by a special class $\mathbf{K}_0$. Since $\mathbf{T} \geqq \mathbf{S}$ every ring in $\mathbf{K}_0$ is also $\mathbf{S}$-semisimple. Thus $\mathbf{K}_0 \leqq \mathbf{K}$ since $\mathbf{K}_0$ and $\mathbf{K}$ consist of prime rings. But then $\mathbf{K}_0 \leqq \mathbf{K}$ obviously implies $\mathbf{T} \geqq \mathbf{S}' \geqq \geqq \mathbf{S}$. □

PROPOSITION 11.10. A supernilpotent radical $\mathbf{S}$ is special if and only if $\mathbf{S} = \mathbf{S}'$, where $\mathbf{S}'$ is the upper radical determined by the class of all $\mathbf{S}$-semisimple prime rings.

PROOF. $\mathbf{S}$ is special if $\mathbf{S} = \mathbf{S}'$. Conversely, if $\mathbf{S}$ is special, then, as before $\mathbf{S} \geqq \mathbf{S}' \geqq \mathbf{S}$ which implies $\mathbf{S}' = \mathbf{S}$. □

PROPOSITION 11.11. If $\mathbf{S}$ is a supernilpotent radical, and B is a non-zero ideal of the $\mathbf{S}$-semisimple ring A with maximum condition for the ideals, contained in B, then there exists an $\mathbf{S}$-semisimple prime ring C such that C is an ideal of A, contained in B.

PROOF. If B is a prime ring, then there is nothing to prove. If B is not a prime ring, then there exist non-zero ideals B_1 and B_2 with $B_1 \cdot B_2 = (0)$ in the ring B. If $\bar{B}_i$ denotes the ideal generated by B_i in A, then we have $(\bar{B}_i)^3 \neq (0)$, $(\bar{B}_i)^3 \leqq B_i$ and thus $\bar{B}_1^3 \cdot \bar{B}_2^3 = (0)$.

If $(\bar{B}_1)^3$ is not a prime ring, then – as above – A has two ideals B_{111} and B_{112}, contained in $\bar{B}_1^3 = B_{11}$, with $B_{111} \cdot B_{112} = (0)$ and $B_{11i} \neq 0$. This procedure gives us a descending chain of ideals of the ring A:

$$B_1 \supset B_{11} \supset B_{111} \supset \dots . \tag{$*$}$$

If I^* denotes the annihilator of I in B, then $I_1 \subset I_2$ obviously implies $I_2^* \leqq I_1^*$. In particular, $B_{11} \supset B_{111}$ gives rise to the strict inclusion $B_{111}^* \supset B_{11}^*$. For, $B_{111}^* = B_{11}^*$ implies $B_{112} \leqq B_{111}^* = B_{11}^*$. But $B_{112} \leqq B_{11}$. Consequently $B_{112}^2 \leqq B_{11}^* B_{11} = (0)$ which is impossible by the $\mathbf{S}$-semisimplicity of B. Consequently, the chain

$$B_1^* \subset B_{11}^* \subset B_{111}^*$$

is strictly ascending. By the maximum condition on ideals contained in B, this chain terminates in a finite number of steps and thus an ideal $B_{11\dots j}$ in ($*$) will be an $\mathbf{S}$-semisimple prime ring. □

THEOREM 11.12. Every supernilpotent radical $\mathbf{S}$ coincides, on every ring A with maximum condition on ideals, with the least special radical $\mathbf{S}'$, for which $\mathbf{S}' \geqq \mathbf{S}$.

PROOF. Let us assume $\mathbf{S}' \neq \mathbf{S}$. If A is a ring with maximum condition on ideals, then we have $\mathbf{S}'(A) \supseteq \mathbf{S}(A)$. Then $B = A/S(A)$ is $\mathbf{S}$-semisimple, and also satisfies the maximum condition on ideals. Then $\mathbf{S}'(A)$ contains an ideal C of B, which is a prime ring by proposition 11.11. Since $\mathbf{S}'$ is hereditary, C is also an $\mathbf{S}'$-ring, and thus C is not an $\mathbf{S}$-semisimple ring by definition of $\mathbf{S}$. On the other hand, C is an $\mathbf{S}$-semisimple ring by theorem 1.8. From this contradiction we have $\mathbf{S}'(A) = \mathbf{S}(A)$. □

Similarly we can prove:

THEOREM 11.13. If $\mathbf{K}$ is the class of all rings such that every homomorphic image contains a non-zero minimal ideal, then on $\mathbf{K}$ every supernilpotent radical $\mathbf{S}$ coincides with the least special radical $\mathbf{S}'$, for which $\mathbf{S}' \supseteq \mathbf{S}$.

PROOF. If $B/\mathbf{S}(A)$ is a minimal ideal of $A/\mathbf{S}(A)$, then $B/\mathbf{S}(A)$ is idempotent since $\mathbf{S}(A)$ is supernilpotent, $B/\mathbf{S}(A)$ is an $\mathbf{S}'$-ideal if $\mathbf{S}'(A) \neq \mathbf{S}(A)$, since $\mathbf{S}'$ is also hereditary. But $B/\mathbf{S}(A)$ is an $\mathbf{S}$-semisimple prime ring, and thus it cannot be an $\mathbf{S}$-ring. Therefore, $\mathbf{S}(A) = \mathbf{S}'(A)$. □

We also have the following as a consequence of theorem 11.5 and proposition 11.10:

PROPOSITION 11.14. The supernilpotent radical $\mathbf{S}$ determined by a weakly special class $\mathbf{K}$ is special if and only if every $\mathbf{S}$-semisimple ring can be homomorphically mapped onto an $\mathbf{S}$-semisimple prime ring.

Next we shall give a supernilpotent radical $\mathbf{S}$ which is not special (cf. Ryabukhin [3]).

As is well known, a ring A is called a *Boolean ring*, if $a^2 = a$ for every element $a \in A$ (see p. 48). Every Boolean ring is obviously commutative, since we have $(2a)^2 = 4a^2 = 4a = 2a$ giving $2a = 0$, and thus $a = -a$ for every $a \in A$. On the other hand we have $ba + ab = 0$ because $(a+b)^2 = a+b$, and $a^2 = a$, $b^2 = b$. Therefore $ba = -ab = ab$. Every homomorphic image and every subring of a Boolean ring is again a Boolean ring. Since in a Boolean ring $ab(a-b) = 0$ holds identically, the prime ring $\{a\}$ with two elements is the only Boolean ring without divisors of zero. Moreover, a commutative prime ring cannot have divisors of zero. For if $a \neq 0$, $b \neq 0$ and $ab = 0$ we have $(a) \cdot (b) = (0)$ for the principal ideals (a) and (b) which is impossible in a prime ring. Therefore, by these preliminaries, the prime field $\{a\}$ of two elements is the only Boolean prime ring.

We now show the existence of a non-zero Boolean ring which does not contain an ideal, which is a prime field of two elements.

EXAMPLE 11.15 (Ryabukhin [3]). Let $\{e_\alpha\}$ be a set of symbols, where the set $\{\alpha\}$ of indices runs over all rational numbers with the natural ordering relation.

Then this set will be a commutative multiplicative semigroup H under the operation $e_\alpha \cdot e_\beta = e_\gamma$, where $\gamma = \max(\alpha, \beta)$. Every element e_α of H is idempotent. Let U be the semigroup algebra of H over the prime field with two elements. Then U is a (commutative)) Boolean ring, since every element u of U has the form

$$u = e_{\alpha_1} + e_{\alpha_2} + \ldots + e_{\alpha_n} \quad (\alpha_1 < \alpha_2 < \ldots < \alpha_n),$$

and a multiplication yields

$$u^2 = e_{\alpha_1} + 3e_{\alpha_2} + 5e_{\alpha_3} + \ldots + (2n-1)\, e_{\alpha_n},$$

and so because $2e_{\alpha_i} = 0$ we have $u^2 = u$.

Moreover, let I be a non-zero ideal of U and $0 \neq i \in I$ with

$$i = e_{\beta_1} + e_{\beta_2} + \ldots + e_{\beta_n} \quad (\beta_1 < \beta_2 < \ldots < \beta_n).$$

If the number n of the non-zero components of i is odd, then we have for every $\gamma \geqq \beta_n$ the relation

$$i \cdot e\gamma = ne_\gamma = e\gamma \in I,$$

and then the ideal I which has been chosen arbitrarily is infinite. But, if n is an even number, then we have for infinitely many γ with $\beta_{n-1} \leqq \gamma < \beta_n$ the relation

$$ie_\gamma = (n-2)\, e_\gamma + e_\gamma + e_{\beta_n} = e_\gamma + e_{\beta_n} \in I,$$

and thus I is again infinite. Since every non-zero ideal I of U is infinite, no ideal of U is a prime field with two elements.

Now we have the following important

THEOREM 11.16 (Ryabukhin [3]). Let **K** be the class of all Boolean rings which do not contain an ideal, which is a prime field with two elements. Then **K** is a weakly special class of rings. The upper radical **S** determined by **K** is supernilpotent but not special.

PROOF. Because of example 11.15, the class **K** is not empty. No ring in **K** has non-zero nilpotent ideal, and thus condition (1) of a weakly special class holds for **K**. Condition (2) is obviously fulfilled. Now we verify condition (3). Let B an ideal of a ring A, where $B \in \mathbf{K}$, and the annihilator B^* of B in A equals zero. If a prime field $\{a\}$ with two elements is an ideal of A, then $\{a\}$ is an ideal direct summand of A. Since $\{a\}^2 = \{a\}$ and $B^* = 0$ we have $\{a\} \cap B = \{a\}$, a contradiction. Consequently, A has no ideal which is a prime field with two elements. Next let a be an arbitrary element of A and b an arbitrary element of B. But $ab \in B$, so we have $(ab)^2 = ab$ and by the commutativity of B we also have $b(ab) = (ab) \cdot b = ab$. Therefore, $a^2 - a \in B^*$ since

$$0 = (ab)^2 - ab = a(bab - b) = (a^2 - a)\, b.$$

As $B^*=(0)$ we must have $a^2=a$. Therefore A is itself a Boolean ring, and thus condition (3) of a weakly special class also holds for **K**.

Now let **S** be the upper radical determined by the class **K**. Since **K** is weakly special, **S** is supernilpotent by theorem 11.5. By theorem 11.5 the **S**-semisimple rings are the subdirect sums of rings in **K**. Since every ring in **K** is a Boolean ring, these subdirect sums are also Boolean rings. Hence no **S**-semisimple prime ring exists, since such a ring would be prime Boolean ring and therefore a prime field of two elements. However, this ring is an **S**-radical ring. Therefore the radical is supernilpotent, but it is not special. □

§ 12. SPECIAL CLASSES OF SUBDIRECTLY IRREDUCIBLE RINGS

For the subject-matter of this section, we refer the reader to Andrunakievich [8, 11, 20].

We recall some important definitions. If B is the complete direct sum of the rings A_α, and A is a subring of B such that for every $a_\alpha \in A_\alpha$, an element $a'= =(\ldots, a_\alpha, \ldots)$ exists, whose component in A_α is the given element $a_\alpha \in A_\alpha$, then A is a subdirect sum of the rings A_α. A is a *subdirect sum* of the rings A_n if and only if there exist ideals I_α of A with $\bigcap_\alpha I_\alpha=(0)$ and $A_\alpha \cong A/I_\alpha$. A ring A is *subdirectly irreducible* if the intersection of all non-zero ideals of A is non-zero.

We have the following:

Proposition 12.1 (Birkhoff [1]). Every ring A is a subdirect sum of subdirectly irreducible rings.

Proof. Let a be an arbitrary non-zero element of A. Then $a \notin (0)$, and by Zorn's lemma there exists a maximal ideal I_a in A such that $a \notin I_a$. Put $D=\bigcap I_a$ for all $a \in A$. If $D \neq (0)$, then there exists an element $d \neq 0$ of D with $d \in I_d$, which contradicts $d \notin I_d$. Therefore $D=(0)$. Consequently, A is a subdirect sum of the rings A/I_a. We shall show that every ring A/I_a is subdirectly irreducible. Every non-zero ideal of A/I_a has the form B/I_a with $B \neq I_a$, and thus $a \in B$. Therefore $((a)+I_a)/I_a$ is the non-zero intersection of all non-zero ideals of A/I_a, and accordingly A/I_a is subdirectly irreducible. □

The uniquely determined minimal ideal of a subdirectly irreducible ring A is called the *heart* of A.

Proposition 12.2. If **R** is a hereditary radical, and A a subdirectly irreducible ring with heart H, then H is either an **R**-radical ring or **R**-semisimple. H is **R**-semisimple if and only if A is **R**-semisimple. The heart of A is either a simple ring or a zero-ring.

PROOF. Since $\mathbf{R}(H)=H\cap\mathbf{R}(A)\subseteq H$, either $\mathbf{R}(H)=H$ or $\mathbf{R}(H)=(0)$. If A is $\mathbf{R}$-semisimple, then because $\mathbf{R}(A)=(0)$ we have $\mathbf{R}(H)=(0)$. But if $\mathbf{R}(H)=(0)$, then $(0)=\mathbf{R}(H)=H\cap\mathbf{R}(A)$ and the subdirect irreducibility of A implies $\mathbf{R}(A)=(0)$. Further we have $H^2\leqq H$, whence $H^2=H$ or $H^2=(0)$. If $H^2=H$, let I be a non-zero ideal of the ring H. Then $I^*=I+AI+IA+AIA$ is a non-zero ideal of A, contained in H, and $(I^*)^3\subseteq I$. The minimality of H yields $I^*=H$ and thus $H^3\subseteq I$. Since $H^2=H$ and $I\subseteq H$ we must have $I=H$. Therefore H is simple if $H^2=H$. □

Let $\mathbf{E}$ be an *algebraic property* (i.e. a property invariant with respect to isomorphisms) of rings. The class of all subdirectly irreducible rings with an idempotent $\mathbf{E}$-heart yields an example of a special class. Namely, we have the following

THEOREM 12.3. The class $\mathbf{K}$ of all subdirectly irreducible rings with an idempotent $\mathbf{E}$-heart is special.

PROOF. First we shall show that every ring in $\mathbf{K}$ is a prime ring. Let B and C be ideals of $A\in\mathbf{K}$. If B and C are non-zero, then $H\subseteq B$ and $H\subseteq C$ for the $\mathbf{E}$-heart H of A, and thus we have $(0)\neq H=H^2\subseteq B\cdot C$. But this means that A is a prime ring. Next we prove that every non-zero ideal B of a subdirectly irreducible ring A with an idempotent $\mathbf{E}$-heart is again subdirectly irreducible with the same heart. If C is a non-zero ideal of the ring B, then we have $BC\neq(0)$, since $BC=(0)$ implies $B\cdot\bar{C}=(0)$ for $\bar{C}=C+CA+AC+ACA$.

But as $B\neq(0)$ and $\bar{C}\neq(0)$ this is impossible since A is a prime ring. Similarly, $BCB\neq(0)$ can also be proved. Since BCB is an ideal of A, $H\subseteq BCB$, and since $BCB\subseteq C$ we have $H\subseteq C$. Accordingly, B is subdirectly irreducible with the same heart H.

Now we prove condition (3) for a special class. Let the non-zero ideal B of A be a subdirectly irreducible ring with an idempotent $\mathbf{E}$-heart H and let B^* be the annihilator of B in A. We shall verify $A/B^*\in\mathbf{K}$. Since $H^2=H$ we have

$$HA = H(HA) \subseteq HB \subseteq H,$$

and similarly $AH\subseteq H$. Therefore, H is an ideal of A, moreover, it is a minimal ideal. Furthermore we have $H\cap B^*=(0)$, because if $H\cap B^*\neq(0)$ then $H=H\cap B^*$ and since $H\subseteq B$ this leads to the contradiction

$$(0) \neq H = H^2 \subseteq B\cdot B^* = (0).$$

Now $H\cap B^*=(0)$ implies the isomorphism

$$(H+B^*)/B^* \cong H/(H\cap B^*) \cong H.$$

Thus A/B^* has the minimal $\mathbf{E}$-ideal $(H+B^*)/B^*$.

If C_1 and C_2 are ideals of A with $C_1\cdot C_2\subseteq B^*$, then $BC_1\cdot C_2B\subseteq BB^*B=(0)$. However, since A is a prime ring, and BC_1 and C_2B are ideals of A in B, we have

$BC_1=(0)$ or $C_2B=(0)$. If $BC_1=(0)$, then $(C_1B)^2=(0)$ and so $C_1B=(0)$ because A is a prime ring. Therefore $C_1 \subseteq B^*$. Similarly, if $C_2B=(0)$ we have $C_2 \subseteq B^*$. Thus B^* is a prime ideal of A and A/B^* is a prime ring. Therefore

$$B^* \not\supseteq C \cdot (H+B^*)$$

for every non-zero ideal C/B^* of A/B^*, and thus $CH \subsetneqq B^*$. In particular this implies $C \cap H \neq (0)$ and thus $H \subseteq C$. But now

$$(H+B^*)/B^* \subseteq C/B^*$$

so that A/B^* is subdirectly irreducible with the **E**-heart $(H+B^*)/B^* \cong H.$ □

The upper radical determined by the special class of all subdirectly irreducible rings with an idempotent **E**-heart will be denoted by $\mathbf{S}_E$. Then $\mathbf{S}_E$ is by definition a special radical, and thus $\mathbf{S}_E$ is also a supernilpotent radical.

If **K** is the class of all subdirectly irreducible rings with idempotent heart, then the corresponding upper radical is called *antisimple*. A ring is called *antisimple,* if it coincides with its antisimple radical.

Hence it follows that:

(1) Every ideal of an antisimple ring is again an antisimple ring.
(2) Every nilpotent ring is antisimple.

PROPOSITION 12.4. For a ring A the following conditions are equivalent:

(a) A is antisimple.

(b) Every homomorphic image of A is a subdirect sum of subdirectly irreducible rings with nilpotent heart.

(c) A does not contain a prime ideal P such that A/P has a minimal ideal.

(d) No ideal of A can be homomorphically mapped onto a (non-trivial) simple ring.

PROOF. (a)⇒(b). If (b) does not hold, then there exists a homomorphic image of A which has a subdirect summand B with idempotent heart. Since B is a homomorphic image of A, (a) does not hold.

(b)⇒(c). If (c) is not valid, then there exists a prime ring A/P containing a minimal ideal M/P. Then we have $M^2 \subsetneqq P$, consequently $(M/P)^2=M/P$. If C/P is a non-zero ideal of A/P then we have $(C/P)(M/P) \neq P/P$ and thus $M/P= =(C/P)(M/P)=C/P$. Therefore A/P is a subdirectly irreducible ring with idempotent heart, in contradiction to (b).

(c)⇒(d). If (d) does not hold, then there exists an ideal I of A such that A/I is subdirectly irreducible with idempotent heart H/I. But then I is a prime ideal of A as shown above. This contradicts (c).

(d)⇒(a). If (a) is not true, then there exists a homomorphic image A/I of A which is subdirectly irreducible and has idempotent heart H/I. But then the ideal

H of A can be homomorphically mapped onto the non-trivial simple ring H/I, which contradicts condition (d). Consequently, conditions (a), (b), (c) and (d) are equivalent. □

PROPOSITION 12.5 (Szász [12, 23]). A ring A is antisimple if and only if the following condition in A is fulfilled:

$(x)=(x+y)$ for every principal ideal (x) and for every element $y\in(x)^2$. $(*)$

PROOF. Both antisimplicity and condition $(*)$ are obviously homomorphically invariant properties.

If A is an antisimple ring, for which $(*)$ does not hold, then there exist $x\in A$ and $y\in(x)^2$ such that $(x)\neq(x+y)$. Then $x\notin(x+y)$. The set of ideals M with $x\notin M\supseteq(x+y)$ is not empty and Zorn's lemma can be applied to this set. So there exists a maximal ideal M_0 in this set. Therefore, if $M_1\supset M_0$, then we have $x\in M_1$, and thus the homomorphic image A/M_0 is subdirectly irreducible. The heart of A/M_0 is $((x)+M_0)/M_0$, and since $x+y\in(x+y)\subseteq M_0$ we have $(x+M_0)=(y+M_0)$. As $y\in(x)^2$, the heart of A/M_0 is idempotent. Therefore A is not an antisimple ring.

Conversely, if A is not antisimple, then A/M_0 is subdirectly irreducible with idempotent heart H/M_0 for some ideal M_0. Then we have $H=(x+M_0)$ and $x\in(x)^2+M$ since $(x+M)^2=(x+M)$.

Therefore one obtains $x=y+m$, where $y\in(x)^2$, $m\in M$. Thus $x-y=m\in M_0$. Hence $(x-y)\subseteq M_0$, but $(x)\nsubseteq M_0$. Consequently $(x)\neq(x-y)$ for a $y\in(x)^2$. Thus condition $(*)$ is not fulfilled. □

PROPOSITION 12.6. $B^2\neq B$ holds for every finitely generated ideal B contained in the antisimple radical $\mathbf{S}(A)$ of the ring A. The antisimple radical $\mathbf{S}(A)$ of a ring with minimum condition for (two-sided) principal ideals is a nil ideal.

PROOF. Let $B=(b_1, \dots, b_n)$ be an ideal of A contained in $\mathbf{S}(A)$, and suppose $B^2=B$. We may assume $b_{i+1}\notin(b_1, \dots, b_i)$ for $i=1, 2, \dots, n-1$. Then

$$b_n\notin C=(b_1, \dots, b_{n-1}).$$

By Zorn's lemma there exists a maximal ideal M of A with $b_n\notin M$ and $M\supseteq C$. Then A/M is a subdirectly irreducible ring with idempotent heart, since every ideal of A properly containing M contains b_n, and thus also contains $B=B^2$. Since B is contained in $\mathbf{S}(A)$ and the antisimple radical is hereditary, B is antisimple. Thus the idempotent simple ring B/M is also antisimple. This is a contradiction. Therefore $B^2\neq B$.

Next let A be a ring with minimum condition on (two-sided) principal ideals, and $\mathbf{S}(A)$ be the antisimple radical of A. For $a\in\mathbf{S}(A)$ we have by the preceding:

$$(a)\supset(a)^2\supseteq(a^2)\supset(a^2)^2\supseteq(a^4)\supset(a^4)^2\supseteq(a^8)\supset\dots.$$

Therefore $(a)\supset(a^2)\supset(a^4)\supset(a^8)\supset\ldots$, and by the minimum condition for principal ideals $a^{2^n}=0$ for some integer n. Since $a\in\mathbf{S}(A)$ is arbitrarily chosen, $\mathbf{S}(A)$ is a nil ideal. □

PROPOSITION 12.7. If $\mathbf{S}(A)$ is the antisimple radical of a ring A and A_n the full matrix ring of type $n\times n$ over the ring A, then $\mathbf{S}(A_n)=(\mathbf{S}(A))_n$.

PROOF. First we assume that the ring A has a unity element. Then there exists a unique correspondence φ between the ideals I of A and the ideals I_n of A_n given by $\varphi: I\to I_n$ (cf. e.g. Jacobson [2], proposition 3.2.1). Here $I_1\subseteqq I_2$ obviously implies $(I_1)_n\subseteqq(I_2)_n$. Hence A is subdirectly irreducible if and only if A_n is subdirectly irreducible. Moreover, $H^2\neq(0)$ for the heart if and only if $(H_n)^2\neq(0)$. Since $(A/I)_n\cong$ $\cong A_n/I_n$ it follows that A/I is subdirectly irreducible with idempotent heart if and only if A_n/I_n is a subdirectly irreducible ring with idempotent heart. By theorem 11.5 $\mathbf{S}(A)$ is the intersection of all ideals I_α of A for which A/I_α is subdirectly irreducible with an idempotent heart. Therefore, we have $\mathbf{S}(A_n)=\bigcap_\alpha(I_\alpha)_n=$ $=(\bigcap_\alpha I_\alpha)_n=(\mathbf{S}(A))_n$.

Next we assume that A does not have unity element. Then A possesses a ring extension A' with unity element such that every ideal of A is an ideal of A'. Since $\mathbf{S}$ is hereditary, we have $\mathbf{S}(A)=A\cap\mathbf{S}(A')$, and thus

$$(\mathbf{S}(A))_n = A_n\cap(\mathbf{S}(A'))_n = A_n\cap\mathbf{S}(A'_n).$$

However, since A_n is an ideal of A'_n, and $\mathbf{S}$ is hereditary, we have

$$A_n\cap\mathbf{S}(A'_n) = \mathbf{S}(A_n).$$

Consequently $(\mathbf{S}(A))_n=\mathbf{S}(A_n)$. □

§ 13. SUPPLEMENTING, DUAL AND SUBIDEMPOTENT RADICALS

For the subject-matter of this section we refer the reader to Andrunakievich [11].

Let $\mathbf{R}$ be a radical in the sense of Amitsur and Kurosh. A radical $\mathbf{S}$ is called the *radical supplementing* $\mathbf{R}$, if $\mathbf{R}(A)\cap\mathbf{S}(A)=(0)$ in every ring A and $\mathbf{T}\leqq\mathbf{S}$ for all radicals $\mathbf{T}$ for which $\mathbf{R}(A)\cap\mathbf{T}(A)=(0)$ in every ring A. If $\mathbf{R}(A)\cap\mathbf{S}(A)=(0)$ for the radicals $\mathbf{R}$ and $\mathbf{S}$ in every ring A, then every $\mathbf{S}$-radical ring is $\mathbf{R}$-semisimple and every $\mathbf{R}$-radical ring is $\mathbf{S}$-semisimple. Now every homomorphic image of an $\mathbf{R}$-radical ring is again an $\mathbf{R}$-radical ring. So every $\mathbf{R}$-radical ring is an $\mathbf{S}$-semisimple ring, whose every homomorphic image is again $\mathbf{S}$-semisimple.

A ring A is called *strongly* **R**-*semisimple* if every homomorphic image of A is an $\mathbf{R}$-semisimple ring.

If **S** is a radical supplementing **R** then every **S**-radical ring is strongly **R**-semisimple and every **R**-radical ring is strongly **S**-semisimple.

By theorem 1.8, the **R**-semisimple rings are hereditary. Therefore every ideal of an **R**-semisimple ring is **R**-semisimple. In order to investigate the question of the existence of a radical **S** supplementing **R** we need to discuss the hereditary radicals.

PROPOSITION 13.1. If **R** is a hereditary radical, then every non-zero **R**-ideal I of a subdirectly irreducible ring A with the heart H can be homomorphically mapped onto a non-zero subdirectly irreducible ring with **R**-radical heart.

PROOF. I is not **R**-semisimple since $I \supseteqq H$. Hence some subdirectly irreducible subdirect summand I_α is not **R**-semisimple in a subdirect sum representation of I by proposition 12.1. (Every subdirect sum of **R**-semisimple rings is **R**-semisimple.) Therefore I_α is a homomorphic image of I, and I_α has the desired property. □

By proposition 13.1 the upper radical $\mathbf{S}_R$ determined by the class **K** of all subdirectly irreducible rings with an **R**-radical heart exists.

PROPOSITION 13.2. For every hereditary radical **R** and for every ring A the following conditions are equivalent:

(a) A is an $\mathbf{S}_R$-radical ring.

(b) Every homomorphic image A' of A is a subdirect sum of subdirectly irreducible rings with **R**-semisimple heart.

(c) A is strongly **R**-semisimple.

(d) Every ideal I of A is the intersection of all those ideals I_α of A such that the factor ring A/I_α is subdirectly irreducible with **R**-semisimple heart.

PROOF. First we shall show the equivalence of (a) and (b). Suppose A satisfies (a). Then we can assume $A' \neq (0)$. A' is a subdirect sum of subdirectly irreducible rings A'_α with hearts H'_α. Since each A'_α is a homomorphic image of the homomorphic image A' of A, every heart H'_α is, by the definition of $\mathbf{S}_R$, **R**-semisimple, and thus (b) holds. But if (a) is not valid, then A has a homomorphic image which is a subdirectly irreducible ring with **R**-radical heart. But then (b) does not hold. Thus (a) and (b) are equivalent.

Moreover, (b)⇒(c). If (b) holds for $A \neq (0)$, and A' is a non-zero homomorphic image of A, then A' is **R**-semisimple, since A' is a subdirect sum of subdirectly irreducible rings A'_α, which are **R**-semisimple (every heart H'_α is also **R**-semisimple) and every subdirect sum of **R**-semisimple rings is **R**-semisimple.

(b) also follows obviously from (c). For, if A' is a non-zero homomorphic image of a strongly **R**-semisimple ring A, and A' is a subdirect sum of the subdirectly irreducible rings A'_α, then every ring A'_α, being a homomorphic image of A, is again **R**-semisimple. Since **R** is hereditary, the heart H'_α of A'_α is **R**-semisimple. Therefore (b) holds.

Now we shall show the equivalence of (b) and (d). Let B be an ideal of a ring satisfying condition (b). Then $A'=A/B$ can be represented as a subdirect sum of subdirectly irreducible rings A'_α with **R**-semisimple hearts. Therefore there exist ideals I'_α in A' with $(0')=\bigcap_\alpha I'_\alpha$ and $A'/I'_\alpha \cong A'_\alpha$. Taking in A the full inverse images, we have

$$B = \bigcap_\alpha I_\alpha \quad \text{and} \quad A/I_\alpha \cong (A/B)/(I_\alpha/B) = A'/I'_\alpha = A'_\alpha.$$

(d)⇒(b) by definition of a subdirect sum. □

We have the following

THEOREM 13.3. If **R** is a hereditary radical, then there always exists a radical **R**′ supplementing to **R**, where the **R**′-radical rings are just the strongly **R**-semisimple rings.

PROOF. By proposition 13.2 the class of all strongly **R**-semisimple rings is the class of all $\mathbf{S}_R$-radical rings. By theorem 1.8 the radical $\mathbf{R}'=\mathbf{S}_R$ is hereditary. Thus $\mathbf{R}(A)\cap\mathbf{R}'(A)=(0)$ in every ring A, since $D=\mathbf{R}(A)\cap\mathbf{R}'(A)$ is both an **R**-radical and **R**-semisimple ring. Now, if $\mathbf{R}(A)\cap\mathbf{Q}(A)=(0)$ for a radical **Q** and for every ring A, then every **Q**-ring is strongly **R**-semisimple, and thus $\mathbf{Q}\leq\mathbf{R}'$. Hence **R**′ is the radical supplementing **R**. □

We say, that **R** and **S** form a *dual pair of supplementing radicals,* if **S** is the radical supplementing **R**, and **R** is the radical supplementing **S**. In other words, for a dual pair of radicals $\mathbf{R}=\mathbf{S}'$ and $\mathbf{S}=\mathbf{R}'$, and thus $\mathbf{R}''=\mathbf{R}$ and $\mathbf{S}''=\mathbf{S}$.

We shall show that for the antisimple radical **S** we have $\mathbf{S}''=\mathbf{S}$. With this aim in mind we determine the radical **S**′ supplementing the antisimple radical **S** and verify that (**S**, **S**′) is a dual pair of radicals. The antisimple radical, as a special radical, is hereditary, and thus **S**′ exists by theorem 13.3. The **S**′-radical rings are the strongly antisimple-semisimple rings which cannot be homomorphically mapped onto a subdirectly irreducible ring with an antisimple heart. A heart H is antisimple only in the case $H^2=(0)$. Thus the **S**′-radical rings cannot be homomorphically mapped onto a zero-ring, that is, onto a ring A with $A^2=(0)$. Therefore $A^2=A$ for every **S**′-radical ring A, and since **S**′ is a hereditary, $I^2=I$ for every ideal of A.

A ring A is called *hereditarily idempotent,* if every ideals I of A is idempotent, i.e. if $I^2=I$. Therefore the radical **S**′ supplementing the antisimple radical **S** is hereditarily idempotent and $\mathbf{S}(A)\cap\mathbf{R}(A)=(0)$ in every ring A for every hereditarily idempotent radical **R** and for the antisimple radical **S**.

Following Blair [1, 2] we call a ring A *f-regular* if $a\in(a)^2$ for every $a\in A$, where (a) is the principal ideal generated by the element a.

Now we have the following

PROPOSITION 13.4. For a ring A the following conditions are equivalent:

(a) A is hereditarily idempotent.

(b) A is f-regular.

(c) $B\cap C=B\cdot C$ for arbitrary ideals B and C of A.

(d) A cannot be homomorphically mapped onto a subdirectly irreducible ring with a nilpotent heart.

(e) Every non-zero homomorphic image of A is a subdirect sum of subdirectly irreducible rings with idempotent heart.

(f) Every ideal I of A is the intersection of all prime ideals P of A such that $P\supseteq I$ and A/P possesses a minimal ideal.

PROOF. (a)$\Rightarrow$(b). If a is an arbitrary element of A and A satisfies condition (a), then $(a)^2=(a)$ obviously implies $a\in(a)^2$ and thus we have the f-regularity of A.

(b)$\Rightarrow$(c). $BC\subseteq B\cap C$ for every pair B, C of ideals in a ring. Now, if $d\in D=B\cap C$ then since $d\in(d)^2$ and $(d)\subseteq B$, $(d)\subseteq C$ we have the relation $d\in(d)^2\subseteq BC\subseteq B\cap C$. Therefore $B\cap C\subseteq BC\subseteq B\cap C$ whence (c) follows.

(c)$\Rightarrow$(d). Together with A every homomorphic image A' of A also obviously fulfils condition (c). Every subdirect summand is a homomorphic image, and since $(0)=H^2=H\cdot H\neq H\cap H$ obviously holds for a nilpotent heart. (d) follows from (c).

(e) is an immediate consequence of (d).

Let I be an ideal of A, where A satisfies condition (e). Then A/I is a subdirect sum of subdirectly irreducible rings A/I_α with $I_\alpha\supseteq I$, $\cap I_\alpha=I$ and with idempotent hearts H_α/I_α. Every ring A/I_α is a prime ring, because $\bar{0}\neq H_\alpha/I_\alpha=(H_\alpha/I_\alpha)^2\subseteq B/I_\alpha\cdot$ $\cdot C/I_\alpha$ and so $BC\not\subseteq I_\alpha$ for the ideals $B\not\subseteq I_\alpha$ and $C\not\subseteq I_\alpha$. Therefore (f) holds.

(f)$\Rightarrow$(a). Let I be an ideal of A. Then I^2 is an intersection of prime ideals P_α of A. Therefore $I^2=\cap P_\alpha\subseteq P_\alpha$. Hence we have $I\subseteq P_n$ for every P_α, since P_α is a prime ideal of A. Thus $I\subseteq\cap P_\alpha=I^2$, whence $I^2=I$ follows. Therefore A is hereditarily idempotent and thus (f) follows from (a). □

A radical **R** is called *subidempotent*, if **R** is hereditary and every **R**-radical ring is idempotent. The ring property characterized in proposition 13.4 is a radical property; moreover it is subidempotent and it is the radical supplementing the antisimple radical.

PROPOSITION 13.5. If **S** is a supernilpotent radical, then every ideal of a strongly S-semisimple ring is again strongly S-semisimple and every strongly S-semisimple ring is a subidempotent radical ring.

If the radical S is either supernilpotent or subidempotent, then every S-semisimple minimal ideal M of an arbitrary ring is a strongly S-semisimple ring.

PROOF. Let S be a supernilpotent radical, B an ideal of A and C an ideal of the ring B. Now, if $\bar{C}=C+AC+CA+ACA$ then we have by the lemma of Andrunakievich [11]

$$(\bar{C})^3 \subseteqq B\bar{C}B = B(C+AC+CA+ACA)B \subseteqq C \subseteqq \bar{C}.$$

Since $A/(\bar{C})^3 \supseteqq \bar{C}/(\bar{C})^3$, and $A/\bar{C}$ is also S-semisimple we have $(\bar{C})^3=(\bar{C})$ and thus $C=\bar{C}$. Accordingly C is an ideal of A. Therefore B/C is an ideal of the S-semisimple ring A/C, and thus B/C is also S-semisimple. But then B is strongly S-semisimple. Since B/B^2 is S-semisimple, we have $B^2=B$ for every ideal B of A.

Now let S be again supernilpotent and M an S-semisimple minimal ideal of the ring A. Then $M^2=M$, and M is a simple ring. For, if I is a non-zero ideal of the ring M, then we have $(\bar{I})^3 \subseteqq I$ for $\bar{I}=I+AI+IA+AIA$. Also $\bar{I}\neq(0)$ and $(\bar{I})^3\neq(0)$. So by the minimality of M in A $\bar{I}=I=M$. Since M is simple and S-semisimple. A is also strongly S-semisimple.

Finally, let S be a subidempotent radical and M an S-semisimple minimal ideal of the ring A. If $M^2=M$, then by a method similar to the preceding M is a simple ring. Thus the simplicity and S-semisimplicity of M imply the strong S-semisimplicity of M. But if $M^2=(0)$ for an S-semisimple ring M, then every homomorphic image M' of M is a zero-ring, and thus M' is again S-semisimple together with M. □

We have the following

THEOREM 13.6. If S is a supernilpotent radical, then the radical S′ supplementing S and the supplementing radical S″ of S′ exist, S′ and S″ form a dual pair of supplementing radicals. S′=S‴, S≦S″ and S′ is subidempotent. Moreover, S=S″ if and only if S is the upper radical determined by the class of all subdirectly irreducible rings with an S-semisimple heart. Therefore S is a special radical when S=S″.

PROOF. Since S is hereditary, S′ exists by theorem 13.3 and by proposition 13.2 the S′-radical rings are just the strongly S-semisimple rings. S′ is also hereditary by the first part of proposition 13.5. S′ is a subidempotent radical because S′ is strongly S-semisimple for a supernilpotent radical S. Now S″ also exists by theorem 13.3 and S″ is an upper radical determined by the class of all subdirectly irreducible rings with $\mathbf{S}'$-radical heart and so with $\mathbf{S}$-semisimple heart. Thus, we have $\mathbf{S} \leqq \mathbf{S}''$ and $\mathbf{S}''$ is also supernilpotent since it is hereditary. Now, if $\mathbf{T}(A) \cap \cap \mathbf{S}''(A) = (0)$ for every ring A, then $\mathbf{T}(A) \cap \mathbf{S}(A) = (0)$ also holds and thus we have $\mathbf{S}''' \leqq \mathbf{S}'$ since $\mathbf{T} \leqq \mathbf{S}'$. Consequently $\mathbf{S}''' = \mathbf{S}'$. Accordingly, $\mathbf{S}'$ and $\mathbf{S}''$ form a dual pair of supplementing radicals because we have $\mathbf{S}'' = (\mathbf{S}')'$ and $\mathbf{S}' = (\mathbf{S}'')'$. The proof of the condition for $\mathbf{S} = \mathbf{S}''$ mentioned in the statement of the theorem is clear. Since an $\mathbf{S}$-semisimple heart of a subdirectly irreducible ring is idempotent for a supernilpotent radical $\mathbf{S}$, the radical $\mathbf{S}''$ is a special radical by its definition and by theorem 12.3. Therefore, $\mathbf{S}$ is a special radical when $\mathbf{S} = \mathbf{S}''$. □

$\mathbf{S} = \mathbf{S}''$ for the antisimple radical $\mathbf{S}$ which is special and thus supernilpotent. However, clearly $\mathbf{S} \neq \mathbf{S}''$ for the supernilpotent radical discussed in example 11.15 because $\mathbf{S}$ is not special but $\mathbf{S}''$ is special.

We now give an example of a radical for which every semisimple ring is strongly semisimple.

EXAMPLE 13.7. For a natural number $n \geqq 2$ let $\mathbf{K}_n$ be the class of rings A in which $a^n = a$ for every element $a \in A$. The rings A from $\mathbf{K}_n$ are commutative by theorem 10.1.1 of Jacobson [2]. Moreover, together with a ring A every homomorphic image of A also belongs to $\mathbf{K}_n$. If $A \in \mathbf{K}_n$ then we also have $B \in \mathbf{K}_n$ for every subring B of A. Every subdirect sum of rings in $\mathbf{K}_n$ is again a ring in $\mathbf{K}_n$. Every subdirectly irreducible ring in $\mathbf{K}_n$ is a field since the heart H of a subdirectly irreducible ring A is idempotent and simple and thus H is a field by commutativity. But then H is a direct summand of A so $H = A$ by the subdirect irreducibility of A. A ring A from $\mathbf{K}_n$ does not have a non-zero nilpotent ideal.

Let B be an ideal of A with $B \in \mathbf{K}_n$. Let the annihilator B^* of B in A be equal to (0). Then A also belongs to $\mathbf{K}_n$. For $ab \in B$ and $ba \in B$ for every $a \in A$ and $b \in B$. Since $(ab)^n = ab$ and B is commutative, we have

$$ab = (ab)^n = ab \cdot (ab)^{n-1} = a(ab)^{n-1} b = a^2 b(ab)^{n-2} b =$$
$$= a^2 (ab)^{n-2} b^2 = \ldots = a^{n-1} ab \cdot b^{n-1} = a^n b^n = a^n b.$$

Thus $(a^n - a)b = 0$. Therefore $a^n - a \in B^*$. Consequently, $a^n = a$ and $A \in \mathbf{K}_n$.

Therefore the class $\mathbf{K}_n$ is weakly special, and the class $\mathbf{K}_n'$ of all fields in $\mathbf{K}_n$ is special. Both classes $\mathbf{K}_n$ and $\mathbf{K}_n'$ determine the same upper radical $\mathbf{S}_n$ and every $\mathbf{S}_n$-semisimple ring is a subdirect sum of rings in $\mathbf{K}_n$, and so a subdirect sum of rings in $\mathbf{K}_n'$. But this means that every $\mathbf{S}_n$-semisimple ring is a ring belonging to $\mathbf{K}_n$ and since $\mathbf{K}_n$ is a homomorphically closed, every $\mathbf{S}_n$-semisimple ring is strongly $\mathbf{S}_n$-semisimple. The radical $\mathbf{S}_n$ is supernilpotent and $\mathbf{S}_n = \mathbf{S}_n''$. Thus, $(\mathbf{S}_n, \mathbf{S}_n')$ is a dual pair of supplementing radicals, and the rings in $\mathbf{K}_n$ coincide with the $\mathbf{S}_n'$-radical rings. Therefore the sum of arbitrarily many ideals I_α of A with $I_\alpha \in \mathbf{K}_n$ is again an ideal in $\mathbf{K}_n$.

§ 14. SUPERNILPOTENT RADICALS AND DIRECT DECOMPOSITIONS OF A RING WITH MINIMUM CONDITION ON PRINCIPAL RIGHT IDEALS

In this section we shall discuss a generalization due to Andrunakievich [15], of a decomposition theorem due to Faith [2].

We need some definitions and preliminaries.

A ring A is called *regular* in the sense of von Neumann if for every element $a \in A$ there exists an element $x \in A$ with $a = axa$. Every ideal and every homomorphic image of a regular ring is again regular.

A ring A is called *weakly regular* if for every element $a \in A$ there exists an element a_1, contained in the principal ideal (a), with $a = aa_1$. A ring is weakly regular if and only if every right ideal R of A is idempotent, i.e. $R^2 = R$. Every ideal I of a weakly regular ring A is weakly regular, because if R is a right ideal of I then

$$(R+RA)^2 \subseteqq (R+RA)I \subseteqq R \subseteqq R+RA = (R+RA)^2.$$

So R is a right ideal of the ring A.

Every regular ring is weakly regular and every weakly regular ring is hereditarily idempotent.

Proposition 14.1 (Andrunakievich [6]). If the factor ring A/Q of a ring is weakly regular and if Q^* is the annihilator of the ideal Q in A, then $Q \cap (Q^*)^2 = (0)$.

Proof. Let $r \in Q \cap (Q^*)^2$. Then there exist elements $s_i \in Q^*$, $t_i \in Q^*$ with

$$r = \sum_{i=1}^{n} s_i t_i \in Q.$$

Under the notation $\bar{t}_i = t_i + Q \in A/Q$ by the weakly regularity of A/Q there exist elements $\bar{e}_i \in (\bar{t}_i)$, such that $\bar{t}_i = \bar{t}_i \cdot \bar{e}_i$. Hence we have $t_i = t_i e_i + r_i$ where $r_i \in Q$, and t_i, $e_i \in (t_i) \subseteqq Q^*$. Therefore we have

$$r = \sum_{i=1}^{n} s_i(t_i e_i + r_i) = \sum_{i=1}^{n} s_i t_i e_i. \tag{$*$}$$

We shall prove $r = 0$ by induction on n. If $n = 1$, then we have $r = s_1 t_1 = s_1 t_1 e_1 = re_1 = 0$ because $r \in Q$ and $e_1 \in Q^*$. Let us assume that the proposition $r = 0$ is already proved for $n-1$, so that $r = \sum_{i=1}^{n-1} u_i v_i$, $u_i \in Q^*$, $v_i \in Q^*$ always implies $r = 0$. We obviously have the equation

$$s_n t_n = r - \sum_{i=1}^{n-1} s_i t_i (s_i,\ t_i \in Q^*).$$

Therefore, ($*$) has the form

$$r = \sum_{i=1}^{n-1} s_i t_i e_i + \left(r - \sum_{i=1}^{n-1} s_i t_i\right) e_n$$

and we have the relation

$$r = \sum_{i=1}^{n-1} u_i v_i$$

with $u_i = s_i t_i \in Q^*$, $v_i = e_i - e_n \in Q^*$ and $re_n \in QQ^* = (0)$. But then we have by the induction assumption $r=0$. Thus $Q \cap (Q^*)^2 = (0)$. □

PROPOSITION 14.2. Let **S** be a supernilpotent radical, A a ring such that $A/\mathbf{S}(A)$ is both strongly **S**-semisimple and weakly regular. Let **S**′ be the radical supplementing **S**. Then $\mathbf{S}'(A) = (0)$ if and only if $\mathbf{S}(A)$ contains its annihilator $(\mathbf{S}(A))^*$ in A.

PROOF. **S**′ exists by theorem 13.3. Assume $(\mathbf{S}(A))^* \subseteq \mathbf{S}(A)$. Since $\mathbf{S}'(A) \cap \mathbf{S}(A) = (0)$ we have $\mathbf{S}'(A) \cdot \mathbf{S}(A) = \mathbf{S}(A) \cdot \mathbf{S}'(A) = (0)$. Therefore $\mathbf{S}'(A) \subseteq (\mathbf{S}(A))^* \subseteq \mathbf{S}(A)$ whence $\mathbf{S}'(A) = 0$. Conversely, suppose $\mathbf{S}'(A) = (0)$. By the assumption on $A/\mathbf{S}(A)$ and by proposition 14.1 we have $\mathbf{S}(A) \cap ((\mathbf{S}(A))^*)^2 = (0)$. This gives the isomorphism $(((\mathbf{S}(A))^*)^2 + \mathbf{S}(A))/\mathbf{S}(A) \cong ((\mathbf{S}(A))^*)^2$. The ring on the left-hand side is a strongly **S**-semisimple ring, i.e. an **S**′-radical ring. Therefore $((\mathbf{S}(A))^*)^2$ is also an **S**′-radical ring. Hence $((\mathbf{S}(A))^*)^2 \subseteq \mathbf{S}'(A)$ holds. Since $\mathbf{S}'(A) = (0)$ we have $((\mathbf{S}(A))^*)^2 = (0)$. But **S** is supernilpotent, so that $(\mathbf{S}(A))^* \subseteq \mathbf{S}(A)$. □

THEOREM 14.3. A ring A is the direct sum of simple rings if and only if $A^2 = A$, and every ideal of A is a direct summand (cf. Cartan and Eilenberg [1]). If **S** is a supernilpotent radical then every **S**-semisimple ring with minimum condition on principal right ideals is the direct sum of simple rings with minimum condition on principal right ideals (cf. Szász [1, 2, 3]).

PROOF. Let A be the direct sum of simple rings A_α, and B an ideal of A. Then $A^2 = A$ since $A_\alpha^2 = A_\alpha$. Let C be an ideal of A, maximal with respect to the condition $B \cap C = (0)$. Then the sum $B + C$ is direct. If there exists an element $a \in A$ with $a \notin B \oplus C$, then there exist a finite number of simple rings A_{α_i}, which are ideals of A, with $a \in A_{\alpha_1} \oplus \ldots \oplus A_{\alpha_n}$. Then $A_{\alpha_i} \not\subseteq B \oplus C$ holds for at least one ideal A_{α_i}. Since $A_{\alpha_i} \not\subseteq B$ and $A_{\alpha_i} \not\subseteq C$, we have $B \cap (C + A_{\alpha_i}) = (0)$ in contradiction to the maximality of C. Therefore $A = B \oplus C$. Conversely, suppose $A^2 = A$ and that every ideal B of A is a direct summand of A. We shall first show that every non-zero ideal B of A contains a minimal ideal M of A. Since $B \neq (0)$ there exists an element $b \in B$ with $b \neq 0$. Let C be an ideal of A maximal with respect to the properties $b \notin C$ and $C \subseteq B$. Since the conditions $b \notin (0)$ and $(0) \subseteq B$ obviously hold for the ideal (0), C exists by Zorn's lemma. We have $B = C \oplus D$ with $D = C_1 \cap B$. If D

contains an ideal D_1 of A with $D_1 \neq D$ and $D_1 \neq (0)$, then we have $D = D_1 \oplus D_2$ with an ideal D_2 of A. But then $B = C \oplus D_1 \oplus D_2$ and $(C \oplus D_1) \cap (C \oplus D_2) = C$, because $c_1 + d_1 = c_2 + d_2 (c_i \in C, d_i \in D)$ implies $c_1 - c_2 = d_2 - d_1 \in C \cap D$ and thus $d_2 = d_1 \in D_1 \cap D_2 = (0)$, $d_2 = d_1 = 0$. Therefore, in fact, $(C \oplus D_1) \cap (C \oplus D_2) = (0)$. On the other hand, since $D_i \oplus C \neq C$ by the maximality of C we have $0 \neq b \in D_i \oplus C$, in contradiction to $(D_1 \oplus C) \cap (D_2 \oplus C) = (C)$ and $b \notin C$. Hence D is a minimal ideal of A contained in B, and thus A has a minimal ideal M in every non-zero ideal B. Let A_0 be the sum of all minimal ideals of A. Then we have $A = A_0 \oplus A_1$. But here $A_1 = (0)$, because if $A_1 \neq (0)$ then A_1 also contains a non-zero minimal ideal of A, which contradicts the definition of A_0. Therefore $A = A_0$, and thus A has a representation $A = \sum \oplus A_\alpha$ where each A_α is a minimal ideal. Since $A^2 = A$ every ring A_α is simple.

Now let **S** be a supernilpotent radical and A a **S**-semisimple ring with minimum condition on principal right ideals. Then A does not contain a non-zero nilpotent ideal. Consequently, it does not contain a non-zero nilpotent right ideal either, because if R is a nilpotent right ideal then so is the ideal $R + AR$. (In every member of $(R + AR)^k$ there are factors contained in R^k.) Every right ideal R of A contains a minimal right ideal R_0 of A, where R_0 is idempotent. Moreover R_0 is of the form $R_0 = eA$ with $e^2 = e$. Thus A is the direct sum of idempotent minimal right ideals $e_\alpha A$ with $e_\alpha^2 = e_\alpha$. The sum of all idempotent minimal right ideals, which are isomorphic as right A-modules, is a minimal two-sided ideal of A. Moreover it is a simple ring, and A is obviously the direct sum of the simple rings obtained in this way. These simple rings obviously satisfy the minimum condition on principal right ideals. □

PROPOSITION 14.4. Let $A = B \oplus C$. Then $C = B^*$ the annihilator of B in A if and only if $B^* \cap B = (0)$.

PROOF. Since $A = B \oplus C$ and $C \subseteq B^*$ we have $B^* = C \oplus (B \cap B^*)$, therefore $B^* = C$ precisely when $B^* \cap B = (0)$. □

We have the following

THEOREM 14.5. Let **S** be a supernilpotent radical, A a ring, for which $A/\mathbf{S}(A)$ satisfies the minimum condition on principal right ideals. Then $A = B \oplus C$, where either $B = (0)$ or B is a strongly **S**-semisimple ring with minimum condition on principal right ideals. C is the annihilator B^* of B in A. Moreover, $\mathbf{S}(C) = \mathbf{S}(A)$ holds, and the annihilator $(\mathbf{S}(C))^*$ of $\mathbf{S}(C)$ in C is contained in $\mathbf{S}(C)$. C is either an **S**-radical ring, or $C/\mathbf{S}(C)$ is an **S**-semisimple ring with minimum condition on principal right ideals. The **S**-semisimple rings are direct sums of simple **S**-semisimple rings with minimum condition on principal right ideals.

PROOF. The radical $\mathbf{S}'$ supplementing $\mathbf{S}$ exists by theorem 13.3. Let $\overline{\mathbf{S}'(A)}=(\mathbf{S}'(A)+\mathbf{S}(A))/\mathbf{S}(A)$ be the image of $\mathbf{S}'(A)$ under the natural homomorphism $x\to\bar{x}=x+\mathbf{S}(A)$. Since $\mathbf{S}'(A)\cap\mathbf{S}(A)=(0)$ we have $\overline{\mathbf{S}'(A)}=\mathbf{S}(A)$. Therefore $\mathbf{S}'(A)$ is strongly $\mathbf{S}$-semisimple. By the definition of the direct sum proposition 14.3 ensures the existence of an ideal T of A with $A=\mathbf{S}'(A)+T$ and $\mathbf{S}'(A)\cap T\subseteq\mathbf{S}(A)$, since $A/\mathbf{S}(A)$ is a direct sum of simple rings. Since $\mathbf{S}'(A)\cap\mathbf{S}(A)=(0)$ we have $\mathbf{S}'(A)\cap T=(0)$, and thus $A=\mathbf{S}'(A)\oplus T$. Therefore we have by proposition 14.4 $T=(\mathbf{S}'(A))^*$ and thus $A=B\oplus C$ with $B=\mathbf{S}'(A)$ and $C=T=(\mathbf{S}'(A))^*$. The relation $\mathbf{S}(A)\subseteq T=(\mathbf{S}'(A))^*$ implies $\mathbf{S}((\mathbf{S}'(A))^*)=(\mathbf{S}'(A))^*\cap\mathbf{S}(A)=\mathbf{S}(A)$.

We still have to prove that the annihilator of $\mathbf{S}(C)$ with respect to $C=T$ is contained in $\mathbf{S}(C)$, i.e. that $\{x;\, x\in C,\, x\mathbf{S}(C)=\mathbf{S}(C)x=0\}\subseteq\mathbf{S}(C)$. If $\mathbf{S}(C)=C$ there is nothing to prove. Now assume $\mathbf{S}(C)\neq C$, consequently $\mathbf{S}(A)\subset(\mathbf{S}'(A))^*=B^*=C$. Putting $\bar{A}=\bar{B}\oplus\bar{C}$, the minimum condition on principal right ideals also holds in $\bar{C}=C/\mathbf{S}(A)$ and thus by theorem 14.3 $C/\mathbf{S}(A)$ is the direct sum of simple $\mathbf{S}$-semisimple rings with minimum condition on principal right ideals. Then $C/\mathbf{S}(A)$ is regular in the sense of von Neumann, and thus it is weakly regular. Morevover, $\mathbf{S}'(C)=C\cap\mathbf{S}'(A)=(\mathbf{S}'(A))^*\cap\mathbf{S}'(A)=(0)$, and $\mathbf{S}'(C)=(0)$ is equivalent to $(0:\mathbf{S}(C))\cap C\subseteq\mathbf{S}(C)$ by proposition 14.2. □

§ 15. SPECIAL MODULES AND SPECIAL RADICALS

To finish this chapter we shall discuss some results of Andrunakievich and Ryabukhin [2]. We shall deal with a particular case of the Baer lower nil radical in Chapter III. The connection between the Jacobson and the Brown–McCoy radical, and the special modules will be discussed in Chapters IV and V.

Let A be a ring and M a right A-module. A subgroup N of M is an A-submodule, if $na\in N$ for every $n\in N$ and $a\in A$.

A right A module M is called an *A-prime module*, if for M the following conditions are fulfilled:

(1) $MA\neq(0)$.

(2) If x is an element of M and B is an ideal of A, then $xB=(0)$ always implies $x=0$ or $B\subseteq(0:M)_A$.

Now, we associate to every ring A a class $\mathbf{M}_A$ of right A-modules. Then the class $\mathbf{M}=\bigcup_A \mathbf{M}_A$ is called a *special class of right modules*, if for M the following conditions are fulfilled:

(1) If $M\in\mathbf{M}_A$, then M is an A-prime module.

(2) If $M\in\mathbf{M}_B$ and B is an ideal of A, then $MB\in\mathbf{M}_A$.

(3) If $M\in\mathbf{M}_A$ and B is an ideal of A with $MB\neq(0)$, then $M\in\mathbf{M}_B$.

(4) If P is an ideal of A and $\bar{A}=A/P$, then $M\in\mathbf{M}_A$ and $P\subseteqq(0:M)_A$ always imply $M\in\mathbf{M}_{\bar{A}}$ (defined by $m(a+P)=ma$ for every $m\in M$) and conversely, $M\in\mathbf{M}_{\bar{A}}$ always implies $M\in\mathbf{M}_A$ (defined by $ma=m(a+P)$ for every $m\in M$). If $\mathbf{M}$ is a special class of modules, then a right A-module $M\in\mathbf{M}_A$ is called a special right A-module.

If A is a ring and $\mathbf{M}$ is a special class of modules, then we mean by the $\mathbf{M}$-radical of the ring A the intersection $\mathbf{M}(A)=\bigcup_{M\in\mathbf{M}_A}(0:M)_A$. If the system $\mathbf{M}_A$ is empty, we define $\mathbf{M}(A)=A$, and then A is called an $\mathbf{M}$-*radical ring*. If $\mathbf{M}(A)=(0)$ then A is said to be $\mathbf{M}$-*semisimple*. An ideal B of the ring A is an $\mathbf{M}$-radical ideal if the ring B is an $\mathbf{M}$-radical ring.

Now we shall give some preliminaries.

Proposition 15.1. Every non-zero A-submodule N of an A-prime module M is again an A-prime module.

Proof. We have $NA\neq(0)$ since $N\neq(0)$, because M is an A-prime module. Now, if we have $xB=(0)$ for $0\neq x\in N$ and for an ideal B of A, then $B\subseteqq(0:M)_A\subseteqq(0:N)_A$, since M is a prime A-module with $N\subseteqq M$. □

Proposition 15.2. If M is an A-prime module, and B is an ideal of A with $MB\neq(0)$, then M is a B-prime module with $(0:M)_B=(0:M)_A\cap B$. Thus, $(0:M)_B$ is an ideal of A.

Proof. If $0\neq x\in M$ and C is an ideal of B with $xC=(0)$, then let $\bar{C}$ be the ideal of A generated by C. Then $(\bar{C})^3\subseteqq C$ (cf. Andrunakievich [6]), and thus $x(\bar{C})^3=(0)$. Let m be the minimal number with $x(\bar{C})^m=(0)$. If $m=1$ then $\bar{C}\subseteqq(0:M)_A$ and since $C\subseteqq\bar{C}\subseteqq B$ we have the relation $C\subseteqq(0:M)_A\cap B=(0:M)_B$. If $m>1$ then take $0\neq y\in x(\bar{C})^{m-1}$. We have $y=xc_1$ with $c_1\in(\bar{C})^{m-1}$. Consequently $y\bar{C}=xc_1\bar{C}\subseteqq\subseteqq x(\bar{C})^m=(0)$. Since $y\neq 0$, $\bar{C}\subseteqq(0:M)_A$, $C\subseteqq\bar{C}$, $C\subseteqq B$ we have $C\subseteqq(0:M)_A\cap B==(0:M)_B$. □

Proposition 15.3. If B is an ideal of the ring A and M a B-prime module, then MB is an A-prime module with

$$(0:M)_B=(0:MB)_A\cap B.$$

Proof. MB can be considered as a right A-module, since if $x\in MB$ then $x==\sum m_ib_i$ where $m_i\in M$ and $b_i\in B$, whence for $a\in A$ we have the relation

$$xa=\sum m_i(b_ia).$$

We shall show that this defines a unique mapping of the product set $MBxA$ into MB. For, let $x=\overline{\sum m_jb_j}$ be another representation of the element $x\in MB$

with $\overline{m_j}\in M$ and $\overline{b_j}\in B$. Then for every $b\in B$ we have

$$(\sum_i m_i(b_i a))b = \sum_i m_i(b_i ab) = \sum_i (m_i b_i)ab =$$

$$= (\sum_j \overline{m}_j \overline{b}_j)ab = \sum_j m_j(\overline{b_j}ab) = (\sum_j m_j(b_j a))b$$

and thus $(\sum m_i(b_i a) - \sum \overline{m_j}(\overline{b_j}a)) = (0)$. Since M is a B-prime module, we have

$$\sum_i m_i(b_i a) = \sum_j \overline{m_j}(\overline{b_j}a).$$

Thus MB is indeed a right A-module.

Now, if $0\neq x\in MB$ and C is an ideal of A with $xC=(0)$, then we have $xBC=(0)$. Moreover, by proposition 15.1 MB is a B-prime module and thus $0\neq x\in MB$ and $xBC=(0)$ imply $BC\subseteq(0:MB)_B$. Thus $MBBC=(0)$. Then M is a faithful $\bar{B}$-prime module where $\bar{B}=P/(0:M)_B$ with $m\bar{b}=mb$. Thus, we have $x\bar{B}\neq(0)$ for every non-zero element $x\in M$. Then $y=x\bar{b}$ with $b\in B$ for $0\neq y\in xB$. Therefore

$$y\overline{BC} = x\bar{b}\cdot\overline{BC} = x(\bar{b}\cdot\overline{BC}\cdot) = (0).$$

Thus $BC\subseteq(0:M)_B$ follows, because M is a faithful $\bar{B}$-prime module. Accordingly we have $MBC=(MB)C=(0)$, and so $C\subseteq(0:MB)_A$. If $b\in(0:M)_B$, then $b\in B$, $Mb=(0)$, $(MB)b=(0)$, and thus $(0:M)_B\subseteq B\cap(0:MB)_A$. Conversely, if $b\in B\cap(0:MB)_A$, then we have $MBb=(0)$, $Bb\subseteq(0:M)_B$, and since $B\nsubseteq(0:M)_B$ it follows that $b\in(0:M)_B$. □

PROPOSITION 15.4. A ring A is a prime ring if and only if there exists a faithful A-prime module.

PROOF. Let A be a prime ring, and M the additive group A^+ of A regarded as a right module over A. If $Aa=(0)\,(a\in A)$ then $A(a)=(0)$ also holds for the principal ideal (a), whence $(a)=(0)$ follows, because A is a prime ring. Thus $(0:A)_A=(0)$. Now, if $xB=(0)$ for an element $x\in A$ and an ideal B, then we have $(x)B=(0)$, whence we obtain $x=0$ or $B\subseteq(0:A)_A$. Accordingly, the right A-module A^+ is a faithful A-prime module.

Conversely, let M be a faithful A-prime module. Suppose B and C are ideals of A with $BC=(0)$ and $B\neq(0)$. If $x\neq0$ and $xB=(0)$ we obtain $B\subseteq(0:M)_A$, because M is a A-prime module. Thus, since $B\neq(0)$, we have $xB\neq(0)$. If y is an arbitrary non-zero element of xB, then we have $y=xb$ with $b\in B$. This yields $yC=xb\cdot C\subseteq \subseteq xBC=(0)$. Therefore $C\subseteq(0:M)_A=(0)$ and thus A is a prime ring. □

PROPOSITION 15.5. An ideal P of a ring A is a prime ideal if and only if there exists an A-prime module M with $P=(0:M)_A$.

PROOF. If P is a prime ideal of A then $\bar{A}=A/P$ is a prime ring, and thus by proposition 15.4 there exists a faithful $\bar{A}$-prime module M such that $(0)=(0:M)_{\bar{A}}=$ $=(0:M)_{A/P}$. Hence $P=(0:M)_A$.

Conversely, if M is an A-prime module with $P=(0:M)_A$, then M is also an A/P-prime module with $(0:M)_{\bar{A}}=(0:M)_{A/P}$, and accordingly M is a faithful A/P-prime module. Then A/P is by proposition 15.4 a prime ring, and therefore P is a prime ideal of A. □

PROPOSITION 15.6. Every homomorphic image $\bar{A}$ of an **M**-radical ring A is again an **M**-radical ring.

PROOF. Put $\bar{A}=A/B$, and assume that A/B is not an **M**-radical ring. Then the class $\mathbf{M}_A$ is not empty, and thus there exists an $\bar{A}$-prime module M in **M**. Then M is also an A-prime module with $B\subseteq(0:M)_A$ by the second part of axiom (4) for a special class **M**. This contradicts the assumption that A is an **M**-radical ring. □

PROPOSITION 15.7. If $\mathbf{M}(A)$ is the **M**-radical of the ring A, then the factor ring $A/\mathbf{M}(A)$ of A is **M**-semisimple.

PROOF. $\mathbf{M}(A)\subseteq(0:M)_A$, since $\mathbf{M}(A)=\bigcap_{M\in\mathbf{M}_A}(0:M)_A$ for every right A-module $M\in\mathbf{M}_{\bar{A}}$. Then M is an $A/\mathbf{M}(A)$-prime module with $M\in\mathbf{M}_{\bar{A}}$ and $\bar{A}=A/\mathbf{M}(A)$. Now if $\bar{a}\in\bigcap(0:M)_{\bar{A}}$ then $a\in\mathbf{M}(A)=\bigcap(0:M)_A$ and thus $\bar{a}=0$. Consequently $\bar{A}=A/\mathbf{M}(A)$ is in fact **M**-semisimple. □

PROPOSITION 15.8. $\mathbf{M}(B)=B\cap\mathbf{M}(A)$ for every ideal B of a ring A, and thus the **M**-radical is hereditary.

PROOF. By propositions 15.2 and 15.3,

$$B\cap\mathbf{M}(A)=B\cap\Big(\bigcap_{M\in\mathbf{M}_A}(0:M)_A\Big)=\bigcap_{M\in\mathbf{M}_B}(0:M)_B\supseteq\mathbf{M}(B)=$$

$$=\bigcap_{M\in\mathbf{M}_B}(0:M)_B=\bigcap\Big(B\bigcap_{M\in\mathbf{M}_A}(0:MB)_A\Big)=B\cap\Big(\bigcap_{M\in\mathbf{M}_A}0:MB)_A\Big)=$$

$$=B\cap\mathbf{M}(A).$$

Hence $\mathbf{M}(B)=B\cap\mathbf{M}(A)$. □

Proposition 15.8 implies

PROPOSITION 15.9. $\mathbf{M}(A)$ is a radical ideal of A and contains every **M**-radical ideal of A. Every ideal of an **M**-radical ring is again an **M**-radical ring.

PROOF. Let I be an ideal of the ring A for which the class $\mathbf{M}_I$ is empty. Then there exists no I-prime module M in **M**. The modules M in $\mathbf{M}_A$ are also I-modules if $MI\neq(0)$, and in every right A-module M in $\mathbf{M}_A$ there exists a non-zero element m and an ideal K of I with $mK=(0)$ and $K\subsetneqq(0:M)_I$. If $\bar{K}$ is the ideal of A generated by K, then $(\bar{K})^3 \subseteq K$ and since $m(\bar{K})^3 = (0)$ we also have $(\bar{K})^3 \subseteq (0:M)_A$. Since $(0:M)_A$ is a prime ideal, we obtain $K \subseteq \bar{K} \subseteq (0:M)_A$. Since $(0:M)_A$ is a prime ideal, we obtain $K \subseteq \bar{K} \subseteq (0:M)_A$ and thus $MK = (0)$ which contradicts $K \nsubseteq (0:M)_A$. Therefore $MI = (0)$ for every $M \in \mathbf{M}$ and thus $I \subseteq \mathbf{M}(A)$. The other statements immediately follow from the above. □

THEOREM 15.10. $\mathbf{M}(A)$ is a radical in the sense of Amitsur and Kurosh. Moreover, $\mathbf{M}(A)$ is hereditary. Every $\mathbf{M}$-semisimple ring is a subdirect sum of rings A_α such that for every ring A_α there exists a faithful A_α-prime module in $\mathbf{M}$.

PROOF is clear by the foregoing propositions. □

We say that a ring A belongs to an $\mathbf{M}$-special class $M(\mathbf{M})$, if there exists a faithful A-module M in $\mathbf{M}_A$.

THEOREM 15.11. Every $\mathbf{M}$-special class $M(\mathbf{M})$ of rings is a special class of rings (cf. § 11). Conversely, every special class K of rings can be considered as an $\mathbf{M}$-special class $M(\mathbf{M})$, where $\mathbf{M}_A = \{M;\ M$ is an A-prime module and $A/(0:M)_A \in K\}$.

PROOF. If a ring A has a faithful A-prime module, then A is a prime ring since $(0:M)_A = (0)$, and $(0:M)_A$ is a prime ideal for an A-prime module M.

If B is a non-zero ideal of a ring $A \in M(\mathbf{M})$, then there exists a faithful A-prime module M, for which $MB \neq (0)$. Then, by the definition of $\mathbf{M}$, M is also a B-prime module (cf. condition (3) in the definition of $\mathbf{M}$). If M is not a faithful B-module, then $K = (0:M)_B$ is a non-zero ideal of B. If K is the ideal generated by K in A, then $(\bar{K})^3 \subseteq K$ implies $M(\bar{K})^3 = (0)$, and thus $(\bar{K})^3 \subseteq (0:M)_A$. However, since $(0:M)_A$ is a prime ideal of A, we obtain $K \subseteq \bar{K} \subseteq (0:M)_A$ and thus $K = (0)$. Accordingly M is a faithful B-prime module.

Let A be a ring with a faithful A-prime module M in $\mathbf{M}_A$, A be an ideal of the ring B and A^* the annihilator of A in the ring B. We shall show that there exists a faithful $\bar{B}$-prime module M in $\mathbf{M}_{\bar{B}}$, where $\bar{B} = B/A^*$. If M is a faithful A-prime module in $\mathbf{M}_A$, then $MA \neq (0)$, and MA is a B-prime module by condition (2) of the definition for MA. $MAb = (0)$ for some $b \in B$ then we have $Ab = (0)$ since $Ab \subseteq (0:M_A) = (0)$. Since $(0:M)_A$ is a prime ideal of A, we obtain $b \in A^*$. Accordingly $(0:MA)_B = A^*$. This implies that MA is a faithful B/A^*-prime module in $\mathbf{M}$. Thus, the $\mathbf{M}$-special class $M(\mathbf{M})$ is a special class.

Conversely, let K be a special class of rings. Put $\mathbf{M} = \bigcup_{A \in K} \mathbf{M}_A$ with $\mathbf{M}_A = \{M;\ M$ is an A-prime module and $A/(0:M)_A \in K\}$. We shall show that $\mathbf{M}$ is a special class of modules.

Every module in $\mathbf{M}_A$ is an A-prime module.

If $M \in \mathbf{M}_A$ and B is an ideal of A with $MB \neq (0)$, then $(0:M)_B$ is an ideal of A, since $(0:M)_B = B \cap (0:M)_A$, and $B/(0:M)_B$ is isomorphic to an ideal of $A/(\bar{0}:M)_A \in$

$K\in$ since

$$B/(0:M)_B \cong (B+(0:M)_A)/(0:M)_A.$$

$MB\neq(0)$ implies $B/(0:M)_B\neq(0)$, and thus we have $B/(0:M)_B\in K$ and $M\in\mathbf{M}_B$.

If $M\in\mathbf{M}_B$ and B is an ideal of A, then we have $B/(0:M)_B\in K$. If $x\in(0:MB)_A$, then we obviously have $Bx=(0:M)_B$ since $MBx=(0)$ and thus $x\in((0:M)_B:B)$. But this right-hand ideal quotient is the right annihilator of $B/(0:M)_B$ in $A/(0:M)_B$. Hence $A/(0:MB)_A\in K$ because $B/(0:M)_B\in K$

$$A/(0:MB)_A \cong (A/(0:M)_B)/(B/(0:M)_B)^*,$$

and K is a special class of rings.

Finally, if $M\in\mathbf{M}_A$, $P\subseteq(0:M)_A$ and $\bar{A}=A/P$, then since $A/(0:M)_A\in K$ and

$$\bar{A}/(0:M)_{\bar{A}} = (A/P)/((0:M)_A/P) \cong A/(0:M)_A$$

we also have $M\in\mathbf{M}_{\bar{A}}$. Similarly it can be shown that $\bar{A}=A/P$ and $M\in\mathbf{M}_{\bar{A}}$ imply also $M\in\mathbf{M}_A$. $\mathbf{M}_A$ contains only faithful modules. □

Problem 16. Find a module-theoretic characterization of supernilpotent radicals!

Problem 17. Does there exist for every natural number n a supernilpotent but non-special radical $\mathbf{S}_n$ for which the following holds: If A_n is an $\mathbf{S}_n$-semisimple non-zero ring, then the isomorphism $A_m\cong A_n$ always implies $m=n$?

Problem 18. Investigate the finitely generated antisimple rings!

Problem 19. Which antisimple rings have a *distributive lattice* of ideals, and of right ideals, respectively?

Problem 20. Investigate non-hereditary radicals $\mathbf{R}$ such that every nilpotent ring is an $\mathbf{R}$-radical ring!

Problem 21. How can we formulate an explicit necessary and sufficient condition for a hereditary radical $\mathbf{R}$ and its supplementing radical $\mathbf{R}'$ to form a dual pair (i.e. $\mathbf{R}=\mathbf{R}''$)?

Problem 22. Let A be a ring, Q^* the two-sided annihilator of the ideal Q of A. Suppose every two-sided ideal of the factor ring A/Q is idempotent. Does $Q\cap(Q^*)^2=(0)$ hold?

Problem 23. Which generalization of Andrunakievich's theorem 14.5 holds for rings with minimum condition on two-sided principal ideals, for linearly compact rings, and for rings with cocompactly generated lattice of ideals? (For the last see also problem 10.)

Problem 24.† Does there exist a simple Jacobson radical ring without divisors of zero?

CHAPTER III

NIL RADICALS

In this chapter we shall investigate the lower nil radical and the upper nil radical of Baer. We shall also discuss the important example of Baer of a nil ring with no non-zero nilpotent ideals. It will be shown that the lower nil radical coincides with the intersection of all prime ideals. Moreover, we shall discuss the hereditary subradicals of the lower nil radical. An important example of Golod will show the existence of a nil ring which is not locally nilpotent. We shall deduce some results connected with the locally nilpotent Levitzki nil radical. Nil rings with certain conditions and their relation to the pivotal monomials will be investigated. We shall also mention in brief the generalized nil radical. Finally, we shall point out some open problems.

§ 16. THE UPPER AND LOWER NIL RADICALS IN THE SENSE OF BAER

For the subject-matter of this section we refer the reader to Baer [1].

We recall the following definitions. An element x of a ring A is called *nilpotent*, if there is a positive integer n with $x^n=0$. A ring A is said to be a *nil ring* if every element of A is nilpotent. A is called *nilpotent*, if $A^n=(0)$ for an exponent $n\geqq 2$. Every nilpotent ring is a nil ring. But the direct sum of the rings B_n with $B_n^{n+1}=$ $=(0)$, $B_n^n\neq(0)$ is a nil ring which is not nilpotent since $(0)\neq B_n^n\subseteqq A^n$. An ideal or a right ideal I of a ring A is called a *nil ideal* (a nilpotent ideal) or a nil right ideal (a nilpotent right ideal), if I is a nil ring (a nilpotent ring).

PROPOSITION 16.1. Every subring and every homomorphic image of a nil ring (nilpotent ring) A is a nil ring (nilpotent ring). If, for an ideal B of A, both B and A/B are nil rings (nilpotent rings), then A is also a nil ring (nilpotent ring).

PROOF is clear. □

PROPOSITION 16.2. The sum of two nil ideals (nilpotent ideals) B_1 and B_2 of a ring A is a nil ideal (nilpotent ideal).

PROOF. Consider the isomorphism $(B_1+B_2)/B_1 \cong B_2/(B_1 \cap B_2)$. Since by proposition 16.1, $B_2/(B_2 \cap B_1)$ is a nil ring, this holds also for $(B_1+B_2)/B_1$ and thus, by a second application of proposition 16.1, also for B_1+B_2. The proof is similar for nilpotent ideals B_1 and B_2. □

PROPOSITION 16.3. The sum U of all nil ideals and the sum N of all nilpotent ideals of a ring are nil ideals, and N contains the sum of all nilpotent right ideals and nilpotent left ideals.

PROOF. If $z \in U$ then there exists a finite number of nil ideals $B_1, B_2, \ldots, B_n$ of A with $z \in B_1 + \ldots + B_n$. But then by proposition 16.2 $B_1 + \ldots + B_n$ is a nil ideal. Consequently z is nilpotent and thus U is a nil ideal of A. Since $N \subseteqq U$, N is also a nil ideal of A. Let R be a nilpotent right-ideal of A. Then $R+AR$ is an ideal of A, for which $(R+AR)^n \subseteqq R^n + AR^n$ can be proved by induction. Accordingly $R+AR$ is nilpotent and $R \subseteqq R+AR \subseteqq N$. Therefore $\sum R \subseteqq N$. □

An ideal I of a ring A is called a *nil radical* if the following conditions are fulfilled:

(1) I is a nil ideal.
(2) A/I does not contain non-zero nilpotent ideals.

By propositions 16.1 and 16.3, the sum U of all nil ideals is a nil radical, and every nil radical is contained in U. But the sum N of all nilpotent ideals is not in general a nil radical. As the ring A considered in the proof of theorem 2.2 (system V) shows, the factor ring A/N of A with respect the sum of all nilpotent ideals can have non-zero nilpotent ideals. But by means of a transfinite procedure we can define another nil radical. This is the *Baer lower nil radical* [1]. We define in a ring A an ideal $N(\alpha)$ for every ordinal number α as follows:

(i) $N(0)=(0)$.

Let us assume that $N(\alpha)$ is already defined for every $\alpha<\beta$.

(ii) If $\beta=\alpha+1$, $N(\beta)/N(\alpha)$ is the sum of all nilpotent ideals of $A/N(\alpha)$.

(iii) If β is a limit ordinal number, $N(\beta)=\sum_{\alpha<\beta} N(\alpha)$.

Since every ring is a set, for every ring there exists an ordinal number γ with $N(\gamma)=N(\gamma+1)$. We denote this ideal $N(\gamma)$ of the ring A by $\mathbf{B}=\mathbf{B}(A)$.

PROPOSITION 16.4. $\mathbf{B}(A)$ is contained in every nil radical of the ring A. Moreover, $\mathbf{B}(A)$ is the intersection of all ideals I of A for which A/I does not have a non-zero nilpotent ideal.

PROOF. Let I be an ideal of A, for which A/I contains no non-zero nilpotent ideal. Then $N(0)=(0)\subseteqq I$. Let us assume that $N(\alpha)\subseteqq I$ is already proved for every ordinal number $\alpha<\beta$. If $\beta=\alpha+1$ then $N(\beta)$ is the sum of all ideals K of A such that K contains $N(\alpha)$ and $K/N(\alpha)$ is nilpotent. Then we have $K^n \subseteq N(\alpha) \subseteq I$ and thus $(K+I)^n \subseteq I$. Hence we have $K+I \subseteq I$ and thus $N(\beta) \subseteq I$. If β is a limit ordinal, then again $N(\beta) \subseteq I$. Therefore we have $\mathbf{B}(A) \subseteq I$ and thus $\mathbf{B}(A) \subseteq \bigcap I$. However, since $A/\mathbf{B}(A)$ does not have non-zero nilpotent ideal, $\mathbf{B}(A)$ occurs among these ideals I and hence we have $\mathbf{B}(A) = \bigcap I$. Thus, the last statement of the proposition is proved. Since A/U contains no non-zero nilpotent ideal, we have $\mathbf{B}(A) = \bigcap I \subseteq U$ and thus $\mathbf{B}(A)$ is a nil ideal. Moreover, it is a nil radical. □

Since every nil radical contains $\mathbf{B}(A)$ and every nil radical is contained in $U = \mathbf{U}(A)$, $\mathbf{B}(A)$ is called the *Baer lower nil radical* and $\mathbf{U}(A)$ is called the *Baer upper nil radical.*

The following example of Baer [1] shows that there exists a ring A such that $A = \mathbf{U}(A)$ and $\mathbf{B}(A) = (0)$.

EXAMPLE 16.5. Let the Abelian group G be the direct sum of countably many infinite cyclic groups

$$\{b(0)\}, \{b(1)\}, \{b(-1)\}, \dots, \{b(n)\}, \{b(-n), \dots\} \quad (n = 0, \pm 1, \pm 2, \pm 3, \dots).$$

Then there exists a unique endomorphism $u(i)$ of G such that

$$b(j)u(i) = \begin{cases} 0 & \text{if } j \equiv 0 \pmod{2^i} \\ b(j-1) & \text{if } j \not\equiv 0 \pmod{2^i}. \end{cases}$$

It can be proved by induction that

$$b(j)u(i_0) \dots u(i_m) = \begin{cases} 0 & \text{if } j \equiv n \pmod{2^{i_n}} \\ & \text{for at least one } n \text{ with } 0 \leqq n \leqq m, \\ b(j-m-1) & \text{if } j \not\equiv n \pmod{2^{i_n}} \\ & \text{for every } n \text{ with } 0 \leqq n \leqq m. \end{cases} \qquad (*)$$

Let A be the ring generated by the endomorphisms $u(1), u(2), u(3), \dots$. We shall show that A is a nil ring, i.e. $A = \mathbf{U}(A)$. Every element $a \in A$ can be represented as a sum $a = \sum_{i=1}^{k} a_i$, where

$$a_i = \pm u((i, 0)), \dots, u((i, m_i))$$

and (i, j) are some positive integers with $m_i \geqq 0$. Let h be the maximum of all numbers $2^{(i,n)}$ for which (i, n) occurs in $\sum a_i$. Then from $(*)$ we obviously have

$$b(j)x_{i_1}x_{i_2} \dots x_{i_h} = 0$$

for every j, because the product $x_{i_1} \dots x_{i_h}$ contains at least h factors $u(s)$ with $2^s \leqq h$. So we obtain $x_{i_1} \dots x_{i_h} = 0$ and thus $a^h = 0$. Therefore A is a nil ring and thus $A = \mathbf{U}(A)$.

Now we shall show that $\mathbf{B}(A)=(0)$, and prove that A does not contain non-zero nilpotent right ideals. With this aim we verify that for every non-zero element $x \in A$ and for every exponent h there exists an element $b=(u(s))^{2^r-m-1} \in A$ with $(xb)^h \neq 0$. Then a principal right ideal $(x)_r$ of A is nilpotent only in the case $x=0$, whence $\mathbf{B}(A)=0$ follows.

Now let x be an arbitrary element of A. Then x can be represented as a sum $x=\sum_{i=0}^{m} y(i)$, where

$$y(i)=\sum_{j=1}^{k(i)} e(i,j)y(i,j),\ e(i,j)=\pm 1,$$

and

$$y(i,j)=u((i,j,0)),\ \ldots,\ u((i,j,i))$$

with $y(n) \neq 0$, $0<(i,j,n)$; $0<k(i)$. Here (i,j,k) are natural numbers.

Let r be a natural number such that $2^r>m+1$ $\left(\text{with } x=\sum_{i=0}^{m} y(i)\right)$ and $r \geqq (i,j,n)$ for every i,j,n with $0 \leqq i \leqq m$; $1 \leqq j \leqq k(i)$; $0 \leqq n \leqq i$.

Put $z(i,j)=y(i,j)u(s)^{2^r-m-1}$. Since $y(m) \neq 0$, there exists an integer t with $b(t)y(m) \neq 0$. Thus, we can assume k to be a number for which

$$b(t)y(m,j)=\begin{cases} b(t-m-1) & \text{if } j=1,2,\ldots,k, \\ 0 & \text{if } j>k. \end{cases}$$

Then we obtain $b(t)y(m)=\sum_{j=1}^{k} e(m,j)b(t-m-1) \neq 0$. Therefore $\sum_{j=1}^{k} e(m,j) \neq 0$. This will be used later on.

If i_v and j_v are integers for $v=1,2,\ldots,h$ with $0 \leqq i_0 \leqq m$ and $1 \leqq j_0 \leqq k(i_0)$ then $(*)$ implies:

$$b(t+eh2^r)z(i_1,j_1)\ldots z(i_h,j_h)=0$$

or

$$b\Big(t+eh2^r-\sum_{v=1}^{h}(i_v+1+2^r-m-1)\Big)=b\Big(t+(e-1)h2^r+mh-\sum_{v=1}^{h} i_v\Big),$$

and this latter element equals to the element $b\big(t+(e-1)h2^r\big)$ if and only if $i_1= =i_2=\ldots=i_h=m$.

If $k<j_v$ for some v, then we have $b(t)y(m,j_v)=0$ and thus by $(*)$ there exists an integer with $0 \leqq n \leqq m$ and $t \equiv n \pmod{2^{(m,j_v,n)}}$. But this implies

$$t+eh2^r \equiv t \equiv n \equiv n+(v-1)2^r \pmod{2^{(m,j_v,n)}}.$$

Hence, by $(*)$ we obtain

$$b(t+eh2^r)z(m,j_1)\ldots z(m,j_h)=0.$$

Therefore, if

$$b(t+eh2^r)\,z(i_1,j_1)\dots z(i_h,j_h) = b\big(t+(e-1)h2^r\big)$$

then we have $i_1=i_2=\dots=i_h=m$ and $1\leqq j_v\leqq k$. Assume that there exist integers j_v, j_v', $1\leqq j_v$, $j_v'\leqq k$ such that

$$b(t)\,z(m,j_1)\dots z(m,j_h) = 0,$$

$$b(t+h2^r)\,z(m,j_i)\dots z(m,j_h') = 0.$$

Then by ($*$), there exist integers n, v with $1\leqq v\leqq h$ and

$$t \equiv \begin{cases} n+(v-1)2^r \pmod{2^{(m,j_v,n,)}} & \text{if } \ 0 \leqq n \leqq m, \\ n+(v-1)2^r \pmod{2^s} & \text{if } \ m < n < 2^r. \end{cases}$$

If $0\leqq n\leqq m$, then we have $t\equiv n \pmod{2^{(m,j_v,n)}}$ and then ($*$) implies $b(t)y(m,j_v)= =0$ which is impossible because $j_v\leqq k$. Therefore there exist integers n, v, and n', v' with $m<n<2^r$, $1\leqq v\leqq h$, $t\equiv n+(v-1)2^r \pmod{2^s}$, and with $m<n'<2^r$, $1\leqq \leqq v'\leqq h$,

$$t+h2^r \equiv n'(v'-1)\,2^r \pmod{2^s},$$

since $r<s$ we have $n\equiv n' \pmod{2^r}$. Moreover, since $m<n$ and $n'<2^r$ even $n=n'$. Thus we have

$$(v-1+h)\,2^r \equiv (v'-1)\,2^r \pmod{2^s}$$

or $h+v-v'\equiv 0 \pmod{2^{s-r}}$. Since $1\leqq v, v'\leqq h$ we have $0<h+v-v'<2h\leqq 2^{s-r}$, which is impossible. Therefore there exists an integer t' such that

$$b(t')\,z(i_1,j_1)\dots z(i_h,j_h) = b(t'-h2^r)$$

holds if and only if $i_1=i_2=\dots=i_h=m$ and $1\leqq j_v\leqq k$ for $1\leqq v\leqq h$. Hence we obtain

$$b(t')\big(xu(s)^{2r-m-1}\big)^h = \sum_{j=1}^{h} e(m,j)^h\, b(t'-h2^r)+\sum_{i\neq t'-h2^r} c(i)\,b(i) \neq 0,$$

because the coefficient $\sum_{j=1}^{h} e(m,j)$ differs from zero by the preceding (here $c(i)$ are integers).

Now, $\big(xu(s)^{2^r-m-1}\big)^h\neq 0$ for arbitrary $x\in A$, and h which proves $\mathbf{B}(A)=(0)$.

Let C denote the ideal generated by $u(2)$, $u(3)$, ..., in A. We can prove that $u(1)\notin C$ and $A^2\leqq C$. Therefore C is an ideal between $\mathbf{B}(A)$ and $\mathbf{U}(A)$ but not a nil radical, because A/C is nilpotent.

§ 17. THE LOWER NIL RADICAL AND PRIME IDEALS

For the subject-matter of this section we refer the reader to Divinsky [1] and Levitzki [12].

We recall that an ideal P of a ring A is said to be a *prime ideal* of A if $BC \subseteqq P$ for ideals B, C implies either $B \subseteqq P$ or $C \subseteqq P$. If A/P has no divisors of zero for an ideal P then P is a prime ideal of A. But a ring A may have a prime ideal P such that A/P contains divisors of zero. For example, in the full $n \times n$ matrix ring $(n \geqq 2)$ over a division ring, (0) is a prime ideal and this ring contains divisors of zero. However, if A is a commutative ring, then P is a prime ideal of A if and only if A/P has no divisors of zero.

Proposition 17.1. An ideal P of a ring A is a prime ideal if and only if $aAb \subseteqq P$ implies $a \in P$ or $b \in P$.

Proof. Let P be a prime ideal of A with $aAb \subseteqq P$. Then we have $AaA \cdot AbA \subseteqq P$. Therefore $AaA \subseteqq P$ or $AbA \subseteqq P$. If $AaA \subseteqq P$ then we have $(a)^3 \subseteqq P$ for the principal ideal (a). Hence $(a) \subseteqq P$ and therefore $a \in P$. Similarly, $AbA \subseteqq P$ implies $b \in P$.

Conversely, let the condition of the statement be fulfilled for P. Let B and C be arbitrary ideals of A with $B \nsubseteqq P$ and $C \nsubseteqq P$. Then there exist elements $b \in B$ and $c \in C$ with $b \notin P$ and $c \notin P$. By the condition for P we have $bAc \nsubseteqq P$. But $bAc \subseteqq$ $\subseteqq BC$ so $BC \nsubseteqq P$ and thus P is a prime ideal. □

From proposition 17.1 it follows that for every prime ideal P and for elements $a \notin P$ and $b \notin P$ there exists an element $x \in A$ with $axb \notin P$. Therefore, the complement $M = A \setminus P$ of a prime ideal P in a ring A has the following property:

$$\text{for every} \quad a, b \in M \quad \text{there exists an} \quad x \in A \quad \text{with} \quad axb \in M. \qquad (*)$$

A subset of a ring is called an *m-system* if it has property (*).

A sequence of elements $b_0, b_1, b_2, \ldots$ of a ring is called an *m-sequence,* if there exists a second sequence $c_0, c_1, c_2, \ldots$ of elements of the ring satisfying the condition $b_{i+1} = b_i c_i b_i$ for $i = 0, 1, 2, 3, \ldots$ Every m-sequence is an m-system. We say that an m-sequence vanishes, if $b_k = 0$ for some k. But in this case $b_m = 0$ for every $m \geqq k$. If b_i, b_j are two members of an m-sequence, then we have $b_k = b_i c_{ij} b_j$ for every $k > i, j$, where $c_{ij} \in A$ is an element such that $b_i c_{ij} b_j$ is contained in the m-sequence.

Proposition 17.2. If B is an ideal of a ring such that A/B does not contain non-zero nilpotent ideals, and C is an ideal of A with $C \supset B$, then there exists a prime ideal P of A with $P \supseteqq B$ and $P \nsupseteqq C$.

Proof. Choose $c_0 \in C$ with $c_0 \notin B$. Then $(c_0)^k \nsubseteqq B$ for the principal ideal (c_0) and for every k. Since $(c_0)^3 \subseteqq Ac_0A$ we have $(Ac_0A)^k \nsubseteqq B$ for every k. Hence we have $c_0Ac_0 \subseteq\!\!\!\!\!/\; B$. Accordingly, there exists an element $a_0 \in A$ with $c_1 = c_0 a_0 c_0 \notin B$. Repeat this procedure with c_1 instead of c_0. We obtain an element $a_1 \in A$ with $c_2 = c_1 a_1 c_1 \notin B$ and $c_2 \in C$. In this way we get an m-sequence $c_0, c_1, c_2, \ldots$, contained in the ideal C and having an empty intersection with the ideal B. Then, by Zorn's lemma there exists an ideal maximal with respect to containing B and having an empty intersection with this m-sequence. We shall show that P is a prime ideal of A. If D and E are ideals of A with $D \nsubseteq P$ and $E \nsubseteq P$, then $D+P$ and $E+P$ have a non-empty intersection with the above m-sequence. Therefore there exist $c_i \in D+P$, $c_j \in E+P$, and then

$$c_k = c_i a_{ij} c_j \in (D+P)(E+P)$$

for a suitable c_k and $a_{ij} \in A$. Thus, we have $(D+P)(E+P) \nsubseteq P$ and so $DE \nsubseteq P$. Therefore P is a prime ideal with $P \supseteq B$ and $P \nsupseteq C$. □

THEOREM 17.3. The lower nil radical of a ring is the intersection of all prime ideals of the ring.

PROOF. Since A/P has no non-zero nilpotent ideals for a prime ideal P, we have $\mathbf{B}(A) \subseteq D$ by proposition 16.4 where D is the intersection of all prime ideals. Now if $D \neq \mathbf{B}(A)$ then by proposition 17.2 there exists a prime ideal such that $P \supseteq \mathbf{B}(A)$ and $P \nsupseteq D$, since $A/\mathbf{B}(A)$ has no non-zero nilpotent ideal. But then $P \nsupseteq D$ is a contradiction, and thus $\mathbf{B}(A) = D$. □

PROPOSITION 17.4. Every ring which does not have non-zero nilpotent ideals, is a subdirect sum of prime rings (i.e. of rings, in which (0) is a prime ideal).

PROOF is clear by theorem 17.3. □

PROPOSITION 17.5. An element a of a ring A is contained in the lower nil radical $\mathbf{B}(A)$ if and only if the following conditions are fulfilled:

(1) Every m-sequence beginning with a vanishes.
(2) Every m-system containing a also contains 0.

PROOF. Suppose $a \notin \mathbf{B}(A)$. Then we have $(a)^k \nsubseteq \mathbf{B}(A)$ for the principal ideal (a) and for every k. Since $(a)^3 \subseteq AaA$ we have $(AaA)^k \nsubseteq \mathbf{B}(A)$ for every k, and thus $aAa \nsubseteq \mathbf{B}(A)$. Therefore there exists an element $c_1 \in A$ with $a_1 = ac_1a \notin \mathbf{B}(A)$. Similarly we obtain $a_i \in A$ with $a_i = a_{i-1}c_{i-1}a_{i-1}$, $a_i \notin \mathbf{B}(A)$, and thus $a, a_1, a_2, \ldots$ is a non-vanishing m-sequence.

Now suppose $a \in \mathbf{B}(A)$, and let $a, a_1, a_2 \ldots$ be an m-sequence. Then we have $a_i \in \mathbf{B}(A)$ for every i. Let α_i be the minimal ordinal number, for which $a_i \in N(\alpha_i)$ (cf. definition in § 16). Then a_i is not a limit ordinal number. Let a_m be the minimum of these ordinal numbers α_i. If $\alpha_m \neq 0$ then $\alpha_m = \beta + 1$. Then we have $(a_m)^{2^k} \subseteq$

$\subseteqq N(\beta)$ for an exponent 2^k. Since $a_{m+1}=a_m c_m a_m \in (a_m)^2$ and $a_{m+2}=a_{m+1}c_{m+1}a_{m+1}\in$ $\in (a_m)^4$ we can show $a_{m+k}\in (a_m)^{2k}\in N(\beta)$. This contradicts the minimality of α_m. Therefore $\alpha_m=0$, and thus $a_m=0$.

Moreover, if $a\notin \mathbf{B}(A)$, then there exists a non-vanishing m-sequence, i.e. an m-system which does not contain 0. Conversely, if $a\in \mathbf{B}(A)$ and M is an m-system with $a\in M$, then we can choose an m-sequence $a, a_1, a_2, \ldots$ from M which vanishes by the preceding, and thus we have $0\in M$. □

§ 18. NIL RADICALS AS AMITSUR–KUROSH RADICALS

For the subject-matter of this section we refer the reader to Divinsky [1].

We will show that the nil radicals, introduced in § 16 are also radicals in the sense of Amitsur and Kurosh. In order to do this we need the following

PROPOSITION 18.1. If B is an ideal of a ring A which has no non-zero nilpotent ideals, then the ring B has no non-zero nilpotent ideals, either.

PROOF. If C is a nilpotent ideal of B with $C^n=(0)$, then we claim that $(ACA)^{2n-1}=$ $=(0)$. We shall show by induction that $(ACA)^{2n-1}\subseteqq AC^nA$. This relation obviously holds for $n=1$. Let us assume that $(ACA)^{2k-1}\subseteqq AC^kA$ is already proved for $k<n$. Then we obtain

$$(ACA)^{2k+1} = (ACA)^{2(k+1)-1} = (ACA)^{2k-1}(ACA)^2 \subseteqq$$
$$\subseteqq AC^kA(ACA)^2 \subseteqq AC^kBCA \subseteqq AC^{k+1}A.$$

Thus $C^n=(0)$ implies $(ACA)^{2n-1}=(0)$. Since A has no non-zero nilpotent ideals, $ACA=0$, because ACA is an ideal of A. If c is an element of C, then $(c)^3\subseteqq C$ for the principal ideal (c) generated in A. Therefore $(c)^5\subseteqq ACA$ and $(c)^5=(0)$. Hence $c=0$, and $C=(0)$. □

A ring A is called a **B**-*ring*, if A is the lower nil radical of a ring C. Every nil ring is the upper nil radical of itself.

THEOREM 18.2. The class of nil rings and the class of **B**-rings are classes of radical rings in the sense of Amitsur and Kurosh.

PROOF. First we discuss the nil rings. Every homomorphic image of a nil ring is a nil ring. Moreover, in every ring A the upper nil radical $\mathbf{U}(A)$ is a nil ideal which contains every other nil ideal of A and $A/\mathbf{U}(A)$ has no non-zero nil ideals. Thus, the class of nil rings is a class of radical rings.

We shall now show that every homomorphic image of a **B**-ring is a **B**-ring. Let A be a **B**-ring, and A/C a homomorphic image of A. Since (0) is a **B**-ring, we may assume $A/C\neq(0)$. If A/C is not a **B**-ring then A/C differs from its lower nil radical B/C. Then $(A/C)/(B/C)\cong A/B$ has no non-zero nilpotent ideals, and thus by the second part of proposition 16.4, B contains the lower nil radical of A. Therefore A differs from its lower nil radical $\mathbf{B}(A)$, and at the same time A is the lower nil radical of a ring D. Accordingly $A=\cup N(\alpha)$, where the ideals $N(\alpha)$ are defined in the ring D. Therefore there exists a minimal ordinal number α with $N(\alpha)\subsetneqq\mathbf{B}(A)$. Then α is not a limit ordinal number, and $N(\alpha-1)\subseteqq\mathbf{B}(A)$ holds. Since $N(\alpha)\subsetneqq\mathbf{B}(A)$, there exists a nilpotent ideal $I/N(\alpha-1)$ of $D/N(\alpha-1)$ such that $I^r=N(\alpha-1)$, $I\subsetneqq B(A)$, and $I^r\subseteqq B(A)$ for an exponent r. The right-hand side of

$$(\mathbf{B}(A)+I)/\mathbf{B}(A)\cong I/(\mathbf{B}(A)\cap I)$$

is nilpotent since $I^r\subseteqq\mathbf{B}(A)$. But since $A/\mathbf{B}(A)$ does not contain non-zero nilpotent ideals, we have $I\subseteqq\mathbf{B}(A)$ and thus $N(\alpha)\subseteqq\mathbf{B}(A)$. Consequently $A=\cup N(\alpha)=\mathbf{B}(A)$ which contradicts our assumption. Accordingly A/C is also a $\mathbf{B}$-ring.

We shall now show that every ring A possesses a $\mathbf{B}$-ideal which contains every other $\mathbf{B}$-ideal of the ring A. Let A be a ring and $\mathbf{B}(A)$ the lower nil radical of A. Then $\mathbf{B}(A)$ is a $\mathbf{B}$-ideal. Let B_1 be an arbitrary $\mathbf{B}$-ideal of A. Then B_1 is the lower nil radical of a ring C. The factor ring $A/\mathbf{B}(A)$ contains no non-zero nilpotent ideals, and by the isomorphism

$$(\mathbf{B}(A)+B_1)/\mathbf{B}(A)\cong B_1/(B_1\cap\mathbf{B}(A))$$

we have $B_1=\mathbf{B}(A)$, because by the above $B_1/(B_1\cap\mathbf{B}(A))$ is a $\mathbf{B}$-ideal. Since $A/\mathbf{B}(A)$ has no non-zero nilpotent ideals nor does $(\mathbf{B}(A)+B_1/\mathbf{B}(A)$ by proposition 18.1, and thus $B_1\subseteqq\mathbf{B}(A)$ holds.

Since $A/\mathbf{B}(A)$ does not contain non-zero nilpotent ideals, we have $\mathbf{B}(A/\mathbf{B}(A))=$ $=(0)$, and thus $\mathbf{B}$ is a radical property in the sense of Amitsur and Kurosh. □

THEOREM 18.3. The lower nil radical $\mathbf{B}$ is the special radical determined by the special class of all prime rings.

PROOF. Let $\mathbf{K}$ be the class of all prime rings. Let B be an ideal of a ring A in $\mathbf{K}$ and C, D ideals of B with $CD=(0)$. Since A is a prime ring, we have $BC\neq(0)$ and $BCB\neq(0)$ if $C\neq(0)$. Since BCB is an ideal in the prime ring A, we obtain $BCB\cdot D\neq(0)$ if $D\neq(0)$. Now $BC\cdot BD\subseteqq CD$ so $CD\neq(0)$. Therefore B is also a prime ring.

Let A be a prime ring which is an ideal of a ring B. Let A^* be the annihilator of A in B. Suppose C and D are ideals of B with $CD\subseteqq A^*$. Then we have $AC=(0)$ or $DA=(0)$ since $AC\cdot DA\subseteqq AA^*A=(0)$, $AC\subseteqq A$, $DA\subseteqq A$, and A is a prime ring. If $AC=(0)$ then $(CA)^2=(0)$ and $CA=(0)$. Thus $C\subseteqq A^*$. Similarly $DA=(0)$ implies $D\subseteqq A^*$. Accordingly $B/A^*\in\mathbf{K}$, and thus $\mathbf{K}$ is a special class of rings.

Let $\mathbf{S}$ be the upper radical, determined by the class $\mathbf{K}$. Then by theorem 11.5, $\mathbf{S}(A)$ is the intersection of all ideals M of the ring A, for which $A/M\in\mathbf{K}$ holds.

Hence $\mathbf{S}(A)$ is precisely the lower nil radical by theorem 17.3. Therefore $\mathbf{S}$ is also special. □

THEOREM 18.4. The upper nil radical $\mathbf{U}(A)$ of A is determined by the special class $\mathbf{K}$ of all those prime rings, which contain no non-zero nil ideals.

PROOF. If B is an ideal of a prime ring A with an upper nil radical (0) then B is a prime ring, as shown in the proof of theorem 18.3. However, since every ideal of a nil ring is again a nil ring, we have $\mathbf{U}(B)=B\cap\mathbf{U}(A)$ by proposition 11.1. Thus $\mathbf{U}(B)=(0)$ since $\mathbf{U}(A)=(0)$. Consequently $B\in\mathbf{K}$.

Now choose $B\in\mathbf{K}$ and let B be an ideal of the ring A, and B^* the annihilator of B in A. Following the proof of theorem 18.3, A/B^* is a prime ring. Assume that C/B^* is a nil ideal of A/B^*. If $C\cap B=(0)$ we have $CB=BC=(0)$ and thus $C\subseteq B^*$ and $C/B^*=(0)$. But $C\cap B$ is an ideal of B if $C\cap B\neq(0)$. For every $x\in C\cap B$ a power x^n belongs to B^*, because C/B^* is a nil ideal and thus we obtain $x^n\in B\cap B^*$. Since $B\cap B^*$ is a nilpotent ideal in the prime ring B, we have $B\cap B^*=(0)$. Thus $x^n=0$. Hence $C\cap B$ is a non-zero nil ideal of B, which is impossible. Therefore $A/B^*\in\mathbf{K}$, and thus $\mathbf{K}$ is a special class of rings.

Let $\mathbf{S}$ be the upper radical determined by the special class $\mathbf{K}$. By theorem 11.5, $\mathbf{S}(A)$ is, in every ring A, the intersection of all those ideals T_α of A, for which $A/T_\alpha\in\mathbf{K}$. Since $\mathbf{U}(A)\subseteq T_\alpha$ for every T_α, we obtain $\mathbf{U}(A)\subseteq\mathbf{S}(A)=\cap T_\alpha$. If $\mathbf{S}(A)\neq$ $\neq\mathbf{U}(A)$, then there exists an element $a\in\mathbf{S}(A)$ with $a\notin\mathbf{U}(A)$. In the principal ideal (a) there exists an element b which is not nilpotent, and thus $\mathbf{U}(A)$ has an empty intersection with the set $\{b, b^2, b^3, \ldots\}$. By Zorn's lemma there exists an ideal P of A maximal with respect to the properties that $P\supseteq\mathbf{U}(A)$ and P has an empty intersection with the set $\{b, b^2, b^3, \ldots\}$. This ideal P is a prime ideal of A. For if D and E are ideals of A with $D\supset P$ and $E\supset P$, then we have $b^m\in D$ and $b^n\in E$ for some m and n, and thus $b^{m+n}\in DE$. Accordingly $DE\nsubseteq P$, which immediately shows that P is a prime ideal. Now, if F/P is a non-zero nil ideal of A/P, then a power b^k of b is contained in F, by the maximality of P. Since $b^k\in F$, and by the condition that F/P is a nil ideal, we have $(b^k)^l=b^{kl}\in P$, a contradiction. Therefore $\mathbf{U}(A/P)=(\bar{0})$, and thus $A/P\in\mathbf{K}$ and $\mathbf{S}(A)\subseteq P$. Hence $a\in\mathbf{S}(A)$ implies $a\in P$, $(a)\subseteq P$ and $b\in(a)=P$, so that P cannot have an empty intersection with the set $\{b, b^2, b^3, \ldots\}$. Therefore $\mathbf{U}(A)=\mathbf{S}(A)$. □

REMARK 18.5. From theorems 18.3 and 18.4 we obtain that the upper nil radical and the lower nil radical are hereditary radicals and also radicals in the sense of Amitsur and Kurosh. Therefore the results 18.1 and 18.2 are consequences of theorems 18.3 and 18.4. However, it was useful to give immediate, more elementary proofs of the important results 18.1 and 18.2, without application of the theory of supernilpotent and special radicals.

§ 19. HEREDITARY SUBRADICALS OF THE LOWER NIL RADICAL

For the subject-matter of this section we refer the reader to Armendariz [2].

For the radicals $\mathbf{S}_1$ and $\mathbf{S}_2$ we say that $\mathbf{S}_1$ is a subradical of $\mathbf{S}_2$, if $\mathbf{S}_1 \leqq \mathbf{S}_2$ holds, i.e. if every $\mathbf{S}_1$-radical ring is also an $\mathbf{S}_2$-radical ring.

Proposition 19.1. The lower nil radical $\mathbf{B}$ coincides with the lower radical, determined by the class of all nilpotent rings. Moreover, it coincides also with the lower radical determined by a zero-ring Z_0 with infinite cyclic additive group.

Proof. Since every nilpotent ring is a radical ring for the lower nil radical $\mathbf{B}$ which is a radical in the sense of Amitsur and Kurosh, we have $\mathbf{R} \leqq \mathbf{B}$, where $\mathbf{R}$ is the lower radical determined by the class of all nilpotent rings. Now, if $\mathbf{R} < \mathbf{B}$, then there exists a ring A with $\mathbf{R}(A) \subset \mathbf{B}(A)$, and then $B_1 = \mathbf{B}(A)/\mathbf{R}(A)$ is a $\mathbf{B}$-radical ring which is $\mathbf{R}$-semisimple. Accordingly, the $\mathbf{B}$-radical ring B_1 contains a non-zero nilpotent ideal which contradicts the $\mathbf{R}$-semisimplicity of B_1. Therefore $\mathbf{R} = \mathbf{B}$. Moreover, every non-zero homomorphic image C of a nilpotent ring A contains a non-zero ideal, which is a homomorphic image of Z_0. This is because if $C^n = (0)$ and $C^{n-1} \neq (0)$ then every non-zero element of C^{n-1} generates an ideal of C which is a homomorphic image of Z_0. Thus A is of degree 2 over Z_0 and therefore every $\mathbf{B}$-radical ring is also of some degree over Z_0. Thus, $\mathbf{B}$ coincides with the lower radical determined by Z_0. □

We denote by Z_p the zero-ring on the cyclic group of order p.

Theorem 19.2 (Armendariz [2]). Let $\mathbf{R}$ be the class of all radical rings where $\mathbf{R}$ is a hereditary radical property with $\mathbf{R} \leqq \mathbf{B}$. Then $\mathbf{R} = \mathbf{B}$ if and only if the class $\mathbf{R}$ contains a zero-ring Z_0 with a non-zero element of infinite additive order. If no zero-ring in $\mathbf{R}$ contains an element of infinite additive order then $\mathbf{R}$ is the lower radical determined by a ring $A = \sum_{p \in Q} \oplus Z_p$ with $Z_p \in \mathbf{R}$. Conversely, every lower radical determined by the class of all homomorphic images of a ring $A = \sum_{p \in Q} \oplus Z_p$, is a hereditary subradical of $\mathbf{B}$, for every set Q of prime numbers.

Proof. If $\mathbf{R} \leqq \mathbf{B}$, then we have $Z_0 \in \mathbf{R}$, because Z_0 is a $\mathbf{B}$-radical ring. Conversely, by proposition 19.1 $Z_0 \in \mathbf{R}$ implies $\mathbf{R} = \mathbf{B}$.

Now we suppose that there exists no zero-ring in $\mathbf{R}$, containing an element of infinite additive order. Since $\mathbf{R} \leqq \mathbf{B}$ and $\mathbf{R}$ is hereditary, $\mathbf{R}$ contains by the preceding a zero-ring Z. Thus the additive group of Z is torsion. Therefore Z contains an ideal Z_p for some prime number p and we have $Z_p \in \mathbf{R}$ since $\mathbf{R}$ is hereditary.

Take a ring from **R** isomorphic to Z_p, and consider $A=\sum_{p\in Q}\oplus Z_p$ where Q is the set of those prime numbers p, for which there exists a ring Z_p in **R**. Then $\mathbf{T}\leqq\mathbf{R}$ for the lower radical **T** determined by the ring A. Conversely, every homomorphic image B' of a ring $B\in\mathbf{R}$ contains an ideal $I\neq(0)$ with $I^2=(0)$, and an ideal J, isomorphic to some Z_p is contained in I. Since Z_p is a homomorphic image of A, we obtain $\mathbf{T}=\mathbf{R}$.

Now let Q be a set of prime numbers p_i with $p_i<p_{i+1}$ and $A=\sum_{p\in Q}\oplus Z_p$. Then it can be shown immediately that every homomorphic image of A is isomorphic to an ideal of A and conversely. Thus the class of all homomorphic images of A is hereditary. Hence it follows that the lower radical **V** determined by the ring A is also hereditary because $\mathbf{V}\leqq\mathbf{B}$, and every ideal I of a **V**-ring is such a **B**-ring whose every homomorphic image contains an ideal which is a homomorphic image of A. Moreover, only the homomorphic images of A occur among the simple zero-rings, which are ideals of a homomorphic image of an ideal of a **V**-radical ring (for details consult Hoffman and Leavitt [2]).

It can easily be shown that if **R** is the lower radical determined by the ring Z_p then for every ring A, every element of $\mathbf{R}(A)$ has an additive order which is a power of a prime number (for this see Armendariz and Leavitt [2]).

Let Q denote a set of prime numbers and $\mathbf{R}_Q$ denote the lower radical determined by the ring

$$A=\sum_{p\in Q}\oplus Z_p$$

and write $\mathbf{R}_p=\mathbf{R}_Q$ for $Q=\{p\}$. Then we obtain

$$\mathbf{R}_p(A)\cap\sum_{q\neq p}\mathbf{R}_q(A)=(0).\ \square$$

We have the following

THEOREM 19.3. $\mathbf{R}_Q(A)=\sum_{p\in Q}\oplus\mathbf{R}_p(A)$ in every ring A.

PROOF. By the preceding remark, the sum of the ideals $\mathbf{R}_p(A)$ is direct in A. Since $\mathbf{R}_p\leqq\mathbf{R}_Q$ we have $I=\sum_{p\in Q}\mathbf{R}_p(A)\subseteqq\mathbf{R}_Q(A)$. Now we prove $I\geqq\mathbf{R}_Q(A)$. Note that the additive group of $\mathbf{R}_Q(A)$ is always torsion. For, if the factor ring T of $\mathbf{R}_Q(A)$, with respect to its maximal torsion ideal differs from zero, then T is a torsion-free, non-zero, $\mathbf{R}_Q$-radical ring which is of degree 3 over A, by a result of Armendariz and Leavitt [2]. Accordingly, T contains a non-zero ideal I of degree 2 over A, and thus I has a non-zero ideal I' which is a homomorphic image of A. But this is a contradiction because I' is torsion-free. Thus $(\mathbf{R}_Q(A))^+$ is torsion. Then $\mathbf{R}_Q(A)$ is the ring direct sum of its p-components A_p. But it is easy to show that $A_p\subseteqq\mathbf{R}_p(A)$, whence $I=\mathbf{R}(A)$ follows. $\square$

§ 20. THE EXISTENCE OF A NIL RING WHICH IS NOT LOCALLY NILPOTENT

For the subject-matter of this section we refer the reader to Golod [1] and Golod and Shafarevich [1].

A ring A is called *locally nilpotent*, if every finite subset $a_1, a_2, \ldots, a_n$ of elements of A generates a nilpotent subring. Every nilpotent ring is locally nilpotent, and every locally nilpotent ring is a nil ring. The ring direct sum A of the rings B_n with $B_n^{n+1}=(0)$, $B_n^n \neq (0)$ is obviously a locally nilpotent, but not nilpotent ring. We shall now discuss Golod's construction of a nil ring which is not locally nilpotent. This ring A is a nil algebra generated by three elements and has infinite rank over a field.

For the discussion of this ring we need some preliminaries, due to Shafarevich and Golod [1]. As for the exposition, we shall follow Herstein [2].

Let F be a field and $T=F[x_1, \ldots, x_d]$ be the polynomial ring in d distinct, non-commutable, transcendental elements x_i over F. Then T, as a vector space over F, can be decomposed into the direct sum $T=\sum_{n=0}^{\infty} \oplus T_n$ where $T_0=F$, and T_n has a basis consisting of d^n distinct monomials $x_{i_1} x_{i_2} \ldots x_{i_n}$. The elements of T_n are called *homogeneous elements* of degree n and thus the *graded algebra* $T=\sum T_n$ has a degree function. Let $\mathfrak{A}=(f_1, f_2, \ldots)$ be an ideal generated by the homogeneous elements $f_1, f_2, \ldots,$ of T where f_i has degree n_i with $n_i \geqq 2$ and $n_1 \leqq n_2 \leqq n_3 \leqq \ldots$. Let r_i denote the number of n_j with $n_j=i$. Then $A=T/\mathfrak{A}$ inherits the graduing of T because $\mathfrak{A}$ is homogeneously generated, and thus we obtain

$$A = \sum_{n=0}^{\infty} A_n \quad \text{and} \quad A_n = T_n/\mathfrak{A} \cap T_n.$$

We put $b_n = \dim_F A_n$.

Proposition 20.1. For the above algebra $A=T/\mathfrak{A}$ we have the inequality

$$b_n \geqq d \cdot b_{n-1} - \sum_{n_i \leqq n} b_{n-n_i}.$$

If $r_i \leqq \left(\frac{d-1}{2}\right)^2$ then A has infinite dimension over F.

We sketch the

Proof as follows. We define linear mappings φ_1 and φ_2 such that the sequence

$$A_{n-n_1} \oplus \ldots \oplus A_{n-n_k} \oplus \ldots \xrightarrow{\varphi_1} \overbrace{A_{n-1} \oplus \ldots \oplus A_{n-1}}^{1 \quad\quad d} \xrightarrow{\varphi_2} A_n \to (0) \tag{1}$$

is exact, where the sum runs over all $n_i \leqq n$. (As is well known, a homological sequence $C_n \xrightarrow{\alpha_n} C_{n+1} \xrightarrow{\alpha_{n+1}} C_{n+2}$ is called *exact*, if the image of α_n coincides with the kernel of α_{n+1} (see e.g. Cartan and Eilenberg [1])). If such mappings φ_1 and φ_2 exist, then we obtain $db_{n-1}=b_n+\dim\ker\varphi_2$. Now $\ker\varphi_2$ is just a homomorphic image of $\sum_{n_i\leqq n} A_{n_i}$, so since $\dim\ker\varphi_2\leqq\sum_{n_i\leqq n} b_{n_i}$ we have the desired inequality $b_n\geqq db_{n-1}-\sum_{n_i\leqq n} b_{n_i}$.

In order to construct φ_1 and φ_2 we first define two mappings ψ_1 and ψ_2 such that the sequence

$$T_{n-n_1}\oplus\ldots\oplus T_{n-n_k}+\ldots\xrightarrow{\psi_1}\overset{1}{\breve{T}}_{n-1}\oplus\ldots\oplus\overset{d}{\breve{T}}_{n-1}\xrightarrow{\psi_2}T_n\to(0) \tag{2}$$

obtained from (1) by substituting A_i by T_i, is exact. We define ψ_2 by

$$\psi_2\colon t_1\oplus\ldots\oplus t_d\to\sum_{i=1}^{d} t_i x_i(t_i\in T_{n-1}).$$

Moreover we have

$$\sum_{n_i\leqq n} s_{n-n_i} f_i=\sum_{i=1}^{d} u_i x_i$$

for the elements $s_i\in T_i$, with uniquely determined elements $u_i\in T_{n-1}$. We define

$$\psi_1\colon s_{n-n_1}+s_{n-n_2}+\ldots\to u_1+u_2+\ldots+u_d.$$

ψ_1 and ψ_2 are obviously linear mappings, and ψ_2 is an epimorphism onto T_n. Hence the sequence (2) is exact in T_i (instead of A_i).

Now we shall show that the exactness of (1) can be deduced from that of (2).

If $\mathfrak{A}_i=\mathfrak{A}\cap T_i$ and $t_1, t_2, \ldots, t_d\in\mathfrak{A}_{n-1}$, then since $\sum t_i x_i\in\mathfrak{A}$ and $\sum t_i x_i\in T_n$ we have $\sum t_i x_i\in\mathfrak{A}_n$. Then φ_2 can be induced from ψ_2:

$$\varphi_2\colon\overset{1}{\breve{A}}_{n-1}\oplus\ldots\oplus\overset{d}{\breve{A}}_{n-1}\xrightarrow{\varphi_2}A_n\to(0).$$

Now we shall show that $\sum s_{n-n_i}f_i=\sum u_i x_i$ for $s_{n-n_i}\in\mathfrak{A}_{n-n_i}$ always implies

$$u_1, u_2, \ldots, u_d\in\mathfrak{A}_{n-1}.$$

Because of the linearity it is sufficient to prove the relation $s_{n-n_i}\in\mathfrak{A}_{n-n_i}$ for a single member. Then we obtain $s_{n-n_i}f_i=\sum_{j=1}^{d} s_{n-n_i}g_{ij}x_j$ with $f_i=\sum_{j=1}^{d} g_{ij}x_j$. If $s_{n-n_i}\in\mathfrak{A}$ we obtain $s_{n-n_i}g_{ij}\in\mathfrak{A}$. Hence,

$$s_{n-n_i}g_{ij}=u_j\in\mathfrak{A}\cap T_{n-1}=\mathfrak{A}_{n-1}.$$

Therefore, φ_1 can be induced from ψ_1, i.e. we have

$$\varphi_1\colon s_{n-n_1}+\ldots\to u_1+u_2+\ldots+u_d\quad(u_i\in\mathfrak{A}_{n-1})$$

with

$$\sum_{n_i\leqq n} s_{n-n_i}f_i=\sum u_i x_i.$$

The proof of the exactness of (1) still remains. For $s_{n-n_k} \in T_{n-n_k}$ we have

$$(s_{n-n_1}+\ldots)\psi_1\psi_2 = \sum u_i x_i$$

with $\sum u_i x_i = \sum s_{n-n_i} f_i \in \mathfrak{A}$, and hence we have

$$(\bar{s}_{n-n_1}+\ldots)\varphi_1\varphi_2 = \bar{0} \quad \text{for} \quad \bar{s}_{n-n_i} \in T_{n-n_i}/\mathfrak{A}_{n-n_i}.$$

Now we verify that if $(t_1+\ldots+t_d)\varphi_2 = \bar{0}$ then there exist elements $u_1, \ldots, u_d \in T_{n-1}$ with $t_i - u_i \in \mathfrak{A}$ and with $\sum u_i x_i = \sum s_{n-n_i} f_i$ for some $s_{n-n_i} \in T_{n-n_i}$.

Suppose $\sum_{i=1}^{d} t_i x_i \in \mathfrak{A}$. Then

$$\sum_{i=1}^{d} t_i x_i = \sum a_{kl} f_l b_{kl} + \sum c_l f_l,$$

where a_{kl}, b_{kl}, c_l are homogeneous with degree $(b_{kl}) \geqq 1$. Since $t_i \in T_{n-1}$ and $\sum t_i x_i \in T_n$ we may suppose that $a_{kl} f_l b_{kl} \in T_n$ and $c_l f_l \in T_n$. Since degree $(b_{kl}) \geqq 1$, we have $b_{kl} = \sum d_{klm} x_m$, and $\sum a_{kl} f_l b_k = \sum d_m x_m$ where $d_m = \sum a_{kl} f_l d_{klm} \in \mathfrak{A}$. If $\sum c_l f_l = \sum_{i=1}^{d} u_i x_i$, then we have $\sum_{i=1}^{d} t_i x_i = \sum_{m=1}^{d} d_m x_m + \sum u_i x_i$. Therefore $t_i - u_i = d_i \in \mathfrak{A}$. Thus we have

$$(c_1+\ldots)\varphi_1 = u_1+\ldots+u_d$$

which proves the exactness of (1). This implies the inequality in proposition 20.1.

Now we prove the second part of proposition 20.1. We consider the formal power series in one indeterminate t. Then

$$\sum_{n=1}^{\infty} b_n t^n = \sum_{n=1}^{\infty} d b_{n-1} t^n - \sum_{n=1}^{\infty} \sum_{n_i \leqq n} b_{n-n_i} t^n.$$

By the definition of r_i we have

$$\sum_{n_i, m} b_m t^{n_i+m} = \sum t^{n_i} \sum_{m=0}^{\infty} b_m t^m = \left(\sum_{i=2}^{\infty} r_i t^i\right)\left(\sum_{m=0}^{\infty} b_m t^m\right).$$

With the notation

$$P_A(t) = \sum_{m=0}^{\infty} b_m t^m$$

the foregoing implies

$$P_A(t) - 1 \geqq d \cdot t \cdot P_A(t) = \left(\sum_{i=2}^{\infty} r_i t^i\right) P_A(t).$$

Therefore

$$P_A(t) \geqq \left(1 - dt - \sum_{i=2}^{\infty} r_i t^i\right)^{-1}.$$

If $r_i \leqq \left(\frac{d-1}{2}\right)^2$ the coefficients of the series expansion of $\left(1 - dt - \sum_{i=2}^{\infty} r_i t^i\right)^{-1}$ are nonnegative, and thus we can show that $b_n \neq 0$ for infinitely many n. Therefore when $r_i \leqq \left(\frac{d-1}{2}\right)^2$ the algebra A has an infinite dimension over F. □

EXAMPLE 20.2 (Golod [1]). Let F be a countable field and $T=F[x_1, x_2, x_3]$ be a polynomial ring in three non-commutable, transcendental elements x_1, x_2, x_3 over F. Then T is a graded countable algebra $T=\sum_{n=0}^{\infty} T_n$. We consider the ideal $\mathfrak{A}=\sum_{n=1}^{\infty} T_n$ of the algebra T. Since $\mathfrak{A}$ is countable, let $s_1, s_2, s_3, \ldots$ be all the elements of $\mathfrak{A}$. Let m_1 be an arbitrary natural number with $m_1 \geqq 2$. Then we have $s_1^{m_1}=s_{12}+\ldots$ $\ldots+s_{1k_1}$ with some $s_{1j} \in T_j$. Moreover, let m_2 be chosen sufficiently large to guarantee $s_2^{m_2} \in T_{k_1+1} \oplus T_{k_1+2} \oplus \ldots$ Then $s_2^{m_2}$ can be represented in the form

$$s_2^{m_2} = s_{2,k_1+1}+\ldots+s_{2,k_2}.$$

By this procedure we obtain natural numbers k_i with $2 \leqq k_1 < k_2 < k_3 < \ldots$ and

$$s_i^{m_i} = s_{i,k_{i-1}+1}+s_{i,k_{i-1}+2}+\ldots+s_{i,k_i} \, (s_{i,j} \in T_j).$$

Let $\mathfrak{B}$ be the ideal generated by all s_{ij} in T. By the construction of $\mathfrak{B}$, $r_i \leqq 1$ for every i. Hence $T/\mathfrak{B}$ has infinite dimension over F by proposition 20.1. Thus, $\mathfrak{A}/\mathfrak{B}$ also has infinite dimension over F. Moreover, $\mathfrak{A}/\mathfrak{B}$ is a nil algebra generated by three elements $x_1+\mathfrak{B}$, $x_2+\mathfrak{B}$ and $x_3+\mathfrak{B}$.

REMARK 20.3. Let $A=\mathfrak{A}/\mathfrak{B}$ be the nil ring, which is not locally nilpotent, discussed in example 20.2. Let F be a prime field of characteristic p, and suppose G is the set of all formal sums $1+a$ with $a \in A$. If $a^n=0$ we have $a^{p^n}=0$, and thus since F has characteristic p, $(1+a)^{p^n}=1$. Hence the elements of G form an infinite p-group generated by three elements. For, if G is finite, then the linear combinations of the elements of G form a finite subalgebra B of A over F. Since $1+a \in G$ and $1 \in G$ we have $B=A$, which contradicts the fact that A is infinite. Therefore, G is an infinite p-group generated by three elements.

§ 21. THE LEVITZKI NIL RADICAL

In this section we shall discuss the Levitzki nil radical defined by means of locally nilpotent ideals (cf. Levitzki [2] and Jacobson [2]).

An ideal I of a ring A is called *locally nilpotent* if the ring I is locally nilpotent, i.e. if every finite set of elements $i_1, i_2, \ldots, i_k \in I$ generates a nilpotent subring.

PROPOSITION 21.1. Every subring and every homomorphic image of a locally nilpotent ring is again locally nilpotent. If B and A/B are locally nilpotent for an ideal B of a ring A, then A is itself locally nilpotent.

PROOF. The first part of the statement is an immediate consequence of the definition. Now, if B and A/B are locally nilpotent, and if $\{x_1, x_2, ..., x_n\}$ is an arbitrary finite set in A, then there exists a natural number m such that $\bar{x}_{i_1} \cdot \bar{x}_{i_2}, ... \cdot ... \cdot \bar{x}_{i_m} = \bar{0}$ for $\bar{x} = x + B$. Then the set of all $x_{i_1}, x_{i_2}, ..., x_{i_m}$ is finite (it is of cardinality at most n^m) and this set is contained in the ideal B. Hence there exists a natural number k such that every product of k factors of the forms $x_{i_1} ... x_{i_m}$ equals 0. Hence $C^{mk} = (0)$ for the subring C generated by $x_1, x_2, ..., x_n$. □

PROPOSITION 21.2. The sum L of all locally nilpotent ideals of a ring A is locally nilpotent, and A/L has no non-zero locally nilpotent ideals.

PROOF. Let $\{x_1, x_2, ..., x_n\}$ be an arbitrary finite set of elements in L. Since every x_j is contained in the sum of a finite number of locally nilpotent ideals $I_{i_1}, I_{i_2}, ..., I_{i_j}$, every x_j $(j = 1, 2, ..., n)$ is contained in the sum $\sum_{i=1}^{n} \sum_{k=1}^{m_i} I_{i_k}$. Thus L is also locally nilpotent, because every finite sum of locally nilpotent ideals is locally nilpotent. For, if I_1 and I_2 are locally nilpotent, then, since $(I_1 + I_2)/I_2 \cong I_1/(I_1 \cap I_2)$ and $I_1 + I_2$ is locally nilpotent by proposition 21.1.

Now, if an ideal T/L of A/L is locally nilpotent, then T itself is locally nilpotent by proposition 21.1, and thus we have by $T \subseteq L$. Hence $T/L = (0)$. □

The ideal L is called the *Levitzki nil radical* of A.

PROPOSITION 21.3. The Levitzki nil radical of a ring A contains every locally nilpotent right (left) ideal.

PROOF. Let R be a locally nilpotent right ideal. Let $\mathbf{L}(A)$ be the Levitzki nil radical of A. Then $(R + \mathbf{L}(A))/\mathbf{L}(A) \cong R/(R \cap \mathbf{L}(A))$ is again a locally nilpotent right ideal of the ring $A/\mathbf{L}(A)$ which is semisimple in the sense of Levitzki. Hence we may assume that $\mathbf{L}(A) = (0)$. Let $x_1, x_2, ..., x_n$ be arbitrary, finitely many elements of the two-sided ideal AR of A. Then every x_i has the form $x_i = \sum_{i,j} a_{ij} r_{ij}$ with $a_{ij} \in A$, $r_{ij} \in R$. Consider the finite system of all possible elements $y_{ijkl} = r_{ij} a_{kl} \in R$. Since R is locally nilpotent, these y_{ijkl} generate a nilpotent subring, and thus there exists a natural number m such that every product of m factors y_{ijkl} equals zero. Now consider the product of $(m+1)$ arbitrary factors x_i. Such a product will begin with some a_{ij}, then m factors y_{ijkl} follow, and at the end there is an r_{ij}. Thus, every product of $(m+1)$ factors from AR equals zero. Hence AR is locally nilpotent. Since $\mathbf{L}(A) = (0)$ we obviously have $AR = 0$, $R^2 = (0)$ and $R = (0)$.

Therefore $R \subseteq \mathbf{L}(A)$. The proof is similar for locally nilpotent left ideals. □

As a consequence of proposition 21.2, it follows that the Levitzki nil radical is a radical in the sense of Amitsur and Kurosh. We shall prove the sharper result that the Levitzki nil radical is a special radical. As a preparation for this, we consider a result of Babich [1].

PROPOSITION 21.4. The Levitzki nil radical $\mathbf{L}(A)$ of a ring A coincides with the intersection D of all those prime ideals P_α of A, for which A/P_α is semisimple in the sense of Levitzki.

PROOF. We have $(\mathbf{L}(A)+P_\alpha)/P_\alpha \cong \mathbf{L}(A)/(\mathbf{L}(A)\cap P_\alpha)$ since A/P_α contains no non-zero locally nilpotent ideals, we must have $\mathbf{L}(A)\subseteq P_\alpha$. Thus $\mathbf{L}(A)\subseteq D=\bigcap P_\alpha$.

Now assume that there exists an element $x\in D$ with $x\notin \mathbf{L}(A)$. Then the principal ideal (x) is not locally nilpotent, and thus there exists a finitely generated subring T of (x) such that $T^n\neq(0)$ for every n. Hence $T^n\nsubseteq \mathbf{L}(A)$ for every n, because if $T^n\subseteq \mathbf{L}(A)$ then both T/T^n and T^n are locally nilpotent and therefore T is also locally nilpotent. Consequently T is nilpotent, which contradicts the fact that $T^n\neq(0)$ for every n. Let $\mathfrak{M}$ be the set of ideals I of A such that $I\supseteq \mathbf{L}(A)$ and $T^n\nsubseteq I$ for every n. This set is not empty since $\mathbf{L}(A)\in\mathfrak{M}$. If $\{t_1, t_2, \ldots, t_k\}$ is a generating system for T, then the set of all possible products $t_{i_1}t_{i_2}\ldots t_{i_m}$ is a generating system for T^m, and thus T^m is finitely generated. Hence the union of every ascending chain of ideals in $\mathfrak{M}$ is again contained in $\mathfrak{M}$. Thus there exists a maximal ideal M in $\mathfrak{M}$ by Zorn's lemma. We shall show that M is a prime ideal of A. If B_1 and B_2 are ideals of A with $B_1\supsetneq M$ and $B_2\supsetneq M$, then we have $T^{m_1}\subseteq B_1$ and $T^{m_2}\subseteq B_2$ for some m_i. Then $T^{m_1+m_2}\subseteq B_1B_2$, so $B_1B_2\nsubseteq M$. Thus M is a prime ideal.

Suppose that A/M contains a locally nilpotent ideal W/M. If $W/M\neq(0)$ we have $T^m\leqq W$ for some m. Since T^m is finitely generated, so is $(T^m+M)/M$. Also since $(T^m+M)/M\leqq W/M$ there exists a natural number n such that $(T^m+M)^n\leqq M$. Therefore $T^{mn}\leqq M$. But this contradicts the definition of M. Hence A/M is a prime ring which is semisimple in the sense of Levitzki. Thus, $D=$ $=\bigcap P_\alpha\leqq M$ and we obtain $T\leqq(x)\leqq D\leqq M$ because $x\in D$ and $T\leqq(x)$. This contradicts the condition $T^m\nleqq M$. So we have $D=\mathbf{L}(A)$. □

THEOREM 21.5. The Levitzki nil radical $\mathbf{L}$ is a special radical which is determined by the special class of all those prime rings which have no non-zero locally nilpotent ideals.

PROOF. Let $\mathbf{K}$ be the class of all prime rings which are semisimple in the sense of Levitzki. If B is an ideal of a ring in K, then B is again semisimple in the sense of Levitzki by theorem 1.8. B is at the same time a prime ring by an argument used earlier in the proof of theorem 18.3.

Now let A be a ring in $\mathbf{K}$ which is an ideal of a ring B. Let A^* be the annihilator of A in B. Then it can be shown by a method similar to the proof of theorem 18.3 that B/A^* is a prime ring. Let C/A^* be a non-zero locally nilpotent ideal of B/A^*. If $C\cap A=(0)$, then we have $CA=AC=(0)$, i.e. $C\leqq A^*$ and $C/A^*=(0)$. Hence,

$C/A^* \neq (0)$ implies $C \cap A \neq (0)$. Let E be an arbitrary finitely generated subring of $C \cap A$. Since $[E+A^*]/A^* \leqq C/A^*$ we obtain $E^n \leqq A^*$ for some n. Now $E \leqq A$, $A \cap A^* = (0)$ imply $E^n = (0)$, and thus $C \cap A$ is locally nilpotent. However, since $A \in \mathbf{K}$, A is semisimple in the sense of Levitzki, and hence $C \cap A = (0)$. Thus $C/A^* = (0)$ and $B/A^* \in \mathbf{K}$.

Accordingly $\mathbf{K}$ is a special class of rings. Let $\mathbf{S}$ be the upper radical determined by this class $\mathbf{K}$. Then $\mathbf{S}$ is a special radical and thus supernilpotent. Hence in every ring A, $\mathbf{S}(A)$ coincides with the intersection D of all ideals I_α of A with $A/I_\alpha \in \mathbf{K}$. But this intersection D is just the Levitzki nil radical $\mathbf{L}(A)$ of A by proposition 21.4. □

§ 22. RESULTS CONCERNING THE LEVITZKI NIL RADICAL

An algebra F is said to be *free* or more precisely freely generated by the elements e_α over a field K, if the following conditions are fulfilled:

(1) the elements e_α are linearly independent over K,
(2) $e_{\alpha_1} e_{\alpha_2} \dots e_{\alpha_m} = e_{\beta_1} \dots e_{\beta_n}$ implies $m=n$ and $\alpha_1 = \beta_1, \dots, \alpha_m = \beta_m$,
(3) the elements e_α generate A as a ring,

Every algebra over K is obviously a homomorphic image of a free algebra F.

An algebra A is a *PI-algebra*, or satisfies a *polynomial identity,* if there exists an element $p(e_1, \dots, e_n)$ of the free algebra F such that p is mapped onto the zero element of A by every homomorphism of F into A. Thus e.g. every commutative algebra is a *PI*-algebra, because $x_1x_2 - x_2x_1 = 0$ identically holds. Therefore *PI*-algebras can be considered as generalizations of the commutative algebras. Boolean rings and nil algebras of bounded nilpotence degree are also *PI*-algebras, since they satisfy the identities $x_1^2 - x_1 = 0$ and $x_1^n = 0$, respectively. It can be shown (proposition 10.4.1 of Jacobson [3]) that every *PI*-algebra satisfies also a *multilinear identity*. Thus, *PI*-algebras can always be considered as algebras satisfying a multilinear identity.

THEOREM 22.1. Every nil *PI*-algebra is locally nilpotent.

PROOF. Let A be a nil *PI*-algebra and $\mathbf{L}(A)$ the Levitzki nil radical of A. We shall show that the assumption $A \neq \mathbf{L}(A)$ leads to a contradiction. If $A \neq \mathbf{L}(A)$ then $B = A/\mathbf{L}(A)$ is a nil *PI*-algebra with Levitzki nil radical (0). Then there exists an element $b \neq 0$ in B with $b^2 = 0$. Now $\mathbf{L}(B) = (0)$ so we have $bB \neq (0)$, because the left annihilator of B is contained in $\mathbf{L}(B)$. Let $f(e_1, e_2, \dots, e_n) = 0$ be the multilinear identity satisfied by B. Then we can write

$$f = f_1(e_2, e_3, \dots, e_n)e_1 + f_2(e_1, e_2, \dots, e_n),$$

where f_2 is the sum of all monomial members of f which do not have e_1 as the last factor. Renumbering the elements if necessary, we may assume that $f_1(e_2, ..., e_n)e_1 \neq$ $\neq 0$. Now, bB is also a PI-algebra. If we substitute b for e_1 and $b_i \in bB$ for e_i in f $(i=2, 3, ..., n)$, then we have $f_2(b_1, b_2, b_3, ..., b_n)=0$ by the definition of f_2 because $b^2=0$. Hence $f_1(b_2, ..., b_n)b=0$. Now, if C is the left annihilator of bB in bB then C is an ideal of bB with $f_1(b_2, ..., b_n) \in C$, because $f_1(b_2, ..., b_n) \in bB$. Hence bB/C satisfies a multilinear identity $f_1(\bar{e}_2, ..., \bar{e}_n)=0$ of a degree $\leqq n-1$, and thus it can be shown by induction on the degree of the multilinear identity that bB/C is locally nilpotent. However, since C is nilpotent, bB is locally nilpotent. This contradicts the assumption that $\mathbf{L}(B)=(0)$. Therefore $A=\mathbf{L}(A)$. □

We mention without proof the sharper result (theorem 10.8.2, Jacobson [2]) that every nil PI-algebra coincides with its Baer lower nil radical. Moreover $A=$ $=N(2)$, where $N(1)$ is the sum of all nilpotent ideals of A and $N(2)/N(1)=$ $=N(1)(A/N(1))$ (that is $N(2)/N(1)$ is the sum of all nilpotent ideals of $A/N(1)$).

The following proposition follows from theorem 22.1 for algebras over a field, but here we formulate it more generally for rings.

Proposition 22.2. If there exists a natural number n such that the polynomial identity $a^n=0$ is satisfied for every element a of a ring A then A is locally nilpotent.

Proof. If $n=2$ we have $yx=-xy$ since $(x+y)^2=x^2=y^2=0$. Now, if T is the subring of A generated by the elements $x_1, x_2, ..., x_n$, then every monomial member of T^{m+1} is the product of $m+1$ elements x_i. But there exist only m distinct x_i's in this product and thus the relation $x_i^2=0$ and a repeated application of $yx=$ $=-xy$ yield $T^{m+1}=(0)$. Hence A is locally nilpotent when $n=2$. We shall prove the theorem by induction on n. Assume that it is already proved for all $m<n$.

Let A be a ring with $x^n=0$ for every element $x \in A$. Let $\mathbf{L}(A)$ be the Levitzki nil radical of A. We shall deduce a contradiction from $A \neq \mathbf{L}(A)$. If $A \neq \mathbf{L}(A)$ then $\mathbf{L}(B)=(0)$ for $B=A/\mathbf{L}(A)$ and $b^n=0$ for every $b \in B$. Let b be a non-zero element of B with $b^2=0$. Then $bB \neq (0)$ since $\mathbf{L}(B)=(0)$. We have $(bx+b)^k=(bx)^k+(bx)^{k-1}b$ for every $x \in B$ and every k, because $(bx+b)^2=(bx)^2+bxb$ holds for $k=2$, and $(bx+b)^{k-1}=(bx)^{k-1}+(bx)^{k-2}b$ implies, if we multiply by $bx+b$, $\big((bx)^{k-1}+$ $+(bx)^{k-2}b\big)(bx+b)=(bx)^k+(bx)^{k-1}b$.

Hence we have $(bx)^{n-1}b=0$ for every $b \in B$ since $y^n=0$ for every y. Now, if S is the left annihilator of bB in bB, then S is an ideal of bB with $S^2=(0)$. Since $(bx)^{n-1}b=0$ we have $(bx)^{n-1} \in S$ for every $x \in B$. Hence bB/S is a ring with $c^{n-1}=0$ for every $c \in bB/S$, and thus bB/S is locally nilpotent by the induction assumption. Since $S^2=0$ bB is also locally nilpotent by proposition 21.1. Thus we obtain a contradiction to $\mathbf{L}(B)=(0)$, since $bB \neq (0)$. Therefore $A=\mathbf{L}(A)$. □

Nagata [2] has proved the following stronger theorem. Its short proof is due to Higgins (cf. Higman [1]).

THEOREM 22.3. Suppose that there exists a natural number n such that $x^n=0$ for every element x of an algebra A over a field of characteristic 0 or $p>n$. Then A is nilpotent. Moreover $A^m=(0)$ where $m=2^n-1$.

PROOF. Let A^* be the smallest algebra with unity element and with $A \leqq A^*$. We define $a^0=1\in A^*$ for every $a\in A^*$. Furthermore, we use the symbol

$$\begin{Bmatrix} a & b \\ n-1 & 1 \end{Bmatrix} = \sum_{i=0}^{n-1} a^i b a^{n-i-1}.$$

In particular we have $\begin{Bmatrix} a & 1 \\ n-1 & 1 \end{Bmatrix} = na^{n-1}$. It can be proved (cf. the proof of theorem 10.2.3 of Jacobson [2]) that $(a+\alpha b)^n=0$ for $\alpha\in K$, $a, b\in A$ implies $\begin{Bmatrix} a & b \\ n-1 & 1 \end{Bmatrix} = 0$. Consider the element

$$f(a, b, c) = \sum_{i,j}^{n-1} a^i c b^j a^{n-i-1} b^{n-j-1}$$

for $a, b, c\in A$. Since

$$f(a, b, c) = \sum_{j=0}^{n-1} \begin{Bmatrix} a & cb^j \\ n-1 & 1 \end{Bmatrix} b^{n-j-1}$$

and $\begin{Bmatrix} a & b \\ n-1 & 1 \end{Bmatrix}=0$ we obtain $f(a, b, c)=0$. On the other hand, we have

$$f(a, b, c) = \sum_{i=0}^{n-1} a^i c \begin{Bmatrix} b & a^{n-i-1} \\ n-1 & 1 \end{Bmatrix} na^{n-1} cb^{n-1},$$

and thus $na^{n-1}cb^{n-1}=0$ for every $a, b, c\in A$. Since $p=0$ or $p>n$, it follows that

$$a^{n-1} cb^{n-1} = 0(a, b, c\in A). \tag{$*$}$$

Now, if B is the ideal of A generated by all elements a^{n-1} $(a\in A)$, then $BAB=(0)$ by $(*)$. Therefore, $\bar{a}^{n-1}=0$ for every element $\bar{a}$ of $\bar{A}=A/B$, and by induction we obtain $A^{2^{n-1}-1}\leqq B$. Hence we have $A^{2^n-1}=A^{2^{n-1}-1}\cdot A\cdot A^{2^{n-1}-1}\leqq BAB=0$. Therefore $A^{2^n-1}=(0)$.

It remains to remark that for $n=2$ $(*)$ implies $acb=0$ for all $a, b, c\in A$ and thus $A^3=(0)$, which yields the starting point for the induction proof. □

The following result can be proved with the help of proposition 22.3:

PROPOSITION 22.4. If there exists a natural number n such that $(xy-yx)^n=0$ for arbitrary elements x and y of a ring A, then the set of all nilpotent elements of A is an ideal which coincides with the Levitzki nil radical $\mathbf{L}(A)$, and thus $A/\mathbf{L}(A)$ is a commutative ring without non-zero nilpotent elements.

PROOF. If $\mathbf{L}(A)$ is the Levitzki nil radical of A, then we shall show that $B=$ $=A/\mathbf{L}(A)$ contains no non-zero nilpotent elements. Let b be an element of B

with $b^2=0$ and $b\neq 0$. Then $(bx-xb)^n=0$ for every $x\in B$. Since $b^2=0$ and $(bx-xb)^n=0$, right multiplication by b yields $(bx)^n b=0$. Hence $(bx)^{n+1}=0$ for every $x\in B$, and thus the right ideal bB is locally nilpotent by proposition 22.2. Since $\mathbf{L}(B)=\mathbf{L}(A/\mathbf{L}(A))$, by proposition 21.3 we have $bB=(0)$. Therefore, as $b\neq 0$, the left annihilator of B differs from zero. This contradicts the assumption that $\mathbf{L}(B)=(0)$. Accordingly, there exists no non-zero nilpotent element in $A/\mathbf{L}(A)$. But $(xy-yx)^n=0$, so $xy-yx\in\mathbf{L}(A)$. Thus $A/\mathbf{L}(A)$ is commutative. □

PROPOSITION 22.5. Every locally nilpotent ring is antisimple (in the sense of § 12).

PROOF. A ring A is antisimple if and only if A cannot be homomorphically mapped onto a subdirectly irreducible ring B with idempotent heart H. Now, if A is a locally nilpotent, but not antisimple ring, then H is a locally nilpotent simple ring. Then $H^2=H$. Moreover, we have $HhH=H$ for every $h\in H$ with $h\neq 0$, because the set of all $x\in H$ for which $HxH=(0)$ is an ideal N with $N^3=(0)$. Since $H^2=H\neq(0)$, and H is simple, $N=(0)$, and thus $HhH=H$ for every non-zero element of H. Hence there exist finitely many elements a_i, $b_i\in H$ such that

$$h=\sum_{i=1}^{n} a_i h b_i.$$

Now, if S is the subring generated by the elements a_i, b_i, then we have $S^k=(0)$, because H is locally nilpotent. By repeated substitution we obtain

$$h=\sum_{i=1}^{n} a_i\left(\sum_{j=1}^{n} a_j h b_j\right) b_i=\sum_{ji,} a_{ij} h b_{ij}$$

with a_{ij}, $b_{ij}\in S^2$, and finally

$$h\in S^k h S^k.$$

Therefore $h=0$. Hence a locally nilpotent ring is antisimple. □

We have the following consequences:

1. The Levitzki nil radical $\mathbf{L}$ is contained in the antisimple radical.
2. There exists no simple, locally nilpotent ring (in the proof of proposition 22.5 we have essentially verified 2).

§ 23. NIL RINGS WITH CERTAIN CONDITIONS

We have seen in the preceding section that every locally nilpotent ring is antisimple. It is still an open question whether every nil ring is antisimple, or whether there exists a non-antisimple nil ring.

If a nil ring A is not antisimple then A can be homomorphically mapped onto a subdirectly irreducible ring with idempotent heart H. Then H is a simple nil ring and conversely, every simple nil ring is a non-antisimple nil ring. But as yet the existence of simple nil rings is neither proved nor disproved.

First we mention some properties of simple nil rings proved by McWorter [1].

PROPOSITION 23.1. Let **P** be a homomorphically invariant property of rings (i.e. every homomorphic image of a **P**-ring A is again a **P**-ring). If there exists a simple ring A containing a non-nilpotent **P**-subring of the form xAy $(x, y \in A)$, then there exists a simple **P**-ring.

PROOF. Let $Q = xAy$ be a non-nilpotent **P**-subring of the simple ring A. Then $yx \neq 0$ since $Q^2 = xAyxAy \neq (0)$. Hence $AyxA = A$, because $AyxA$ is an ideal in the simple ring A, and if $AzA = 0$, $z \neq 0$ we obtain $A^3 = (0)$, in contradiction to $A^2 = A \neq 0$. Therefore, we have $Q^2 = xAy = Q$. If

$$R = \{z;\ z \in Q,\ yzx = 0\},$$

then R is an additive group. Moreover, it is an ideal of Q, because we obviously have $y(xayz)x = yxa \cdot yzx = 0$ for every $xay \in Q$ and $z \in R$. Thus $xAyR \subseteq R$. Also $y(zxay)x = yzx \cdot ayx = 0$, and thus $R \cdot xAy \subseteq R$, consequently $QR \subseteq R$ and $RQ \subseteq R$. If $R = Q$, we have $yQx = (0)$. Therefore $y(xAy)x = (0)$ and thus $(yxA)^2 = (0)$; but then A contains a non-zero nilpotent ideal, which contradicts the simplicity of A. Therefore $R \neq Q$, and then Q/R is an idempotent **P**-ring. If $q \in Q$ is not contained in R, we have $yqx \neq 0$, and so $AyqxA = A$. Hence

$$QqQ = (xAy)q(xAy) = xAy = Q.$$

Thus Q/R is a simple **P**-ring. □

COROLLARY 23.2. If there exists a simple ring A containing a non-nilpotent nil right ideal, a non-nilpotent nil left ideal or a non-nilpotent nil subring of the form xAy, then there exists a simple nil ring.

PROOF. The property of a ring to be a nil ring is homomorphically invariant, and so the statement follows from proposition 23.1 if xAy is a non-nilpotent nil subring of A. Now, if R is a non-nilpotent nil right ideal of the simple ring A, then rA is a non-zero nil right ideal of A for every non-zero element $r \in R$, because $rA \subseteq R$, and A has no non-zero left annihilator. Since $r \neq 0$, $Ar \neq (0)$ also holds, and thus there exists an element s with $sr \neq 0$. Consequently, $srA \neq 0$ and $(srA)^k \neq (0)$ for every k. Then $rAs \neq (0)$, because A is a prime ring. Moreover $rAs \subseteq rA$ and thus rAs is a nil ring. If $(rAs)^n = (0)$ we obtain $(srA)^{n+1} = s(rAs)^n rA = (0)$, a contradiction with respect to $(srA)^k \neq (0)$ for every k. Hence $rAS \neq (0)$ is a non-nilpotent nil subring, and thus proposition 23.1 can be applied. The proof is similar in the case of a non-nilpotent nil left ideal. □

PROPOSITION 23.3. If there exists a simple nil ring A then there also exists a simple nil ring B which is the sum of two proper right ideals.

PROOF. First we shall show that in the simple nil ring A there exist elements x, y for which neither $xA \leqq yA$ nor $yA \leqq xA$ holds. In the contrary case we have either $xA \leqq yA$ or $yA \leqq xA$ for all x, y, and thus the right ideals of the form zA $(z \in A)$ form a chain. If $xA \nleqq yA$, then $x \neq 0$, and we have $xA \nleqq yA$. Since x is nilpotent and $xA > x^2A > x^3A > \ldots > x^nA = (0)$ there exists a k with $x^kA \geqq yA > x^{k+1}A$. Hence we have $yA > x^{k+1}A \geqq xyA$. But we get the same relation if $xA \leqq yA$, because then $x(yA) \leqq xA \leqq yA$. Therefore, whether $yA \leqq xA$ or $xA \leqq yA$, we have $x(ya) \leqq yA$, and thus yA is an ideal of A. By the simplicity of A we obtain either $yA = A$ or $yA = (0)$. Since the left annihilator of A is zero, $yA = (0)$ is impossible. If $yA = A$, then $ya = y$ for some $a \in A$. Therefore $ya^2 = ya = y$ and by the nilpotence of a we have $y = 0$. Thus, the right ideals zA of a simple nil ring do not form a chain. Therefore there exist elements x, y in the simple nil ring such that neither $xA \leqq yA$ nor $yA \leqq xA$ holds.

Put $B = C/D$ with $C = xA + yA$, where x and y are elements of A with $xA \nleqq yA$ and $yA \nleqq xA$. Here D is the left annihilator of the right ideal C in C. Then D is an ideal of the ring C, and thus $B = C/D$ exists. We shall show that B is a simple nil ring. $AxA = AyA = A$, therefore $(xA)^2 = xA$ and $(yA)^2 = yA$ because $x \neq 0$, $y \neq 0$ and A is simple. Hence $C = xA + yA$ is idempotent and, consequently, so is $B = C/D$. If $c \in C$ is an element with $c \notin D$ then $cx \neq 0$ or $cy \neq 0$. Obviously, if $cx \neq 0$ then

$$CcC = (xA + yA)c(xA + yA) = xAcxA + yAcxA +$$

$$+ xAcyA + yAcyA = xA + yA + (xAcyA + yAcyA)$$

since $AcxA = A$.

Hence it follows that $CcC = xA + yA = C$ for every $c \in C$ with $c \notin D$ and $cx \neq 0$. If $c \in C$, $c \notin D$, and $cy \neq 0$, we can similarly prove $CcC = C$. Thus C/D is a simple nil ring. If $(xA + D)/D) \leqq (yA + D)/D$ then for every $a \in A$ there exists an element $a' \in A$ with $xa - ya' \in D$. Hence $(xa - ya')xA = (0)$ and $(xa - ya')yA = (0)$, because $xA = C$ and $yA \leqq C$. Since A has zero left annihilator, we have $xax = ya'x$ and $xay = ya'y$. Therefore $xAx \leqq yAx$ and $xAy \leqq yAy$. Hence we obtain $(xA)^2 = xA \leqq$ $\leqq yAxA \leqq yA$. However $xA \nleqq yA$. Consequently, $(xA + D)/D = (yA + D)/D$. Similarly $(yA + D)/D = (xA + D)/D$. Therefore $B = C/D$ is a simple nil ring which is the sum of two right ideals. □

The following theorem is a result of Levitzki [10]. The very short proof given here is due to Herstein [3].

THEOREM 23.4. Every nil right ideal R of a ring with maximum condition for right ideals is nilpotent.

PROOF. Let A be a ring with *maximum condition for right ideals*, i.e. in A every ascending chain of right ideals is finite. Then the sum N of all nilpotent ideals of A is nilpotent by the maximum condition and by proposition 16.2. If the assertion of our theorem is false, then there exists a nil right ideal R of A with $R \subsetneqq N$. Then A/N is a ring with maximum condition. A/N contains no non-zero nilpotent ideal. However, it contains the non-nilpotent nil right ideal $(R+N)/N$. This right ideal is non-nilpotent because $R^n \subsetneqq N$ for every n. By the preceding it may be assumed that the sum N of all nilpotent ideals of A equals (0). Let R be a non-zero nil right ideal of A, r a non-zero element of R, M the set of all non-zero elements of the left ideal Ar of A, and R_{xr} the right annihilating right ideal of the element $xr \in M \subset Ar$. Then $R_{xr} = \{y; y \in A, xry = 0\}$. By the maximum condition for right ideals there exist maximal right ideals R_{x_ir} among the right ideals R_{xr}. If $t \in A$ with $tx_ir \neq 0$, then $R_{x_ir} \leqq R_{tx_ir}$ and $tx_ir \in M$. So $R_{x_ir} = R_{tx_ir}$ by the maximality of R_{x_ir}. We shall show that for every finite set $x_1r, x_2r, \ldots, x_nr$ such that each R_{x_ir} is maximal there exists a non-zero element $u \in Ax_1rA$. Assume that no non-zero element u in Ax_1rA for which $ux_1r = \ldots = ux_{n-1}r = 0$, annihilates x_nr from the left. There exists a non-zero element $u \in Ax_1rA$ with $x_nru \neq 0$, because if $x_nru = 0$ holds for every $u \in Ax_1rA$ with $ux_1r = \ldots = ux_{n-1}r = 0$, we have $x_nrAu = 0$. Therefore $(ux_nrA)^2 = (0)$. Since $N = (0)$, we obtain $ux_nrA = (0)$ and $ux_nr = 0$, wich contradicts the last condition. Therefore there exists an element $u \in Ax_1rA$ with $x_nru \neq 0$, $ux_1r = \ldots = ux_{n-1}r = 0$, $ux_nr \neq 0$. Since $rux_n \in R$; rux_n is nilpotent. Therefore $(x_nru)^k = 0$ for some k, because $(rux_n)^{n-1} = 0$ implies

$$(x_n ru)^n = x_n (rux_n)^{n-1} ru = 0.$$

If $(x_nru)^m = 0$, $(x_nru)^{m-1} \neq 0$, then since $(x_nru)^{m-1} \cdot x_nru = 0$ we have $u \in R_{(x_nru)^{m-1}x_nr}$. But $u \notin R_{x_nr}$ so $(x_nru)^{m-1}x_nr = 0$ by the remark made about t earlier in the proof. But then the non-zero element $(x_nru)^{m-1} \in Ax_1rA$ annihilates both x_nr and the elements $x_1r, \ldots, x_{n-1}r$.

Consider the ascending chain of right ideals

$$x_1rA \leqq x_1rA + x_2rA \leqq x_1rA + x_2rA + x_3rA \leqq \ldots .$$

There exists an integer n such that

$$x_irA \leqq x_1rA + \ldots + x_nrA$$

for all x_ir for which R_{x_ir} is maximal. Moreover, either $bx_1r = 0$ or $R_{bx_1r} = R_{x_ir}$ for an i, therefore,

$$Ax_1r \leqq x_1rA + \ldots + x_nrA.$$

Since there exists a $u \in Ax_1rA$ with $ux_1r = \ldots = ux_nr = 0$, we have $uAx_1r = (0)$. Since $(0) \neq uA \leqq Ax_1rA$ it follows that

$$(uA)^2 \leqq uA(Ax_1rA) \leqq uAx_1rA = (0).$$

Consequently uA is a non-zero nilpotent right ideal, which contradicts the assumption $N=(0)$. □

Next we give a result of Herstein and Small [1], which is similar, but neither more general nor more special than theorem 23.4. We need some preparation.

If A is a ring and S is a subset of A, we denote

$$l(S) = \{y;\ y\in A,\ yS = 0\},$$

$$r(S) = \{y;\ y\in A,\ Sy = 0\}.$$

$l(S)$ and $r(S)$ are called respectively, the *left annihilator* and the *right annihilator* of S in A. Then $l(S)$ is a left ideal and $r(S)$ is a right ideal of A. The maximum condition for the right annihilators $r(S)$ holds in A if and only if the minimum condition is satisfied by the left annihilators $l(S)$ in A.

PROPOSITION 23.5. Every nil ring A with maximum condition for right annihilators of the form $r(x)$ $(x\in A)$ coincides with its Baer lower nil radical $\mathbf{B}(A)$. If the maximum condition holds for right annihilators $r(S)$ in a nil ring A then A contains a non-zero element $x\in l(A)$.

PROOF. If A has a non-zero lower nil radical $\mathbf{B}(A)$, then we put $\bar{A}=A/\mathbf{B}(A)$ and $\mathfrak{M}=\{r(x);\ x\in A,\ x\notin\mathbf{B}(A)\}$. Let $r(a)$ be a right ideal maximal in $\mathfrak{M}$. Since $r(xa)\supseteqq r(a)$ for every $x\in A$, we have either $xa\in\mathbf{B}(A)$ or $r(xa)=r(a)$. Let y be an arbitrary element such that $ya\notin\mathbf{B}(A)$. Then there exist natural numbers n, k such that $(ya)^n=0$, $(ya)^{n-1}\neq 0$, $(ya)^k\in\mathbf{B}(A)$, $(ya)^{k-1}\notin\mathbf{B}(A)$. Since $r((ya)^{k-1})\geqq r(a)$ and $(ya)^{k-1}\notin\mathbf{B}(A)$ we have $r((ya)^{k-1})=r(a)$. Thus since $(ya)^{n-k+1}\in r((ya)^{k-1})$ it follows that $a(ya)^{n-k+1}=0$, whence $(ya)^{n-k+2}=0$ and $n\leqq n-k+2$. Hence $k=2$. Therefore $(ya)^2\in\mathbf{B}(A)$ for every $y\in A$ with $ya\notin\mathbf{B}(A)$. For every $z\in A$ we have $a+za\notin\mathbf{B}(A)$, because $a+za\in\mathbf{B}(A)$ and $z^t=0$ imply $a+za+(a+za)(-z)+\ldots+(a+za)(-z)^{t-1}= =a\in\mathbf{B}(A)$. Since $r(a+za)\geqq r(a)$ and $a+za\notin\mathbf{B}(A)$ we have $r(a)=r(a+za)$. Consequently $(a+za)^2\in\mathbf{B}(A)$, whence we obtain $aza\in\mathbf{B}(A)$. But then $(Aa)^2\leqq\mathbf{B}(A)$ and thus $Aa\leqq\mathbf{B}(A)$ by the definition of $\mathbf{B}(A)$. Since $(a)+\mathbf{B}(A)$ is contained in the right annihilator of $A/\mathbf{B}(A)$, we obtain the contradiction that $a\in\mathbf{B}(A)$. Thus $A=\mathbf{B}(A)$.

Now, let $\mathfrak{N}$ be the set of all $l(S)$, where S is a finitely generated subring of A. Since the maximum condition holds for all $r(T)$, there exists a minimal left ideal $l(S_0)$ of A in the set $\mathfrak{N}$.

Let $S_1=\{S_0, x\}$ where x is an arbitrary element of A. We have $l(S_1)=l(S_0)$ and thus $l(S_0)=l(S_1)=l(S_0)\cap l(x)\leqq l(x)$. Hence, $l(S_0)x=(0)$ for every $x\in A$ and so $l(S_0)A=(0)$. Since A is locally nilpotent $A=\mathbf{B}(A)$. Therefore S_0 is nilpotent. Hence there exists a k with $S_0^k=(0)$ and $S_0^{k-1}\neq 0$. Then $(0)\neq S_0^{k-1}\leqq \leqq l(S_0)$ and $S_0^{k-1}A=(0)$. □

THEOREM 23.6. If A is a nil ring with maximum condition for right annihilators $r(S)$ and for left annihilators of the form $T_k=\{x; x\in A, xA^k=(0)\}$, then A is nilpotent.

PROOF. We have $T_1\leqq T_2\leqq T_3\leqq\ldots$. By assumption there exists an n such that $T_n=T_{n+1}=T_{m+2}=\ldots$. If $T_n=A$, we have $A^{n+1}=(0)$. If $T_n\neq A$, $\bar{A}=A/T_n$ has a non-zero left annihilator $\bar{x}$ by proposition 23.5. Thus $x\cdot\bar{A}=(0)$ for $\bar{A}=A/T_n$ and $\bar{x}=x+T_n$. Consequently $xA\leqq T_n$, $xAA^n=(0)$, $xA^{n+1}=(0)$ and $x\in T_{n+1}$. Since $T_{n+1}=T_n$ we have $x\in T_n$, whence the contradiction $\bar{x}=0$ follows. Therefore $A=T_n$, and thus $A^{n+1}=(0)$. □

We remark in connection with the second part of proposition 23.5 that every nilpotent ring contains a non-zero left annihilator. Namely, if $A^k=(0)$, and $A^{k-1}\neq(0)$ we have $A^{k-1}\cdot A=(0)$. Proposition 23.5 asserts the existence of non-zero left annihilators in nil rings with maximum condition for right annihilators $r(S)$. These rings coincide with their lower nil radicals. However, there exists a nil ring (naturally without maximum condition for right annihilators) which does not have a non-zero left annihilator. Nevertheless this ring is the sum of its nilpotent ideals.

EXAMPLE 23.7 (Baer [2]). Let G be a countable Abelian group with $x+x=0$ for every $x\in G$. Then G is a direct sum of cyclic subgroups $\{a_i\}$ or order 2, i.e.

$$G=\sum_{i=0}^{\infty}\{a_i\}.$$

If $G_0=G$ and $G_i=\sum_{j\geqq i}\{a_j\}$, then we have

$$\bigcap_{i=1}^{\infty}G_i=(0),\quad G_{i+1}\subsetneqq G_i\quad\text{and rank}\quad(G_i/G_{i+1})=1.$$

Let $\varepsilon(G)$ be the ring of all endomorphisms of the group G, and let A be a subset of all those endomorphisms α in $\varepsilon(G)$ for which the following conditions are fulfilled:

(1) $G_i\alpha\leqq G_{i+1}$ for all i;
(2) $G_j\alpha=(0)$ for a sufficiently large j, depending on α.

Then A is a right ideal of $\varepsilon(G)$. Let R_i be the set of all elements α of $\varepsilon(G)$, for which $G_i\alpha=0$. We obviously have $R_i\leqq A$, and since $G_i(AR_i)=(G_iA)R_i\leqq G_{i+1}R_i=(0)$ and $G_i(R_iA)=(G_iR_i)A=(0)$ every R_i is an ideal in A. These ideals R_i form an ascending chain

$$R_1\leqq R_2\leqq R_3\leqq\ldots,$$

and by (2) the union of this chain is A itself.

Now, $G=\{a_0\}\oplus G_1$. Let C be the set of all elements α of A for which $a_0\alpha=0$. Then C is a right ideal of A with $C\cap R_1=(0)$. For if $\beta\in R_1\cap C$ then obviously $G\beta=0$, and so $\beta=0$. Hence we have $CR_1=(0)$, because CR_1 is contained in $C\cap R_1$.

Assume that the idealizer of the right ideal C is larger than C. Then there exists an element $\alpha\in A$ with $\alpha\notin C$ and $\alpha C\leqq C$. We obviously have $\{a_0\}\alpha\neq 0$ and $\{a_0\}\alpha$ is a cyclic subgroup in G_1 by definition of A. There exists an element $\beta\in A$ with $a_0\beta=0$ and $a_0\alpha\beta\neq 0$, because $G=\{a_0\}\oplus\{a_0\}\alpha+G^*$ with $G^*=G_1$. Hence we have $\beta\in C$, $\alpha\beta\notin C$, so $\alpha C\not\leqq C$ which contradicts the assumption $\alpha C\leqq C$. Therefore the idealizer D of C in A coincides with C.

Now, if F is the sum of all ideals of A contained in the right ideal C, then F is an ideal of A with $F\leqq C$. Let $B=A/F$. Since A is a nil ring by (1) and (2), B is again a nil ring. Moreover, $GR_i^{i+1}=(0)$, and thus A is the union of the nilpotent ideals R_i. Consequently B is the union of the nilpotent ideals $(R_i+F)/F$.

If b is a non-zero element of B with $bB=(0)$, then we have $b\notin F$ and $bA\leqq F$. If $T=\{x, x\in A, xA\leqq F\}$ then T is an ideal of A with $b\in T$, $T\not\leqq F$. Then $C+T$ is a subring of A with $C\neq C+T$, because obviously $T\not\leqq C$ since $T\not\leqq F$. Moreover, we have

$$(C+T)C\leqq C+TC\leqq C+TA\leqq C+F=C,$$

and thus the subring $C+T$, for which $C+T\neq C$, is contained in the idealizer of the right ideal C, which contradicts the fact that C is its own idealizer. Therefore $A/F=B$ is a sum of its nilpotent ideals. It does not have non-zero left annihilator and thus the maximum condition does not hold in B by proposition 23.5.

§ 24. PIVOTAL MONOMIALS

We shall now discuss a general result of Amitsur [13] which implies both the local nilpotence of nil PI-rings and the nilpotence of nil subrings of rings with minimum condition for right ideals. For this we need some preparation.

The notion of a pivotal monomial was introduced by Drazin [2]. Let $\lambda_1, \lambda_2, \ldots, \lambda_t$ be non-commutating indeterminates and $\pi(\lambda)=\lambda_{i_1}\lambda_{i_2}\ldots\lambda_{i_d}$ be a monomial of degree d in the polynomial ring $A[\lambda_1, \lambda_2, \ldots, \lambda_t, \ldots]$ over a ring A. Let P_π denote the set of all monomials $\sigma(\lambda)=\lambda_{j_1}\lambda_{j_2}\ldots\lambda_{j_q}$, for which either $q>d$ or $i_h\neq j_h$ holds for some $h\leqq q$, if $q\leqq d$. Let S_π denote the set of all elements of P_π, for which the degree q of $\sigma(\lambda)=\lambda_{j_1}\ldots\lambda_{j_q}$ is at least d. Then $\pi(\lambda)$ is called a *right pivotal monomial*, if $\pi(x)=x_{i_1}\ldots x_{i_d}$ is contained in the right ideal, generated by all $\sigma(\lambda)\in P$, for every substitution $\lambda_i=x_i\in A$ $(i=1, 2, 3, \ldots,)$. If $\pi(x)=x_{i_1}\ldots x_{i_d}$ is contained in the right ideal, generated by all $\sigma(\lambda)$, contained in S_π, then $\pi(\lambda)$ is called *right strongly pivotal*.

A ring A is called a *right PM-ring* of degree d if there exists a monomial $\pi(\lambda)=$ $=\lambda_{i_1}\dots\lambda_{i_d}$ with

$$\pi(x)A \leqq \sum_{\sigma\in S_\pi} \sigma(x)A$$

for every substitution $\lambda_i=x_i\in A$. A left PM-ring A is analogously defined. A is a PM-ring of degree d if there exists a monomial $\pi(\lambda)=\lambda_{i_1}\lambda_{i_2}\dots\lambda_{i_d}$ with

$$A\cdot\pi(x)\cdot A \leqq \sum_{\sigma\in S_\pi} A\sigma(x)A$$

for every substitution $\lambda_i=x_i\in A$.

We may assume (by theorem 2 of Drazin [2]) that $\pi(\lambda)$ is linear in every λ_i. Let $N(A)$ denote the sum of all nilpotent ideals of A.

THEOREM 24.1. If A is a PM-ring of degree d, and T a nil subring of A, then $T^d\leqq N(A)$, and thus T is locally nilpotent.

PROOF. Let $n\geqq d+1$, moreover $A_i=T^{n-i}AT^i$ for $1\leqq i\leqq d$ and $A_i=AT^nA$ for $i>d$. Then

$$A_i\cdot A_j \leqq AT^n A. \tag{1}$$

Let $x_i\in A_i$ $(i\geqq 1)$. Then

$$x_{i_1}\dots x_{i_p}\in AT^n A, \tag{2}$$

if $p>d$ or if $p=d$ and $(i_1, i_2, \dots, i_d)=(1, 2, \dots, d)$. In fact, if $p>d$ either $i_j\geqq i_{j+1}$ holds for a j, and then (1) implies (2), or $i_p>d$, and then (2) follows from the definition of A_i. In the second case, if $(i_1, \dots, i_d)\neq(1, 2, \dots, d)$, obviously either $i_j\geqq i_{j+1}$ or $i_d>d$ holds.

Now A is a two-sided strongly PM-ring of degree d if

$$Ax_1x_2\dots x_dA \leqq \sum_{\sigma\in S_\pi} A\sigma(x)A.$$

By relation (2) we obtain $A\sigma(x)A\leqq AT^nA$, if $\sigma(x)\in S_\pi$. On the other hand, the elements $x_1\, x_2\, x_3 \dots x_d$ run over all elements of

$$(T^{n-1}A)^dT^d = (T^{n-1}AT)(T^{n-2}AT^2)\dots(T^{n-d}AT^d),$$

if x_i runs over all elements of A_i. Hence it follows that

$$(AT^{n-1}A)^{d+1} \leqq A(T^{n-1}A)^dT^dA \leqq AT^nA. \tag{3}$$

First let us assume that T is nilpotent. Let k be the minimal natural number such that AT^kA is nilpotent. If $k>d$ the nilpotence of $AT^{k-1}A$ follows from (3), which contradicts the minimality of k. Therefore $k\leqq d$. Now

$$T^k+AT^k+T^kA+AT^kA \leqq N(A),$$

since

$$(T^k+AT^k+T^kA+AT^kA)^3 \leqq AT^kA.$$

Hence $T^k\subseteq N(A)$.

Let T be an arbitrary nil subring of A and $t\in T$. Since the subring $\{t\}$ generated by t is nilpotent, $t^d \in N(A)$ by the preceding. Hence $(T+N(A))/N(A)$ is a nil ring with $\bar{t}^d = \bar{0}$ for every $t \in T$, whence by proposition 22.2 it follows that $(T+ + N(A))/N(A)$ is locally nilpotent. But then T is locally nilpotent, because $N(A)$ is also locally nilpotent. If $t_1, t_2, \ldots, t_d$ are arbitrary elements in T, then the subring $\{t_1, t_2, \ldots, t_d\}$ is nilpotent, and thus we have $\{t_1, t_2, \ldots, t_d\}^d \leqq N(A)$ by the preceding. In particular $t_1, t_2, \ldots t_d \in N(A)$. Therefore $T^d \leqq N(A)$. But then T is locally nilpotent, because T is the extension of a locally nilpotent ring by the nilpotent ring T/T^d. □

We only consider division-rings D such that every matrix-ring over D is a *PI*-algebra. If A is a ring with minimum condition for right ideals, or if A is a *PI*-algebra, then there exists a natural number m such that for every element x of A there exists an $a \in A$ with $x^m = x^{m+1}a$. Then λ^m is a pivotal monomial since $x^m = x(x^m)a = \ldots = x^k x^m a^k$ for every k.

For, if A is a ring with minimum condition for right ideals and N is the nil radical of A, then A/N is the direct sum of full matrix rings M_i over division rings D_i by the well-known Wedderburn–Artin structure theorem (cf. Chapter IV, theorem 33.4). Let $n_i \times n_i$ be the type of M_i over D_i. If $k_i = n_i^2 + 1$, then M_i is a *PI*-algebra satisfying the multilinear polynomial $f(x_1, \ldots, x_{k_i}) = \sum (-1)^\alpha x_{j_1} x_{j_2} \ldots \ldots, x_{j_{k_i}}$, where α is the number of inversions of $j_1, j_2, \ldots, j_{k_i}$ and the sum is taken over all possible permutations $j_1, j_2, \ldots, j_{k_i}$, since the rank of M_i over D_i is less than k_i. Since M_i is a *PI*-algebra, there exists a natural number m_i such that for every $x \in M_i$ there exists an $a \in M_i$ with $x^{m_i} = x^{m_i+1}a$. If m is the maximum of the m_i's, then corresponding to every element $\bar{x} \in A/N$ one can find an element $\bar{a} \in A/N$ with $\bar{x}^m = x^{m+1}\bar{a}$. Therefore $x^m - x^{m+1}a \in N$. Hence, $x^m \equiv x^{m+r}a^r \pmod N$ for every natural number r. If $N^s = (0)$ (cf. proposition 33.5), then since $(x^m - x^{m+r}a^r)^s = 0$ we have the desired relation $x^{ms} = x^{ms+1}y$ with $r = ms - m + 1$, $y \in A$ and with an s, independent of x. Hence λ^{ms} is a pivotal monomial for the ring A with minimum condition for right ideals, as claimed.

PROPOSITION 24.2. Every nil subring B of a *PI*-algebra A is locally nilpotent.

PROOF is clear by the foregoing. □

§ 25. THE GENERALIZED NIL RADICAL

Let **K** be the class of all non-zero rings without divisors of zero. Every ring in **K** is a prime ring, and every non-zero ideal of a ring in **K** belongs to **K**. Let $B \in \mathbf{K}$, B an ideal of A and B^* the annihilator of B in A. If $BR = (0)$, then $(RB)^2 = RBRB$ and $RB \leqq B \in \mathbf{K}$ imply $R \leqq B^*$. Similarly, $L \leqq B^*$ if $LB = (0)$. Here and R and L denote the right and left annihilator of the ideal B, respectively.

For $a_1 \notin B^*$ and $a_2 \notin B^*$ we have by the preceding $Ba_1 \neq (0)$ and $a_2B \neq (0)$. Since $Ba_1 \leqq B \in \mathbf{K}$ and $a_2B \leqq B \in \mathbf{K}$ we have $(0) \neq Ba_1a_2B$. Therefore $a_1a_2 \notin B^*$, and thus A/B^* is also a ring without divisors of zero.

Therefore, $\mathbf{K}$ is a special class of rings. If $\mathbf{S}$ is the upper radical, determined by the class $\mathbf{K}$, then $\mathbf{S} > \mathbf{U}$, where $\mathbf{U}$ is the upper nil radical, because a ring from $\mathbf{U}$, certainly cannot be homomorphically mapped onto a ring in $\mathbf{K}$. Then $\mathbf{N}_g = \mathbf{S}$ is called the *generalized nil radical* (cf. Thierrin [1]).

Therefore we have:

THEOREM 25.1. The class of all non-zero rings without divisors of zero is a special class $\mathbf{K}$, and the upper radical $\mathbf{N}_g$ determined by $\mathbf{K}$, is strictly larger than the upper nil radical. Every $\mathbf{N}_g$-semisimple ring is a subdirect sum of rings without divisors of zero. For a commutative ring $\mathbf{N}_g$ coincides with the lower nil radical.

PROOF. $\mathbf{U} \leqq \mathbf{N}_g$. Since a full matrix ring A of type $n \times n$ over a division ring is simple and $\mathbf{U}$-semisimple, and since A contains nilpotent elements, A is an $\mathbf{N}_g$-radical ring, and thus $\mathbf{U} \neq \mathbf{N}_g$.

If A is commutative, and $\mathbf{B}(A)$ is the lower nil radical of A, then $A/\mathbf{B}(A)$ is a subdirect sum of prime rings A_α. Now, if x, y are elements of A_α with $xy=0$, then we have by commutativity $xA_\alpha y = (0)$. Hence we obtain $x=0$ or $y=0$, because A_α is a prime ring. Therefore A_α has no divisors of zero, and thus $A/\mathbf{B}(A)$ is $\mathbf{N}_g$-semisimple. Hence $\mathbf{N}_g \leqq B$. Since, on the other hand $\mathbf{B} \leqq \mathbf{U} \leqq \mathbf{N}_g$, we have, $\mathbf{B} = \mathbf{N}_g$ for commutative rings. □

The upper radical determined by the class of all division rings is again special and it contains the generalized nil radical (cf. Thierrin [2]).

PROBLEM 25 (Levitzki). Does there exist an idempotent simple nil ring?

PROBLEM 26 (Koethe). Let R be a nil right ideal of a ring A. Is $R + AR$ necessarily a nil ideal?

PROBLEM 27.† Does there exist a nil ring A with zero Baer lower nil radical such that the structure of A is less complicated than that of the important example 16.5 of Baer?

PROBLEM 28. Investigate the various concrete radicals of the full matrix rings over an arbitrary nil ring!

PROBLEM 29. By an MHR-ring (an MH_1R-ring) we mean a ring with minimum condition on principal right ideals (on right ideals, contained in a principal right ideal) (cf. Szász [1, 2, 3]). Does there exist a nilpotent MHR-ring, or a nil MHR-ring (a nil MH_1R-ring), in which an ideal or a subring does not inherit the same chain condition?

PROBLEM 30. Does there exist a nil MHR-ring without minimum condition on principal left ideals?

PROBLEM 31. Does there exist a nil MHR-ring (or an arbitrary MHR-ring) without minimum condition for finitely generated right ideals?

PROBLEM 32. Does there exist a canonical representation of the lower or of the upper Loewy system of operator modules over an arbitrary MHR-ring with a non-zero, non-nilpotent upper nil radical (cf. Szász [2]).

PROBLEM 33. For which ordinal numbers γ does there exist a nil MHR-ring A with $A^{\gamma}=(0)$ and $A^{\beta}\neq(0)$ for $\beta<\gamma$?

PROBLEM 34. Describe all MHR-rings A, for which the right annihilator of the α-th transfinite right hypersocle A_{α} of A is just the α-th transfinite power $(\mathbf{U}(A))^{\alpha}$ of the upper nil radical of A for every α.

PROBLEM 35. Does there exist a nil MHR-ring without minimum condition for the powers of a fixed principal right ideal?

PROBLEM 36.† Find a (ring-theoretical) necessary and sufficient condition in order that the adjoint group (with $x \circ y = x + y - xy$) of a nil MHR-ring should possess a *transfinite solvable invariant system* (cf. Szász [2]).

PROBLEM 37. Let $\mathbf{Z}$ be the lower radical determined by the class of all zero-rings of prime number order. Is every nil MHR-ring $\mathbf{Z}$-radical ring?

PROBLEM 38.† Let $\mathbf{M}$ be the lower radical determined by the class of all those nil rings which are lower nil radicals of an MHR-ring. Let $\mathbf{D}$ be the (Divinsky) lower radical determined by the class of all those nilpotent rings, which are (lower) nil radicals of a ring with minimum condition for right ideals. Is $\mathbf{M}\neq\mathbf{D}$?

PROBLEM 39.† Does the radical $\mathbf{M}$ (defined in problem 38) differ from the Baer lower nil radical?

PROBLEM 40.† Investigate the (nilpotent) rings with minimum condition for right ideals which have an infinite distributive lattice of left ideals (cf. problem 2.13 of Szász [3]).

PROBLEM 41. Does there exist for every natural number $n \geqq 4$ a finite ring A with the following three conditions:

(a) $(\mathbf{U}(A))^{n}=(0)$ and $(\mathbf{U}(A))^{n-1}\neq(0)$ for the (upper) nil radical $\mathbf{U}(A)$ of A;

(b) A has two modular right ideals R_1 and R_2 such that $R_1 \cap R_2$ is not modular in A (R is modular in A, if there exists an element $a \in A$ with $x - ax \in R$ for every $x \in A$).

(c) The adjoint semigroup H (with the operation $x \circ y = x + y - xy$) of the ring A contains a subsemigroup F such that the Frattini subsemigroup Φ of F is empty but not every subset of F is a subsemigroup (Szász [20] has shown the existence of such rings for $n=3$, cf. problem 2 of Kertész [1], and problem of Rédei [1], p. 90).

PROBLEM 42. Investigate the upper radical determined by the class of all subdirectly irreducible rings without non-zero nil radical.

PROBLEM 43. Is every nil subring of an arbitrary MHR-ring locally nilpotent?

PROBLEM 44. Investigate the upper radical determined by the class of all subdirectly irreducible rings without divisors of zero (cf. Thierrin radical, Thierrin [1]).

CHAPTER IV

THE JACOBSON RADICAL

In this chapter the Jacobson radical is defined as the annihilator of all irreducible right modules over a ring. Right primitive ideals, modular right ideals and quasi-modular right ideals are introduced. The Jacobson radical coincides both with the sum of all quasi-regular right ideals and with the intersection of all modular maximal right ideals. It also coincides with the intersection of all quasi-modular maximal right ideals. We shall show by an example, that a quasi-modular maximal right ideal is not necessarily modular, and that a right primitive ring is not necessarily left primitive. We shall discuss in detail the Jacobson radical of a full matrix ring, that of further rings related to a ring, and that of an algebra over a field. We shall explicitly give examples of a right primitive ring, whose proper homomorphic images are nilpotent, and of a simple Jacobson radical ring. Moreover, we shall prove the density theorem for right primitive rings, and a criterion for a radical to coincide with the Jacobson radical on the class of rings with minimum condition on right ideals. Finally, some open problems are raised.

§ 26. IRREDUCIBLE RIGHT MODULES, RIGHT PRIMITIVE IDEALS, MODULAR AND QUASI-MODULAR RIGHT IDEALS

Let A be a ring and M a right A-module. M is said to be *irreducible*, if $MA \neq (0)$, and M contains only the trivial submodules (0) and M. Every irreducible right A-module M is a prime module (in the sense of § 15). For, if m is a non-zero element of M and B is an ideal of A with $mB=(0)$, then the set $N= \{x;\ x\in M,\ xB=(0)\}$ is a submodule of M with $N\neq(0)$. Therefore, $N=M$ and $B\leqq(0:M)_A$. For a ring A, let $\mathbf{M}_A$ denote the class of all irreducible right A-modules, and set $\mathbf{M}=U\mathbf{M}_A$.

We have the following

THEOREM 26.1. $\mathbf{M}$ is a special class of prime modules.

PROOF. As we have shown, every right A-module M of $\mathbf{M}_A$ is an A-prime module. Let B an ideal of A and $M\in\mathbf{M}_A$ with $MB\neq(0)$. If N is a B-submodule of M, then we have $NB\leqq N$. If $N\neq M$ then NB is a right A-module, therefore $NB=(0)$.

Since B is an ideal with $MB \neq (0)$, by the A-irreducibility of M we have $N=(0)$. Thus M is also a B-irreducible right module. Hence $M \in \mathbf{M}_B$.

If B is an ideal of A, and $M \in \mathbf{M}_B$, then we have $MB \in \mathbf{M}_A$. For, if N is a non-zero A-submodule of MB, then since $MB=M$ and M is B-irreducible it follows that $N=M$. Thus $MB=M \in \mathbf{M}_A$.

Finally, if $M \in \mathbf{M}_A$ and P is an ideal of A contained in $(0:M)_A$, then we have $M \in \mathbf{M}_{\bar{A}}$ where $\bar{A}=A/P$. Conversely, if P is an arbitrary ideal of the ring A, $\bar{A}=A/P$ and $M \in \mathbf{M}_A$, then we have $M \in \mathbf{M}_A$ with $P \leqq (0:M)_A$. □

Let A be a ring. Then $\mathbf{J}(A)=M \in \bigcap_{M \in \mathbf{M}_A} (0:M)_A$ is called the *Jacobson radical* of the ring A.

An ideal P of a ring A is called *right primitive* (in short *primitive*) if there exists an irreducible right A-module $M \in \mathbf{M}_A$ with $P=(0:M)_A$. A ring A is called *right primitive* (in short *primitive*), if the ideal (0) is a right primitive ideal. Therefore, A is a right primitive ring if and only if there exists a *faithful* irreducible right A-module $M \in \mathbf{M}_A$. By theorem 15.1 and by the definition we have

THEOREM 26.2. The Jacobson radical $\mathbf{J}(A)$ of the ring A is the intersection of all right primitive ideals of A. Every $\mathbf{J}$-semisimple ring is a subdirect sum of right primitive rings. The class of all right primitive rings is a special class of rings, and thus the Jacobson radical is a special radical. For every ideal B of a ring A $\mathbf{J}(B)=B \cap \mathbf{J}(A)$.

PROOF is clear by the results of § 15. □

A right A-module M is said to be *strongly cyclic,* if there exists an element m of M with $M=mA$. The element m is called a *strong generator* of M. A right ideal R of a ring A is called *modular* in A if there exists an element $b \in A$ such that $a-ba \in R$ for every $a \in A$.

PROPOSITION 26.3. A right A-module M is strongly cyclic if and only if there exists a modular right ideal R of A such that the module isomorphism $M \cong A/R$ holds. A right ideal R of A is modular in A if and only if $R=(0:m)_A$ for a strong generator m of a strongly cyclic right A-module M.

PROOF. If $M=mA$, there exists an element B with $m=mb$. Then $a-ba \in (0:m)$ since $ma=mba$. The mapping $a \to ma$ is a module isomorphism of A onto $M=$ $=mA$ with kernel $(0:m)_A=R$, where R is modular since $a-ba \in R$. Conversely, if R is a modular right ideal of the ring A with $a-ba \in R$ for every $a \in A$ then A/R is a strongly cyclic right A-module with strong generator element $b+R$ since $a+R=(b+R)a$. □

PROPOSITION 26.4. If R is a modular right ideal of A, then $R \geqq (R:A)$ where $R:A=\{x;\, x \in A,\, Ax \leqq R\}$.

PROOF. $R=(0:m)_A \geqq (0:M)_A=(0:A/R)_A=R:A$.

A right ideal R of a ring A is called *quasi-modular* in A if $R:A \leqq R$. By proposition 26.4 every modular right ideal of A is quasi-modular in A. But there exists a ring A containing a quasi-modular maximal right ideal which is not modular in A. We construct such a ring in the following

EXAMPLE 26.5 (cf. Szász [17]). Let p be zero or a prime number, K_p a prime field, $\mathfrak{m}$ an infinite cardinal number, Γ an index set of cardinality $\mathfrak{m}$, $\delta_{\alpha\beta}$ the Kronecker-symbol, A an algebra over K_p with the basis elements $a_\alpha, r_{\alpha\beta}, s_{\alpha\beta\gamma}$ $(\alpha, \beta, \gamma \in \Gamma)$, and R the subalgebra of A generated over K_p by all $r_{\beta\gamma}$ and $s_{\varepsilon\eta\vartheta}$. The multiplication of the basis elements is defined by the following table:

	a_ε	$r_{\varepsilon\eta}$	$s_{\varepsilon\eta\vartheta}$
a_α	a_ε	$\delta_{\alpha\varepsilon}\cdot a_\eta$	$\delta_{\alpha\varepsilon}\cdot a_\vartheta$
$r_{\alpha\beta}$	$s_{\alpha\beta\varepsilon}$	$\delta_{\beta\varepsilon}\cdot r_{\alpha\eta}$	$\delta_{\beta\varepsilon}\cdot s_{\alpha\eta\vartheta}$
$s_{\alpha\beta\gamma}$	$s_{\alpha\beta\varepsilon}$	$\delta_{\gamma\varepsilon}\cdot s_{\alpha\beta\eta}$	$\delta_{\gamma\varepsilon}\cdot s_{\alpha\beta\vartheta}$

Every element of this monomial algebra can be represented in the form

$$a = \sum\nolimits^* \pi_i a_{\alpha_i} + \sum_{i,j}{}^* \varrho_{ij} r_{\alpha_i\beta_j} + \sum\nolimits^* \delta_{ijk} s_{\alpha_i\beta_j\gamma_k} \tag{$*$}$$

where π_i, ϱ_{ij}, $\sigma_{ijk}\in K_p$ and all sums $\sum^*$ are finite. $a\in R$ if and only if in $(*)$ $\pi_i=0$ for every i.

On the basis of the multiplication table it can be proved that A is an associative algebra.

R is obviously a right ideal of A. Moreover, R is a maximal right ideal of A. For, if $a\notin R$ then there exists $\pi_i \in K_p$ with $\pi_i \neq 0$ in $(*)$. Therefore we have a $(\pi_i^{-1}\varrho r_{\alpha_i\beta})=\varrho a_\beta + r'$ $(r'\in R)$ for every $\beta\in\Gamma$ and $\varrho\in K_p$. Consequently $aR+R=A$.

We claim that R is not modular in A. If $a\in R$, then we certainly have $(1-a)A \not\subseteq R$ since $(1-a)a_\alpha = a_\alpha - aa_\alpha \notin R$. However, if $a\notin R$ then there exists an i with $\pi_i\neq 0$ in $(*)$, whence we obtain $(1-a)(-\pi_i^{-1}r_{\alpha_i\beta})=a_\beta+r''\in R$ $(r''\in R)$. Consequently $(1-a)A\not\subseteq R$. Therefore R is not modular in A.

We shall now show that R is quasi-modular in A. Let a be an element of A with $a\notin R$. Assume that $a_\alpha a = r^* \in R$ for an index α. Then a obviously has the form $a=\sum \pi_i a_{\alpha_i} + r$ with a finite sum $\sum$ and $r\in R$. Then since $a_\alpha r = -\sum \pi_i a_{\alpha_i} + r^*$, the element r has the form

$$r = \sum_i{}^* \sigma_i r_{\alpha\alpha_i} + \sum_{i,j}{}^* \tau_{ij} s_{\alpha\beta_j\alpha_i} + r_\alpha$$

with σ_i, $\tau_{ij} \in K_p$ and with finite sums $\sum^*$. Also, $r_\alpha \in R$, $a_\alpha r_\alpha \in R$. Consequently, r_α has a representation of the form

$$r_\alpha = \sum_{i,j} \pi_{ij}(r_{\alpha\gamma_i} - s_{\alpha\beta_j\gamma_i}) + \sum_{i,j} \sigma'_{ij} r_{\beta_i\gamma_j} + \sum_{i,j,k} \tau'_{i,j,k} s_{\varepsilon_i\eta_j\vartheta_K} \qquad (**)$$

with π_{ij}, σ'_{ij}, $\tau'_{ijk} \in K_p$, and all three sums $\sum^*$ are finite. Since $|\Gamma| = \mathfrak{m}$, there exists an index $\varepsilon \in \Gamma$ different from α and from all indices β_i and ε_i occurring in $(**)$ on the left-hand side. Hence, since $a_\varepsilon r_\alpha = a_\varepsilon r = 0$, we obviously have

$$a_\alpha a = \sum_i{}^* \pi_i a_{\alpha_i} \notin R.$$

Therefore, $R : A \leqq R$. Thus R is quasi-modular in A.

Proposition 26.6. Every proper modular right ideal R of a ring A can be embedded in a modular maximal right ideal.

Proof. Let $b \in A$ be an element with $a - ba \in R$ for every $a \in A$. Since $R \neq A$ we have $b \notin R$ because $b \in R$ would imply $ba + (a - ba) = A \in R$ for all $a \in A$ and therefore would imply $R = A$. Let $\mathfrak{M}$ be the set of all right ideals R' of A with $b \notin R'$. The set $\mathfrak{M}$ is not empty since $R \in \mathfrak{M}$. Therefore by Zorn's lemma there exists a right ideal M which is maximal in $\mathfrak{M}$. By the above argument $R'' > M$ implies $b \in R''$ and thus $R'' = A$. Thus M is a maximal right ideal of A, M is modular since $(1-b)A \leqq R \leqq M$. □

Proposition 26.7. A right A-module M is irreducible if and only if $MA \neq (0)$ and M is strongly cyclic with every non-zero element a strong generator. A right A-module M is irreducible if and only if there exists a modular maximal right ideal R with $M \cong A/R$.

Proof. If M is irreducible then $mA = M$ for every non-zero element m of M, because the set $N = \{m; m \in M, mA = (0)\}$ is an A-submodule with $NA = (0)$. We have $N = (0)$ since $MA \neq (0)$ and M is irreducible. Hence $mA \neq (0)$ for every element $m \neq 0$. Thus $mA = M$ for every non-zero $m \in M$.

Conversely, if N is a proper A-submodule of M, then $nA \leqq N$ $(0 \neq n \in N)$ implies that the element n is not a strong generator of M. Therefore, if $MA \neq (0)$, and if every non-zero element is a strong generator of M, then M is irreducible.

If M is an irreducible right A-module, then $M \cong A/(0:m)_A$ for every $m \in M$ with $m \neq 0$, $R = (0:m)_A$ is modular since $m = mb\,(b \in A)$, $ma = mba$, $a - ba \in R$. R is a maximal right ideal of A by the irreducibility of M. Conversely, if R is a modular maximal right ideal of A with $(1-b)A \subseteq R$ then A/R is a right A-module without non-trivial submodules. If $A^2 \leqq R$, we have $A \subseteq R$ since $(1-b)A \subseteq R$. This is a contradiction, and thus $(A/R)A = A/R$. Hence A/R is irreducible. □

PROPOSITION 26.8. An ideal P of a ring A is right primitive in A if and only if there exists a modular maximal right ideal R of A with $P=(R:A)$.

PROOF. If P is a primitive ideal of A then $P=(0:M)_A$ with an irreducible right A-module M. But we have $M\cong A/R$ for some modular maximal right ideal R of A. It follows that $P=(0:A/R)_A=(R:A)$. □

PROPOSITION 26.9 (cf. Szász [18]). If R is a quasi-modular maximal right ideal of a ring A, then $R:A$ coincides with the primitive ideal $(R_x:A)$, where

$$R_x = \{x\}^{-1}A = \{z;\ z\in A,\ xz\in R\}$$

is a modular maximal right ideal of A for *every* element $x\in A$ with $x\notin R$.

PROOF. R_x is a right ideal of A. We have $A^2+R=A$ by the quasi-modularity of R. Thus $RA^{-1}=\{z; z\in A, zA\subseteq R\}\subseteq R$. Consequently $xA+R=A$ for every $x\in A$ with $x\notin R$. Therefore, for every $x\in A$ with $x\notin R$ there exists $y\in A$ with $xy\notin R$. Thus $y\notin R_x$. Hence $R_x\neq A$. For every element $z\in A$ with $z\notin R_x$ we obviously have $xz\notin R$, and thus $xzA+R=A$. Hence there exist an $a\in A$ and an $r\in R$ with $xza+r=$ $=xb$. Therefore we have $x(za-b)=-r\in R$, $za-b\in R_x$ and $b\in zA+R_x$ for every $b\in A$. Consequently $A=zA+R_x$, because b has been chosen arbitrarily. Hence R_x is a maximal right ideal of A. Moreover, because $xzA+R=A$, there exist elements $a_1\in A$, $r_1\in R$ with $xza_1+r_1=x$, and thus we have $x(1-za_1)A=r_1A\subseteq R$. Therefore, $(1-za_1)\subseteq R_x$, hence R_x is modular in A.

Now, on the one hand we have $(xA)^{-1}R\subseteq A^{-1}R$ for every $x\notin R(x\in A)$ since $xA+R=A$ and $A((xA)^{-1}R)=(xA+R)(xA)^{-1}R\subseteq R$. On the other hand, if $y\in A^{-1}R$ then $xAy\subseteq R$ and thus $y\in(xA)^{-1}R$ since $A^{-1}R=(xA+R)^{-1}R$. Hence we have $A^{-1}R=(xA)^{-1}R$ for every $x\in R(x\in A)$. We also have trivially that $(xA)^{-1}R=$ $=A^{-1}\{x\}^{-1}R=A^{-1}R_x$. Therefore, the modularity of the maximal right ideal R_x implies that $Q=A^{-1}R=(xA)^{-1}R$ coincides with the primitive ideal $P_x=A^{-1}R_x$ for every $x\notin R$, $x\in A$. □

REMARK 26.10. In the paper [16] of Szász an ideal $Q=A^{-1}R=R:A$ was called *quasi-primitive*, if R is a quasi-modular maximal right ideal. Proposition 26.9 also says that the Jacobson radical $\mathbf{J}(A)$ of a ring coincides with the intersection of all quasi-primitive ideals (for a fully elementary proof of this fact we refer the reader to Szász [16]). But by Steinfeld [7] and by proposition 26.9 every quasi-primitive ideal is a primitive ideal.

PROPOSITION 26.11. Every commutative primitive ring is a field.

PROOF. In a primitive ring A there exists a modular maximal right ideal R with $(0)=(R:A)=A^{-1}R$. Therefore $Ax\subseteq R$ implies $x=0$. By the commutativity we have $RA=AR\subseteq R$, and thus $R=(0)$. Therefore A is a ring with a left unity element and without non-trivial right ideals, and hence A, being commutative, is a field. □

PROPOSITION 26.12. Every primitive ideal of a ring A is a prime ideal of A.

PROOF. This follows from § 15; but we give here another proof. If P is a primitive ideal of the ring A, then there exists an irreducible right A-module M with $P=(0:M)_A$. If B and C are ideals of A with $B\subsetneqq P$ and $C\subsetneqq P$, then we have $MB\neq$ $\neq(0)$ and $MC\neq(0)$. Consequently $MB=MC=M$. Hence $M=MC=(MB)C=$ $=M(BC)$. Thus $BC\nsubseteqq P$. Hence P is a prime ideal of A. □

REMARK 26.13. For a proof of the fact that every semisimple ring (in the sense of Jacobson) with minimum condition on right ideals has a unity element and is the direct sum of finitely many idempotent minimal right ideals, the following theorem proves useful:

The intersection of finitely many modular maximal right ideals of a ring is again modular (cf. proposition 3.6.1 (2) of Jacobson [2]). Following the methods of Jacobson [2] and a proposition of Kertész [1] (Satz 5.2, p. 121) it is possible to prove more generally that:

If $A=R_1+R_2$ for the modular right ideals R_1 and R_2 of a ring A, then the intersection $R_1\cap R_2$ is again modular in the ring A.

For if $(1-e_i)A\leq R$ with $e_i=e_{i1}+e_{i2}$ and $e_{ij}\in R_j$, then it is easy to see that $(1-e)A\leq R_1\cap R_2$ where $e=e_{12}+e_{21}$.

In the book [1] (p. 123) of Kertész, the following problem is raised:

Is the intersection of two modular right ideals of a ring always modular?

Szász [19, 20] has solved this problem by giving examples in the negative.

EXAMPLE 26.14. Consider the algebra A generated by the elements a, b, c, d over the prime field K_2 (with elements 0 and 1), and with the following table of multiplication

	a	b	c	d
a	a	$a+b+c$	a	d
b	$a+b+d$	b	c	b
c	c	b	c	$a+c+d$
d	a	d	$b+c+d$	d

An easy calculation shows that this algebra, consisting of 16 elements, is associative.

The radical J of A is a modular maximal right ideal with $J^3=(0)$ and $J^2\neq(0)$, and J contains eight elements.

Moreover, we have $A/J\cong K_2$.

The right ideals $(1-a)A=R_1$ and $(1-b)A=R_2$ are modular in A. Since A obviously has no left unity element, and $(1-a)A\cap(1-b)A=(0)$, the intersection $D=(1-a)A\cap(1-b)A$ is certainly not modular in A. Here $K=(1-a)A+(1-b)A$ is a two-sided ideal for which the factor ring, consisting of four elements, is non-commutative and non-semisimple (for a more detailed discussion we refer the reader to Szász [20]).

Szász [19] has obtained the following stronger result:

For every infinite cardinal $\aleph_\alpha$ there exists a ring A containing $\aleph_\alpha$ modular right ideals such that every pair of modular right ideals has a non-modular intersection in A.

§ 27. THE EXISTENCE OF A RIGHT PRIMITIVE RING WHICH IS NOT LEFT PRIMITIVE

The following example of a ring with the properties mentioned in the title is due to Bergman [1]; we sketch it following Herstein [2].

EXAMPLE 27.1. Let Q be the field of rational numbers, $Q(x)$ the field of all rational functions of the indeterminate x over Q, and α the monomorphism α: $r(x)\to r(x^2)$. Let $A=Q(x)[y]$ be the polynomial ring in an indeterminate y over $Q(x)$ with the ordinary definition of the equality, addition and subtraction and with multiplication defined by $yr(x)=r(x^2)y=\alpha(r(x))y$. Then A is a non-commutative ring with a unity element and without divisors of zero. Moreover, it is a principal left ideal ring, because A has a *Euclidean division algorithm on the left*.

We shall show that every subalgebra B of A over Q with $x\in B$ and $y\in B$, is a right primitive ring. We shall prove this by explicitly constructing a faithful irreducible right B-module M.

We have $r(x)\in Q(x)$. Define $r^*(x)\in Q(x)$ by the equality

$$r^*(x^2)=\frac{r(x)+r(-x)}{2}.$$

Then $Q(x)$ is a right B-module with the operations $r(x)\cdot q(x)=r(x)q(x)$ and $r(x)y=r^*(x)$. For we have $(ry)q=r(yq)=r(\alpha(q))y=(r_\alpha(q))y$. Consequently, $r^*q==(r\alpha(q))^*$, i.e. in detail

$$(r^*q)(x^2)=r^*(x^2)\,q(x^2)=\frac{1}{2}(r(x)+r(-x))\,q(x^2)=(r\alpha(q))^*(x^2).$$

Moreover,

$$x^n y^m=\begin{cases} x^{2m-n} & \text{if } \ 2^m \mid n \\ 0 & \text{if } \ 2^m \nmid n. \end{cases}$$

Let M be the right B-module generated by x in $Q(x)$. We assert that M is a faithful irreducible right B-module. Consider $0\neq p(x)\cdot(q(x))^{-1}$ where $p(x)$, $q(x)$ are polynomials over Q. Suppose $n=$degree $p(x)$, and let m be a natural number with $2^m>n$. Then we have

$$\frac{p(x)}{q(x)}(q(x)\cdot x^{2m-n}y^m)=(p(x)x^{2m-n})y^m=cx,$$

where $c\neq 0$ is the principal coefficient of $p(x)$. Therefore, x is contained in the right B-module, generated by the arbitrary element $p(x)\cdot(q(x))^{-1}$. Thus M is an irreducible right B-module.

Let $0\neq b=\sum_i r_i(x)y^i\in(0:M)_B$ and let $xp(x)$ be a non-zero polynomial such that all rational functions $xp(x)\cdot r_i(x)$ are polynomials. Let

$$xp(x)\cdot b=\sum s_i(x)\,y^i \quad (\neq 0)$$

where each $s_i(x)$ is a polynomial. Then we have $\sum s_i(x)y^i\in(0:M)_B$. Let j be the minimal number with $s_j(x)\neq 0$, $d=$max degree $(s_i(x)-s_j(x))$, $m=$degree $s_j(x)$ and n a natural number such that

$$2^n\geqq \max\{\text{degree } s_j(x),\ 2^{j^i+1}\cdot d\}.$$

Let $F(x,y)=x^{2^n-m}(\sum s_i(x)y^i)$. Since x^{2^n} is the highest power of x with non-zero coefficient in $x^{2^n-m}\cdot s_j(x)$, x^{2^n-j} has a non-zero coefficient in $x^{2^n-m}s_j(x)y^j$. The highest power of x in $x^{2^n-m}s_i(x)$ is obviously $x^{2^n-m}\cdot x^{\text{degree } s_i(x)}$, and 2^n-m+ $+$degree $s_i(x)\leqq 2^n+d$. Hence x occurs in $x^{2^n-m}s_i(x)y^i$ at most with exponent $\dfrac{2^n+d}{2^i}$. Since

$$\frac{2^n+d}{2^i}\leqq 2^{n-j-1}+d\leqq 2\cdot 2^{n-j-1}=2^{n-j},$$

$x^{2^{n-j}}$ has a non-zero coefficient in $F(x,y)$. This contradicts the definition of

$\sum_i s_i(x)y^i$. Therefore, M is a faithful, irreducible right B-module, and thus B is right primitive.

Using a certain *valuation* we shall now explicitly construct a subalgebra B of A with $x \in B$ and $y \in B$ which is not left primitive.

Let $p_n(x)$ be the n-th cyclotomic polynomial over Q and y_n the valuation of $Q(x)$ induced by $p_n(x)$.

First we prove that the equation $v_n(r(x)) = v_n(r(x^2))$ holds for every odd natural number n and for every $r(x) \in Q(x)$. If $p_n(x)$ and $p(x)$ have a common root then we have $p_n(x)|p(x)$, because $p_n(x)$ is irreducible over Q. For odd n the roots of $p_n(x^2)$ are just the elements $\sqrt{\vartheta}$ and $-\sqrt{\vartheta}$ with $p_n(\vartheta) = 0$. Therefore we have $p_n(x)|$ $|p_n(x^2)$ and $p_n(-x)|p_n(x^2)$. By comparison of the degrees we have $p_n(x^2) = p_n(x) \cdot$ $\cdot p_n(-x)$. If $(p_n(x), q(x)) = 1$ then $(p_n(x), q(x^2)) = 1$ and thus $v_n(q(x)) = 0$. Hence we have $v_n(f(x)) = v_n(f(x^2))$ for every polynomial. Therefore, the statement is proved for rational functions. Now we define the function w_n on A by $w_n(\sum r_i(x)y^i) =$ $= \min_i v_n(r_i(x))$. Then w_n is a valuation on the ring A since $v_n(r(x)) = v_n(r(x)^2)$. Let $B_0 = \{z; z \in A, w_n(z) \geqq 0$ for every odd $n\}$. Then B_0 is a subalgebra of A with $x \in B_0$ and $y \in B_0$, and thus B_0 is right primitive by the above. The fact that B_0 is not left primitive, follows from the stronger proposition that every maximal left ideal L of B_0 contains a non-zero ideal of B_0.

If L is a maximal left ideal of B_0, then AL is a left ideal of A. Now, A is a principal left ideal ring. Therefore, if $AL \neq A$ then AL can be generated by an element $g = a_0(x) + \ldots + a_m(x) \cdot y^m$ with $m > 0$ and $a_m(x) \neq 0$, because we have $(g)_l = A$ if $m = 0$. Let n be an odd natural number such that $w_n(g) = 0$. For an arbitrary element $a = \sum b_i y^i$ there exists an i by the definition of w_n such that $w_n(a) = w_n(b_i)$. Let $\delta(a)$ be the largest number i with $w_n(a) = w_n(b_i)$. Then we have $\delta(ac) = \delta(a) + \delta(c)$ for every $a \in A$ and $c \in A$. Then since $w_n(g) = 0$, we have $\delta(g) = m$, where m denotes the degree of g in the indeterminate y.

If $u \in AL$, then we have $\delta(u) \geqq \delta(g) > 0$. For the n-th cyclotomic polynomial $p_n(x)$ we have $w_n(p_n(x)) = 1$ and $p_n(x) \in B_0$. Since $p_n(x)$ has an inverse element in $Q(x)$, $p_n(x) \in AL$, and since L is a maximal left ideal of B, it follows that $B_0 = L +$ $+ B_0 p_n(x)$. Thus we have $1 = l + bp_n(x)$ with $l \in L$ and $b \in B_0$. Hence, on the one hand, $\delta(1 - bp_n(x)) = 0$ and, on the other, $\delta(l) > 0$. This is a contradiction. So we have $AL = A$ for every maximal left ideal L of B_0. Hence there exist elements $a_j \in A$ and $l_j \in L$ such that $1 = \sum_{j=1}^{s} a_j l_j$. Moreover, there exists a non-zero element $z \in Q(x)$ such that $w_n(za_j) \geqq 0$ for every n and for every j with $1 \leqq j \leqq s$. Thus we have $za_j \in B_0$. Hence $z = z(\sum a_j l_j) = \sum (za_j)\, l_j \in B_0 L \leqq L$. Therefore $B_0 z \leqq L$. Since $z \in Q(x)$, z^{-1} exists in A, and since $w_n(z^{-1}tz) = w_n(t)$ we have $z^{-1}B_0 z = B_0$. Thus $I = zB_0 = B_0 z$ is an ideal of B_0 contained in the left ideal L. Hence B_0 is not left primitive. However B_0 is right primitive.

§ 28. QUASI-REGULAR ELEMENTS

One can associate to every element a of a ring A a modular right ideal R with $x-ax\in R$ for every $x\in A$. For example, one such modular right ideal is the set of all elements $x-ax$. This set is denoted by $(1-a)A$, even if A possesses no left unity element. If $(1-a)A=A$, then the element a is called *right quasi-regular*. $(1-a)A=$ $=A$ if and only if $a\in(1-a)A$. For, $(1-a)A=A$ obviously implies $a\in(1-a)A$, and if $a\in(1-a)A$ then we have $x=(x-ax)+ax\in(1-a)A$ for every $x\in A$ and so $(1-a)A=A$. Now, if $-a\in(1-a)A$, then there exists an element $b\in A$ such that $-a=b-ab$; i.e. $a+b-ab=0$. This element b is called *right quasi-inverse* of a.

For the discussion of the quasi-regular elements, it is advantageous to introduce the *circle operation* $a\circ b=a+b-ab$. Let 1 denote the unity element of a ring extension of A. For the mapping

$$\delta\colon\; a\to 1-a$$

we have $(a\circ b)\sigma=1-a\circ b=(1-a)(1-b)=a\sigma\cdot b\sigma$. Therefore $a\circ b=(a\sigma\cdot b\sigma)\sigma^{-1}$. Thus

$$a\circ(b\circ c)=(a\sigma(b\sigma\cdot c\sigma)\sigma^{-1})\sigma^{-1}=(a\sigma b\sigma c\sigma)\sigma^{-1}=$$

$$=((a\sigma b\sigma)\sigma^{-1}\sigma c\sigma)\sigma^{-1}=(a\circ b)\circ c,$$

i.e. the circle operation is associative. Hence, since $a\circ 0=0\circ a=a$, the elements of every ring A form a semigroup with respect to the circle operation. The element 0 is the unity element of this semigroup.

An element a is *quasi-regular* in A, if there exists an element b of A such that $a\circ b=b\circ a=0$. The inverse element b is uniquely determined because $a\circ c=c\circ a=$ $=0$ implies $b=b\circ 0=b\circ(a\circ c)=(b\circ a)\circ c=0\circ c=c$. The quasi-regular elements of the semigroup $(A,\circ)$ form a group. A right ideal R of A is called right quasi-regular (quasi-regular) if every element r of R is right quasi-regular (quasi-regular) in A.

PROPOSITION 28.1. Every right quasi-regular right ideal R of a ring A is quasi-regular.

PROOF. For every $r\in R$ there exists an $r'\in A$ with $r+r'=rr'=0$. Therefore, $r'=rr'-r\in R$. Hence there exists an element $r''\in A$ with $r'+r''-r'r''=0$. Hence again $r''\in R$. Therefore

$$r=r\circ 0=r\circ(r'\circ r'')=(r\circ r')\circ r''=r''.$$

Thus $r\circ r'=r'\circ r=0$. □

PROPOSITION 28.2. Every nil right ideal of a ring A is quasi-regular.

PROOF. If $a^n=0$ for an element a of R, then we have $a\circ b=b\circ a=0$ where

$$b=-a-a^2-\ldots-a^{n-1}.\square$$

Now our aim is to give a comprehensive characterization of the Jacobson radical $\mathbf{J}(A)$ of a ring A by means of certain subsets H_i $(i=1, 2, \ldots, 10)$ of A. In particular, in order to define H_6, we need some preparation.

Let $D_r(A)$ be the intersection of all maximal right ideals of the ring A if there exist any, and $D_r(A)=A$ if A has no maximal right ideals. As is well known, $D_r(A)$ is the set of all elements x of A which are redundant in every right generating system of A, i.e. such that $A=(B, x)_r$ implies $A=(B)_r$ for every subset B. Obviously $D_r(A)$ is a right ideal. It will be shown later that $D_r(A)$ is an ideal, moreover, a quasi-regular ideal of A.

We define

$$H_6=A^{-1}D_r(A)=\{x;\ x\in A,\ Ax\subseteq D_r(A)\}.$$

H_6 is obviously an ideal of A. With the methods of Hille [1] (theorem 22.15.3) and Kertész [5] it can be proved that H_6 is also a quasi-regular ideal, and H_6 coincides with the Jacobson radical $\mathbf{J}(A)$ of A.

Moreover, we show

THEOREM 28.3 (cf. Jacobson [2], Kertész [1]). The following subsets of an arbitrary ring A coincide with the Jacobson radical $\mathbf{J}(A)$ of A:

(1) the union H_1 of those subgroups of the maximal quasi-regular group of the semigroup $(A, \circ)$ which are also right ideals of the ring A,

(2) the union H_2 of all quasi-regular right ideals,

(3) the set H_3 of all elements $x\in A$ such that the product xy is right quasi-regular in A for every $y\in A$,

(4) the set H_4 of all elements $x\in A$ such that the product zxy is right quasi-regular in A for every $y\in A$ and $z\in A$,

(5) the set H_5 of all elements $x\in A$ such that the product zxy is quasi-regular in A for every $y\in A$, $z\in A$,

(6) the ideal $H_6=A^{-1}D_r(A)$,

(7) the intersection H_7 of all quasi-modular maximal right ideals of A,

(8) the intersection H_8 of all modular maximal right ideals of A,

(9) the intersection H_9 of all primitive ideals of A,

(10) the intersection H_{10} of all quasi-primitive ideals of A.

PROOF. $\mathbf{J}(A)=H_9=H_{10}$ follows essentially from the definition of $\mathbf{J}(A)$. Therefore, we only need to prove that the subsets H_1, H_2, H_3, H_4, H_5, H_6, H_7, H_8, and H_9 coincide.

We claim $H_1\subseteq H_2$. Every subgroup B of the maximal quasi-regular subgroup of the semigroup $(A, \circ)$ which is at the same time a right ideal of the ring A, is a quasi-regular right ideal of A, and thus we have $B \leqq H_2$ and $H_1 \leqq H_2$.

$H_2 \leqq H_3$ holds, because every element x of H_2 is contained in a quasi-regular right ideal and thus we have $xy \in H_2$ for every $y \in A$. Therefore, $x \in H_3$ and $H_2 \leqq H_3$.

We shall now show that $H_3 \leqq H_4$. Let $x \in H_3$. Then $x(yz)$ is right quasi-regular in A for every $y \in A$ and $z \in A$. Thus there exists an element $u \in A$ with $xyz+u-xyzu=0$. But then since

$$zxy+(zuxy-zxy)-zxy(zuxy-zxy) = z(xyz+u-xyzu)xy = 0,$$

we have that zxy is right quasi-regular in A for every $y \in A$ and $z \in A$.

We claim $H_4 \leqq H_5$. For, $x \in H_4$, $y \in A$, $z \in A$ imply the existence of an element $w \in A$ with

$$zxy+w-zxyw = 0. \tag{$*$}$$

Hence we have $w=zx(yw-y)$. Therefore, the element w is not only left quasi-regular in A, as $(*)$ shows, but also right quasi-regular by the definition of H_4 and because $w=zx(yw-y)$. Since the quasi-inverse is uniquely determined, w is the quasi-inverse of zxy. Thus $H_4 \leqq H_5$.

Also $H_5 \leqq H_6$. For, if $x \in A$, $x \notin H_6$, then we have $Ax \not\equiv D_r(A)$. Accordingly, there exists an element $a \in A$ and a maximal right ideal R of A with $ax \notin R$. Since $\bar{a}x \neq \bar{0}$ the right A-module A/R is irreducible and thus there exists an element $b \in A$ with $\bar{a}xb=\bar{a}$. Therefore, $axb-a \in R$. Suppose $x \in H_5$. We shall deduce a contradiction. Since $x \in H_5$ there exists a $y \in A$ such that

$$xbxb+y-xbxby = 0.$$

But we have $\bar{a}xbxb=\bar{a}xb=\bar{a}$. Consequently $\bar{a}=\bar{a}(xbxb)=\bar{a}(xbxby-y)= =\bar{a}(y-y)=0$, which contradicts the condition $ax \notin R$. Since $xbxb$ is not quasi-regular, we have $x \notin H_5$. Thus $H_5 \leqq H_6$.

Now we prove $H_6 \leqq H_7$. If $x \notin H_7$, then there exists a quasi-modular maximal right ideal R of A with $x \notin R$, and thus since $A^{-1}R \subseteqq R$ we can find an element $y \in A$ such that $yx \notin R$. Hence we have $Ax \not\equiv D_r = D_r(A)$ and thus $x \notin H_6$. Therefore $H_6 \leqq H_7$.

$H_7 \leqq H_8$, because every modular maximal right ideal of A is also quasi-modular in A.

Now we shall show $H_8 \leqq H_9$. If $x \in A$ and $x \notin H_9$, then there exists a primitive ideal P of A with $x \notin P$. Then there exists a modular maximal right ideal R of A such that $P=A^{-1}R=\{y;\, y \in A,\, Ay \leqq R\}$. Moreover, we have $Ax \subsetneqq R$. Consequently $x \notin A^{-1}D_r(A)=H_6$. Since $H_6 \geqq H_5 \geqq H_4 \geqq H_3$ we also have $x \notin H_3$. Hence there exists an element $z \in A$ such that xz is not right quasi-regular. Therefore $xz \notin (1- -xz)A$. Thus $x \notin (1-xz)A$. By Zorn's lemma a maximal right ideal R_1 such that

$x\notin R_1$ $(1-xz)A\subseteq R_1$, and R_1 is a modular maximal right ideal of A. Since $x\notin R_1$ we have $x\notin H_8$. Thus $x\in H_8$ implies $x\in H_9$. This proves that $H_8\subseteq H_9$.

The proof of $H_9\subseteq H_1$ is similar to the proof of $H_8\subseteq H_9$. We shall show that every element x of H_9 is right quasi-regular in A. If $x\in H_9$, and x is not right quasi-regular, then $x\notin(1-x)A$. Let R be a maximal right ideal of A with $x\notin R$ and $R\supseteq(1-x)A$. Such a right ideal exists by Zorn's lemma, R is a modular maximal right ideal. Then $P=A^{-1}R=\{z;\, z\in A,\, Az\subseteq R\}$ is a primitive ideal of A, and since $x\in H_9$ we have $x\in P$. Therefore, $Ax\subseteq R$. In particular $x^2\in R$. Since $x-x^2\in(1-x)A\subseteq \subseteq R$ it follows that $x\in R$, which contradicts condition $x\notin R$. Therefore, x is right quasi-regular in A if $x\in H_9$. Since H_9 is an ideal, x is quasi-regular in A by proposition 28.1. Thus $H_9\subseteq H_1$. □

This proof of theorem 28.3 is due to Kertész [1].

Proposition 28.4. If we assign to every set H_i $(i=1, 2, \ldots, 9)$ the left-right dual set H_i' (i.e. for 'right' we substitute 'left' and vice versa), then these left-right dual sets H_i' also coincide with the Jacobson radical $\mathbf{J}(A)$ of the ring.

Proof. The set H_5 is obviously left-right self dual. Hence we have $H_5=H_5'$. Since $H_i'=H_5'$ and $H_5'=\mathbf{J}(A)$ we also have $\mathbf{J}(A)=H_i'$ for $i=1, 2, \ldots, 9$. □

Remark 28.5. By remark 26.10 every right quasi-primitive ring is also a right primitive ring, and thus the Jacobson radical $\mathbf{J}(A)$ coincides with the intersection of all right quasi-primitive ideals. Moreover, $\mathbf{J}(A)$ coincides with the intersection of all left primitive ideals, but there exists a right primitive ring which is not left primitive, as in example 27.1.

Moreover, the Jacobson radical $\mathbf{J}(A)$ of A is the sum of all quasi-regular right ideals, and it contains every quasi-regular left idal of A. Hence $\mathbf{J}(A)$ contains every nil right ideal, every nil left ideal and every nil ideal of A. But it is still an open question whether the upper nil radical contains every nil right ideal and every nil left ideal of the ring.

§ 29. RIGHT PRIMITIVE MATRIX RINGS

For the subject-matter of this section we refer the reader to Posner [3].

Let A be a ring, $n(\geq 1)$ a natural number, A_n the ring of all matrices of type $n\times n$, and M a right A_n-module with $MA_n=M$ and without trivial submodules (i.e. $mA_n=(0)$ for $m\in M$ implies $m=0$).

We have the following

Theorem 29.1 (Posner [3]). If M is considered as an right A-module over the ring of the scalar matrices of A_n, then there exist a right A-module $\overline{M}$ and n

isomorphisms $f_i:\overline{M}\to\overline{M}_i\subset M$ such that $M=\overline{M}_1\oplus\ldots\oplus\overline{M}_n$. If a_{ij} here is the matrix with $a\in A$ in the (i,j)-th place and with 0's elsewhere then a_{ij} maps the submodule $\overline{M}_k$ into 0 for $k\neq i$ and $\overline{M}_i$ into $\overline{M}_i\, f_i^{-1}\, rf_j\subseteq\overline{M}_j$ for $k=i$ with a suitable element $r\in A$. Moreover, every $\overline{M}_i$ is uniquely determined and it coincides with MI_i, where I_i denotes the left ideal of all those matrices of A_n in which at most the i-th column differs from zero. $\overline{M}A=\overline{M}$, and $\overline{M}$ has no trivial submodule. $\overline{M}$ is a faithful right A_n-module if and only if $\overline{M}$ is a faithful right A-module. M is an irreducible right A_n-module if and only if $\overline{M}$ is an irreducible right A-module.

PROOF. Since $MA_n=M$, every element $m\in M$ has the form

$$m = m_1+\ldots+m_n \quad \text{with} \quad m_i\in MI_i.$$

If for $1\leqq k\leqq n$, J_k is the right ideal of all those matrices of A_n in which at most the k-th row differs from zero, then we have $I_i\cdot J_k=0$ for $i\neq k$, and thus

$$m_1'=m_2'+\ldots+m_n' \quad (m_i'\in MI_i)$$

obviously implies $m_2'J_1=\ldots=m_n'J_1=(0)$ and $m_1'J_1=(0)$. Since $m_1'J_2=\ldots=m_1'J_n=$ $=(0)$ we have $m_1'A_n=(0)$ and hence $m_1'=0$. This fact shows that the sum

$$M = MI_1+\ldots+MI_n$$

is direct. Now let $\overline{M}$ be an arbitrary (abstract) right A-module, which is isomorphic to the right A-module $\overline{M}_1=MI_1$. Moreover, let f_1 be an A-isomorphism of $\overline{M}$ onto $\overline{M}_1$. We shall define isomorphisms f_{ij}: $\overline{M}_i\to\overline{M}_j=MI_j$ such that f_{ii} is the identity mapping on $\overline{M}_i$ and $f_{ij}\cdot f_{jk}=f_{ik}$ holds. Then we define $f_j:\overline{M}\to\overline{M}_j$ by $f_j=f_1\cdot f_{1j}$. Thus the mapping property, mentioned in the theorem, for a_{ij} is automatically fulfilled. If $m_i\in\overline{M}_i=\overline{M}I_i$, then we have

$$m_i = m_{i1}a_1+\ldots+m_{in}a_n \quad \text{with} \quad m_{ij}\in M,\quad a_k\in I_k,\quad 1\leqq j\leqq n,\quad 1\leqq k\leqq n.$$

Let $a_k^{(j)}$ denote the matrix obtained from a_k by interchanging the i-th column with the j-th, and let

$$m_i f_{ij} = m_{i1}a_1^{(j)}+\ldots+m_{in}a_n^{(j)}.$$

We shall show the uniqueness of f_{ij} by proving that $m_{i1}a_1+\ldots+m_{in}a_n=0$ implies $m_{i1}a_1^{(j)}+\ldots+m_{in}a_n^{(j)}=0$. The image of $m_{in}a_n^{(j)}+\ldots+m_{i1}a_1^{(j)}$ under a_{jk} coincides with the image of $m_{i1}a_1+\ldots m_{in}a_n$ under a_{ik} for $1\leqq k\leqq n$, and thus it is zero. But $a_{lk}(l\neq j)$ also annihilates this element which is thus annihilated by the whole A_n. Hence this element is 0. Therefore f_{ij} is unique. Moreover, it is compatible with the operation of A. It is easy to show that $\overline{M}_1$ contains no trivial right A-module. Moreover, $\overline{M}_1A=\overline{M}$ because

$$\overline{M}_1\subset M = MA_n = MA_n^2 = M(I_1+\ldots+I_n)(J_1+\ldots+J_n) =$$
$$= MI_1J_1+\ldots+MI_nJ_1 = \overline{M}_1A+\ldots+\overline{M}_nA.$$

Therefore, $M_iA\leqq\overline{M}_i$ and the directness of the decomposition yield $\overline{M}_1=\overline{M}_1A$.

We shall now show that the submodules $\overline{M}_i$ are uniquely determined. For, if $M=N_1\oplus\ldots\oplus N_n$ is another direct decomposition of M, where the effect of A_n is given by the multiplication of the matrices, then $N_j a_{ji}\leqq N_i\cap\overline{M}_i$ for all $a\in A$ and $N_j a_{ki}=(0)$ for $j\neq k$. Hence we have

$$MA_n = (N_1+\ldots+N_n)A_n \leqq (N_1\cap\overline{M}_1)+\ldots+(N_n\cap\overline{M}_n),$$

$$\overline{M}_1+\ldots+\overline{M}_n = (N_1\cap\overline{M}_1)+\ldots+(N_n\cap\overline{M}_n),$$

and thus $N_i\cap\overline{M}_i=\overline{M}_i\leqq N_i$. Therefore $N_i=\overline{M}_i$.

Next we prove that M is a faithful right A_n-module if and only if $\overline{M}_1$ is a faithful right A-module. For, if $a\in A$ acts as a zero operator on $\overline{M}_1$, then a_{11} also acts as a zero operator on M. Conversely, if the matrix $a=(a_{ij})$ annihilates the A_n-module M, then the j_0-th row of a annihilates the submodule $\overline{M}_{j_0}$, which is also annihilated by the other rows. The i_0-th component of the image of $\overline{M}_{j_0}$ coincides with $(\overline{M}_{j_0})f_{j_0}a_{j_0i_0}$ under the mapping induced by the matrix, in which only the j_0-th row differs from zero, and this acts in the same way as a does. Hence $a_{j_0i_0}$ annihilates the whole $\overline{M}_1$, and thus $a_{j_0i_0}=0$ if $\overline{M}_1$ is faithful. Since i_0, j_0 are arbitrarily chosen, $a=0$.

Moreover, M is an irreducible right A_n-module if and only if $\overline{M}_1$ is an irreducible right A-module. For, if $\overline{N}$ is a non-zero A-submodule of $\overline{M}_1$, then we have

$$M = \overline{N}\oplus\overline{N}f_{12}\oplus\ldots\oplus\overline{N}f_{1n}$$

and thus $\overline{N}=\overline{M}_1$. Conversely, if $\overline{M}_1$ is an irreducible right A-module, then we have $mA_n=M$ for every non-zero element m of M. For, let $p_1+\ldots+p_n$ be an arbitrary element of mA_n with $p_i\in\overline{M}_i$. Assume that the first component m_1 of m differs from zero. Then put $a_{ij}=0$ for $i\neq 1$ and $m_1a_{ij}=p_j$. This is possible since $m_1A= =\overline{M}_1$. Now, if $a=(a_{ij})\in A_n$ then the i-th component of ma is obviously $m_1a_{1i}+\ldots \ldots+m_na_{ni}=p_i$. □

Theorem 29.1 implies

THEOREM 29.2. Let A be a ring and $n(\geqq 1)$ a natural number. Then the ring A_n of all matrices of type $n\times n$ over A is right primitive if and only if A is a right primitive ring.

We recall that an ideal Q of a ring A is called *semi-prime* in A if $B^2\leqq Q$ implies $B\leqq Q$ for an ideal B of A.

PROPOSITION 29.3. An ideal Q of a full matrix ring A_n over A is semi-prime if and only if there exists a semi-prime ideal P of A with $Q=P_n$.

PROOF. If the ideal P is semi-prime in the ring A and I is an ideal of A_n with $I^k\leqq P_n$ for some k, then let J be the ideal in A generated by the elements of the matrices of I. Then we have $J_n \geqq I$ and $J_n^3 \leqq A_n I_n A_n \leqq I$. Hence we have

$$J_n^{3k} = (J_n^3)^k \leqq I^k \leqq P_n.$$

Consequently, $J^{3k} \leqq P$, $J \leqq P$, $J_n \leqq P_n$ and $I \leqq P_n$. Therefore P_n is semi-prime in A_n. Conversely, if the ideal Q is semi-prime in A_n then let P be the ideal in A generated by the elements of the matrices of Q. Then we have $P_n \geqq Q$ and $P_n^3 \leqq Q$. Therefore it follows that $P_n \leqq Q$ and thus $Q = P_n$. □

As we shall see later, proposition 29.3 remains valid if we replace 'semi-prime' by 'right primitive'. For, we have

THEOREM 29.4. An ideal Q of the full matrix ring A_n is right primitive in A_n if and only if there exists a right primitive ideal P of A with $Q = P_n$.

PROOF. If P is a right primitive ideal of the ring A, then there exists a faithful irreducible right A/P-module $\overline{M}$. Then $\overset{1}{\breve{\overline{M}}} \oplus \ldots \oplus \overset{n}{\breve{\overline{M}}}$ is a faithful irreducible right $(A/P)_n$-module with $\overset{i}{\breve{\overline{M}}} \cong \overline{M}$. But $(A/P)_n \cong A_n/P_n$. Therefore P_n is a right primitive ideal of A_n. Conversely, if Q is a right primitive ideal of A_n, then Q is semiprime in A_n, and thus there exists an ideal P of A with $Q = P_n$ by proposition 29.3. Since A_n/P_n is right primitive, and $A_n/P_n \cong (A/P)_n$ A/P is also right primitive, and hence P is a right primitive ideal of A. □

§ 30. THE JACOBSON RADICAL OF RINGS RELATED TO A RING

We have the following:

THEOREM 30.1. The Jacobson radical $\mathbf{J}(A_n)$ of a full matrix ring A_n over A coincides with the full matrix ring $(\mathbf{J}(A))_n$ over the Jacobson radical $\mathbf{J}(A)$ of the ring A.

PROOF I. By theorem 29.4 $\mathbf{J}(A_n) = \bigcap_P P_n$, where P runs over all right primitive ideals of the ring A. Hence a matrix of A_n is contained in $\mathbf{J}(A_n)$ if and only if every element of the matrix is contained in every right primitive ideal of A, and thus in $\mathbf{J}(A)$. Hence $\mathbf{J}(A_n) = (\mathbf{J}(A))_n$. □

PROOF II. (The original proof of Jacobson (cf. [2]) uses quasi-regular elements.) If

$$Z = \begin{pmatrix} z_{11} & z_{12} & \ldots & z_{1n} \\ 0 & 0 & \ldots & 0 \\ \vdots & \vdots & & \vdots \\ 0 & 0 & \ldots & 0 \end{pmatrix}$$

is a matrix where z_{11} is a right quasi-regular element of A, then since $(1-z_{11})A=$ $=A$, there exist elements $z'_{1i}\in A$ with $z'_{1i}-z_{11}z'_{1i}=-z_{1i}$. Now $z_{11}\circ z'_{11}=0$. Moreover $Z\circ Z'=0$ where

$$Z' = \begin{pmatrix} z'_{11} & z'_{12} & \dots & z'_{1n} \\ 0 & 0 & \dots & 0 \\ \vdots & \vdots & & \vdots \\ 0 & 0 & \dots & 0 \end{pmatrix}$$

Accordingly Z is a right quasi-regular matrix in A_n. Let I_j be the set of all matrices Z of A_n, whose j-th row consists of elements of $\mathbf{J}(A)$, and whose k-th row equals zero for $k\neq j$. Then I_j is a right ideal of A_n, and I_j is quasi-regular by an argument similar to that given for $j=1$. Hence $I_j\leqq\mathbf{J}(A_n)$, and we have $(\mathbf{J}(A))_n\subseteqq\mathbf{J}(A_n)$ since $(\mathbf{J}(A))_n = I_1+I_2+\dots+I_n$.

On the other hand, let

$$W = \begin{pmatrix} w_{11} & \dots & w_{1n} \\ \vdots & & \vdots \\ w_{n1} & \dots & w_{nn} \end{pmatrix}$$

be an arbitrary matrix of $\mathbf{J}(A_n)$ and A_{pq} the matrix with an arbitrary fixed element a in the (p, q)-th place and 0's elsewhere. Then we have

$$\sum_{k=1}^{n} A_{kp}WB_{qk} = \text{diag}\,(aw_{pq}b, \dots, aw_{pq}b).$$

Since W is contained in $\mathbf{J}(A_n)$, there exists a quasi-inverse matrix (w'_{ij}) for W. Hence $(aw_{pq}b)\circ w'_{11}=w'_{11}\circ(aw_{pq}b)=0$. Therefore, $aw_{pq}b$ is quasi-regular for every element $a\in A$, $b\in A$. Hence $w_{pq}\in\mathbf{J}(A)$ by theorem 28.3, and thus $\mathbf{J}(A_n)=(\mathbf{J}(A))_n$. Therefore, we have $\mathbf{J}(A_n)=(\mathbf{J}(A))_\alpha$. □

THEOREM 30.2. If $\mathbf{R}$ is a general radical containing every $\mathbf{R}$-right ideal (e.g. the Jacobson radical, the Levitzki radical or the Baer lower nil radical), and if K is a right ideal of the $\mathbf{R}$-semisimple ring A, then the radical $\mathbf{R}(K)$ coincides with the left annihilator of the ring K.

PROOF. We have $\mathbf{R}(A)=(0)$. Now, $\mathbf{R}(K)\cdot K$ is a right ideal of A which is an $\mathbf{R}$-right ideal since $\mathbf{R}(K)K\leqq\mathbf{R}(K)$ and thus it is contained in $\mathbf{R}(A)$ by the assumption on $\mathbf{R}$. Hence $\mathbf{R}(K)K=(0)$. □

PROPOSITION 30.3. If A is an arbitrary ring, I the ring of the rational integers, $A^*=A+I$ with $A\cap I=(0)$, and if the unity element of I is also the unity element of A^*, then $\mathbf{J}(A)^*=\mathbf{J}(A)$.

PROOF. First we shall show that I is **J**-semi-simple. Since the intersection of all principal ideals (p) is zero, where p runs over all prime numbers, I is a subdirect sum of the fields $I/(p)$ and thus $\mathbf{J}(I)=(0)$.

Since every ideal of A is also an ideal of A^*, $A^*/A\cong I$ and $\mathbf{J}(I)=(0)$ obviously imply $\mathbf{J}(A^*)\leqq A$. But the Jacobson radical **J** is hereditary by the last statement of theorem 26.3, and hence $\mathbf{J}(A)=A\cap\mathbf{J}(A^*)=\mathbf{J}(A^*)$. □

For the discussion of a result of Amitsur on the Jacobson radical of a polynomial ring $A[\lambda]$ we need some preparation. The ring $A[\lambda]$ consists of polynomials $a_0+a_1\lambda+\ldots+a_n\lambda^n$ in an indeterminate λ with coefficients $a_i\in A$ and with $a_i\lambda=$ $=\lambda a_i$ for every a_i. The determination of $\mathbf{J}(A[\lambda])$ is still an open question.

PROPOSITION 30.4. If B is an ideal of $A[\lambda]$, $p(\lambda)=a_0+a_1\lambda+\ldots+a_n\lambda^n$ with $a_n\neq 0$ is a non-zero polynomial of the lowest degree in the ideal B and $r(\lambda)$ is a polynomial in $A[\lambda]$ satisfying the condition $a_n^k r(\lambda)=0$ for an exponent $k\geqq 1$, then $a_n^{k-1}p(\lambda)r(\lambda)=0$.

PROOF. If $r(\lambda)=\sum_{i=0}^{m} r_i\lambda^i$, then $a_m^k r(\lambda)=0$ is equivalent to $a_n^k r_i=0$ for $i=0, 1, 2, \ldots$ $\ldots, m$. Hence it is sufficient to prove the proposition only for polynomials $r(\lambda)=r$ of degree 0 in λ. Now $a_n^k r$ is the coefficient of λ^n in the polynomial $a_n^{k-1}p(\lambda)r$, and $a_n^k r=0$. Hence the degree of $a_n^{k-1}p(\lambda)r$ is less than the degree of $p(\lambda)$, and since $a_n^{k-1}p(\lambda)r\in B$ we have $a_n^{k-1}p(\lambda)r=0$ by the definition of $p(\lambda)$. □

THEOREM 30.5 (Amitsur [10]). If the ring A has no non-zero nil ideal, then the polynomial ring $A[\lambda]$ is semi-simple in the sense of Jacobson.

PROOF. Suppose $\mathbf{J}(A[\lambda])\neq(0)$. If M is the set of non-zero polynomials of the lowest degree in $\mathbf{J}(A[\lambda])\neq(0)$, and N is the set of coefficients of the terms of highest degree in the polynomials of M, then N is an ideal of A. We shall show that N is a nil ideal of A. Let $p(\lambda)=a_0+a_1\lambda+\ldots+a_n\lambda$. Then $p(\lambda)\lambda a_n$ is quasi-regular in $A[\lambda]$, and so there exists a polynomial $p(\lambda)$ with

$$p(\lambda)\lambda a_n+q(\lambda)-p(\lambda)\lambda a_n q(\lambda)=0 \tag{1}$$

and

$$p(\lambda)\lambda a_n+q(\lambda)-q(\lambda)p(\lambda)\lambda a_n=0. \tag{2}$$

These equations show that $q(\lambda)\in\lambda A[\lambda]$. Let $s(\lambda)=p(\lambda)a_n$ and $q(\lambda)=t(\lambda)$. Then (1) and (2) imply

$$s(\lambda)+t(\lambda)-\lambda s(\lambda)t(\lambda)=0 \tag{3}$$

and

$$s(\lambda)+t(\lambda)-\lambda t(\lambda)s(\lambda)=0. \tag{4}$$

We shall show that $a_n^k t(\lambda)=0$ for a sufficiently large k. For, suppose $a_n^k t(\lambda)\neq 0$

for every k. If f is the smallest among the degrees of the polynomials $a_n^k t(\lambda)$ and

$$t(\lambda) = t_1(\lambda) + \lambda^{f+1} t_2(\lambda)$$

with

$$t_1(\lambda) = b_0 + b_1\lambda + b_2\lambda^2 + \ldots + b_f\lambda^f,$$

then $a_n^k t_2(\lambda) = 0$ for a sufficienty large number k, but $a_n^k b_f \neq 0$ for every k. The coefficient of the term of highest degree in $s(\lambda)$ is a_n^2, and since $a_n^k t_2(\lambda)=0$ we have $a_n^l s(\lambda) t_2(\lambda) = 0$ for some l by proposition 30.4. Now (3) implies

$$a_n^l s(\lambda) + a_n^l t_1(\lambda) - a_n^l \lambda s(\lambda) t_1(\lambda) = 0. \tag{5}$$

The coefficient of λ^{n+f+1} is obviously $-a_n^{l+b+1} b_f$ on the left-hand side of (5), and hence $a_n^{l+f+1} b_f = 0$, which contradicts the condition $a_n^k b_f \neq 0$ for every k. Hence we have $a_n^k t(\lambda) = 0$ for a suitable number k. From (4) we have $a_n^k s(\lambda) = 0$. Consequently $a_n^{k+2} = 0$. Therefore N is a nil ideal. □

There exist commutative radical rings in the sense of Jacobson without divisors of zero:

EXAMPLE 30.6. Let A be the set of rational numbers, numerator is an even number, and denominator is an odd number. Then A is a commutative ring without divisors of zero under the ordinary addition and multiplication. The solution of the equation $x+y-xy=0$ is $y = \dfrac{x}{x-1}$. Now, if x has the form $\dfrac{2k}{2l-1}$, then

$$y = \frac{2k}{2l-1} \cdot \left(\frac{2k}{2l-1} - 1\right)^{-1} = \frac{2k}{2l-1} \cdot \frac{2l-1}{2k-2l-1} = \frac{2k}{2(k-l)-1}.$$

Thus, y is also contained in the ring A, i.e. every element of A has a quasi-inverse in A. Therefore $\mathbf{J}(A) = A$. Since A does not contain a non-zero nil ideal, $A[\lambda]$ is semisimple. The radical ring A contains the semisimple ring of the even integers.

§ 31. THE JACOBSON RADICAL OF AN ALGEBRA OVER A FIELD

For the subject matter of this section we refer the reader to Amitsur [15].

In Chapter I it was shown that every general radical of an algebra A coincides with the radical of the ring A considered without operators. Now we discuss the Jacobson radical of an algebra A over a field K. Then A is a vector space over K. In every algebra A we can classify the elements by the following definition:

An element a of an algebra A is said to be *algebraic* (*transcendental*), if the subalgebra $\{a\}$, generated by a over K has a finite (infinite) dimension over K. The subalgebra $\{a\}$ consists of all polynomials $\sum \alpha_i a^i$ with $\alpha_i \in K$.

We have the following

THEOREM 31.1. Every element of the Jacobson radical $\mathbf{J}(A)$ of an algebra A over a field is either nilpotent or transcendental.

PROOF. If $a \in \mathbf{J}(A)$ is algebraic, then $\{a\}$ has a finite dimension over K, and thus since $a^{k-1}\{a\} \supseteq a^k\{a\}$ there exists an integer m with $a^{m-1}\{a\} = a^m a$. Hence there exists an element $b \in \{a\}$ with $a^m = a^m b$. Since $b \in \mathbf{J}(A)$ there exists a quasi-inverse element c for b such that $b + c - bc = 0$. Thus

$$a^m = a^m - a^m(b+c-bc) = a^m(1-b)(1-c).$$

Hence $a^m = 0$, because $a^m - a^m b = 0$. Therefore a is nilpotent. □

An algebra A over a field K is called *algebraic* if every element of A is algebraic. From theorem 31.1 we have immediately

PROPOSITION 31.2. The Jacobson radical $\mathbf{J}(A)$ of an algebraic algebra A over a field K is a nil ideal which contains every nil right ideal and every nil left ideal of A.

Now we shall discuss an important result of Amitsur, for which we need some preparation.

If the algebra A over K has no unity element, then let A^* be its least (canonical) algebra extension with unity element. Then A^* is also an algebra over K. Let $\{a\}^*$ be the subalgebra in A generated by a and 1. Then a is obviously algebraic if and only if $\{a\}^*$ has a finite dimension over K. If $K[\lambda]$ is the polynomial ring over K in an indeterminate λ then by $1 \to 1$, $\lambda \to a$ defines a homomorphism of $K(\lambda)$ onto $\{a\}^*$. The kernel of this homomorphism is a principal ideal $(\mu(\lambda))$, because $K[\lambda]$ is a principal ideal ring. The element a is algebraic if and only if $\mu(\lambda) \neq 0$ holds. We have $\mu(a) = 0$ and $\mu(\lambda)$ is said to be the *minimal polynomial* of a. It may be assumed that the coefficient of the term of the highest degree in $\mu(\lambda)$ equals 1. If $\nu(\lambda) \in K[\lambda]$ and $\nu(a) \in \{a\}^*$ then $\nu(a)$ has an inverse element in $\{a\}^*$ if and only if $(\nu(\lambda), \mu(\lambda)) = 1$, i.e. if the polynomials $\nu(\lambda)$ and $\mu(\lambda)$ are relatively prime.

We say that an element $\alpha \in K$ belongs to the *spectrum* $\sigma(a)$ of the element $a \in A$, if the element $a - \alpha 1$ has no inverse element in A. Here $a - \alpha 1$ has no inverse element in A if and only if the polynomials $\lambda - \alpha$ and $\mu(\lambda)$ are not relatively prime, i.e. if $\lambda - \alpha$ is a divisor of $\mu(\lambda)$. But this is the case if and only if $\mu(d) = 0$. Therefore every element of the spectrum of a is a root of the minimal polynomials $\mu(\lambda)$ of A. The complement $\varrho(a)$ of the spectrum $\sigma(a)$ of A in K is called the *resolvent set* of A.

PROPOSITION 31.3. If $\alpha_1, \alpha_2, \dots, \alpha_r$ are distinct elements of the resolvent set $\varrho(a)$ of a, and $(a - \alpha_1 1)^{-1}, \dots, (a - \alpha_r 1)^{-1}$ are the corresponding inverse elements, then either $(a-\alpha_1 1)^{-1}, \ldots, (a-\alpha_r 1)^{-1}$ are linearly independent over K, or a is an algebraic element with a minimal polynomial of degree $\leqq r-1$.

PROOF. If the elements $(a-\alpha_i 1)^{-1}$ $(i=1, 2, \ldots, r)$ are linearly dependent, then there exist elements $\beta_i \in K$ which are not all zero, such that

$$\sum_{i=1}^{r} \beta_i (a-\alpha_i 1)^{-1} = 0.$$

Multiplying by $\prod_{i=1}^{r} (a-\alpha_i 1)$ we obtain

$$\sum_{i=1}^{r} \beta_i \mu_i(a) = 0$$

where $\mu_i(\lambda) = \prod_{j\neq i} (\lambda-\alpha_j)$. Then the polynomial $\nu(\lambda) = \sum_{i=1}^{r} \beta_i \mu_i(\lambda)$ differs from zero, because the substitution of α_j obviously gives $\mu_j(\alpha_j) \neq 0$ since $\alpha_i \neq \alpha_k$ for $i \neq k$, and $\beta_j \neq 0$ for some j. Since $\nu(a)=0$, a is algebraic, and since the minimal polynomial $\mu(\lambda)$ is a divisor of $\nu(\lambda)$, the degree of $\mu(\lambda) \leqq r-1$. □

We have the following

THEOREM 31.4. Let A be an algebra over an infinite field K, and let the cardinality of K be strictly larger than the dimension of A over K. Then the Jacobson radical $\mathbf{J}(A)$ of A is a nil ideal.

PROOF. First suppose that there exists a unity element in A. If $a \in \mathbf{J}(A)$, then $(1-\alpha a)^{-1}$ exists for every $\alpha \in K$, and thus α^{-1} is contained in the resolvent set $\varrho(a)$ of A for every non-zero element α. If a is not an algebraic element, then the elements $(1-\alpha a)^{-1}$ are linearly independent by proposition 31.3, which is impossible by the condition imposed on the cardinality of K and the dimension of A over K. Hence a is algebraic, and thus a is nilpotent by theorem 31.3. If no unity element exists in A, then we form the canonical extension $A^* = A + K1$ of A, which has the unity element 1. Then $\mathbf{J}(A) = \mathbf{J}(A^*)$ and $\dim A^* = 1 + \dim A$. Thus $\mathbf{J}(A^*) = \mathbf{J}(A)$ is a nil ideal by the preceding consideration. □

Hence we have

PROPOSITION 31.5. The Jacobson radical $\mathbf{J}(A)$ of every finitely generated algebra over an uncountable field K is a nil ideal.

PROOF. The dimension of A over K is countable. □

PROPOSITION 31.6. If N is a nil algebra over an uncountable field K, then the full matrix ring N_n is also a nil algebra.

PROOF. Let $Z=(z_{ij})$ be an arbitrary element of N_n and P the subalgebra of N generated by the elements z_{ij}. Then P has a countable basis over K which can be obtained by choosing for basis elements of P_n all matrices which have a basis element of P in the (i, j)-th place and 0 elsewhere. Moreover, P is a nil algebra and thus it is a radical algebra in the sense of Jacobson. Since P_n is a radical algebra whose dimension is less than the cardinality of K, P_n is a nil algebra. Now Z is nilpotent since $Z \in P_n$. Thus N_n is a nil algebra. □

§ 32. EXAMPLE OF A RIGHT PRIMITIVE RING WHOSE PROPER HOMOMORPHIC IMAGES ARE NILPOTENT, AND EXAMPLE OF A SIMPLE RING WHICH IS A RADICAL RING IN THE SENSE OF JACOBSON

The following problem is due to Kurosh [3]: Does the Jacobson radical coincide with the upper radical determined by the class of all right primitive simple rings?

The following example of Sąsiada and Suliński [1] provides a negative answer to this question.

EXAMPLE 32.1. Let K be a field of characteristic zero, and S an automorphism of infinite order of K. For instance, let K be a purely transcendental field extension of the field of real numbers with infinitely many algebraically independent transcendental elements

$$\dots, x_{-3}, x_{-2}, x_{-1}, x_0, x_1, x_2, x_3, \dots$$

and S be an automorphism of K with $x_i^S = x_{i+1}$ $(i=0, \pm 1, \pm 2, \pm 3, \dots)$ and with $r^S = r$ for every real number. Then $S^n \neq 1$ for every natural number where 1 denotes the identity automorphism of K.

Let R be the set of all polynomials $a_0 + za_1 + \dots + z^n a_n$ in an indeterminate z over K with coefficients a_i from the field K. Let equality and addition of these polynomials be defined as usual. Let the multiplication be given by $az = za^S$ for every $a \in K$, together with the distributive and associative laws. Thus, in particular, we have $az^m = z^m a^{S^m}$, and R is a ring which we shall denote by $K[z, S]$.

The degree of a polynomial in R is defined as usual. Since degree $f_1 f_2 =$ degree $f_1 +$ degree f_2, R is a ring without divisors of zero in which only the non-zero elements of K have an inverse element.

The polynomials of R have the Euclidean algorithm on both the left-hand side and the right-hand side. If e.g.

$$f(z) = a_0 + za_1 + \dots + z^m a_m \quad \text{and} \quad g(z) = b_0 + zb_1 + \dots + z^r b_r$$

with $m \geqq r$, then

$$f(z) - g(z)\, z^{m-r} (b_r^{-1})^{S^{m-r}} a_m$$

is a polynomial of degree less than m. Continuing this procedure, we have $f(z)= =g(z)q(z)+r(z)$ holds, where either degree $r(z)<$degree $g(z)$, or $r(z)=0$. Similarly, $f(z)=q_1(z)g(z)+r_1(z)$, where either degree $r_1(z)<$degree $g(z)$, or $r_1(z)=0$.

From these facts it follows that every right ideal and every left ideal of R is principal.

If I is an ideal of R, then $I=f(z)R=Rg(z)$. Then we have $f(z)=h(z)g(z)$ and $g(z)=f(z)\cdot k(z)$. Therefore, $g(z)=h(z)g(z)k(z)$. Hence $h(z)$ and $k(z)$ are elements of K, and since $f(z)k=h^{-1}f(z)=g(z)$ and $Rh^{-1}=R$ we have $Rg(z)=g(z)R$ and thus

$$I=f(z)R=Rf(z).$$

We shall now show that $f(z)R=Rf(z)$ holds only if $f(z)=z^na$. For, if $f(z)= =a_0+za_1+\ldots+z^n$ with $a_0\neq 0$, we have $f(z)b\in Rf(z)$ for every $b\in K$. Hence there exists an element $c\in K$ with $f(z)b=cf(z)$. Therefore by comparison of coefficients we have:

$$b=c^{S^n},\ a_{n-1}b=a_{n-1}c^{S^{n-1}},\ \ldots,\ a_0b=a_0c.$$

Since $a_0\neq 0$, and $a_0b=a_0c$ implies $b=c$, and thus $b=b^{S^n}$ for every $b\in K$. This is impossible by the assumption on S. Hence $f(z)$ has the form $f(z)=z^kg(z)$ with $k\geqq 1$, where $g(z)$ has a non-zero constant term.

Since

$$(z^rd_r+\ldots+zd_1+d_0)z^k=z^k(z^rd_r^{S^k}+\ldots+zd_1S^k+d_0^{S^k})$$

we have $Rz^k=z^kR$, and since $Rf(z)=f(z)R$ we have $z^kg(z)R=Rz^kg(z)$. Hence it follows that $g(z)R=Rg(z)$ as R has no divisors of zero. By the preceding this condition does not lead to a contradiction if and only if $g(z)$ is constant. Hence $f(z)$ has the form z^ka, and thus we have $I=z^kR=Rz^k$.

We shall now show that R is right primitive. First we prove that $(1-z)R$ is a modular maximal right ideal of R.

$(1-z)R$ is a right ideal, but it is not an ideal of R, because $a(1-z)b=ab-za^Sb\notin \notin R$ for $a\neq 0$. Hence $(1-z)R\neq R$. Let M be a maximal right ideal of R with $(1-z)R\leqq M$. Then $M=f(z)R$ where $f(z)$ is a polynomial. Since $f(z)$ is a non-zero polynomial of the lowest degree in M, and $(1-z)\in(1-z)R\leqq M, f(z)$ has the degree 1 because $M\neq R$. If $f(z)=za_1+a_0$ then $a_0+a_1=f(z)+(1-z)a_1$ and therefore $a_0+a_1\in M$. Hence $a_0+a_1=0$ and thus $f(z)=(1-z)a_0$. Therefore, $M= =(1-z)R$. Hence $M=(1-z)R$ is a modular maximal right ideal of R.

Now $(M:R)\subseteqq M$. If $(M:R)\neq 0$ then $(M:R)=z^kR=Rz^k$. Then $z^k\in(M:R)\subseteqq M$ and so we have $z^k=(1-z)g(z)$ with a polynomial $g(z)=z^{k-1}d_{k-1}+\ldots+zd_1+d_0$. Therefore

$$z^k=z^kd_{k-1}+z^{k-1}(d_{k-2}-d_{k-1})+\ldots+z(d_0-d_1)+d_0.$$

This leads to the contradiction $0=d_0=d_1=\ldots=d_{k-1}=1$. Hence $(M:R)=(0)$. Thus R is right primitive.

The ideal $T = xR$ of R is semisimple, even right primitive in the sense of Jacobson. Furthermore, T is not a simple ring, because T contains the proper ideals $z^k R$ for $k = 2, 3, 4\ldots$. If S is a non-zero ideal of T and we put

$$S^* = S + RS + SR + RSR,$$

then $(S^*)^3 \leqq S$ and $S^* = z^k R$ hold for some k. Hence $(S^*)^3 = z^{3k} R \subseteqq S$. The ring T/S is nilpotent because

$$T/S \cong (zR/z^{3k} R)/(S/z^{3k} R),$$

and $zR/z^{3k}R$ is nilpotent. Hence T is a semisimple ring in the sense of Jacobson and a radical ring for the upper radical determined by the class of all right-primitive simple rings.

For simple radical rings we first mention the following

PROPOSITION 32.2. A ring A without non-trivial ideals is a radical ring in the sense of Jacobson if and only if A contains no maximal right ideal.

PROOF. If A is a ring without non-trivial ideals, and $\mathbf{J}(A)$ is the Jacobson radical of A, then we have either $\mathbf{J}(A) = (0)$ or $\mathbf{J}(A) = A$. If $\mathbf{J}(A) = (0)$, A has modular maximal right ideals. Conversely, if R is an arbitrary maximal right ideal of A, then the intersection $D_r(A)$ of all maximal right ideals of the ring A differs from A. Then $A \cdot \mathbf{J}(A) \leqq D_r(A)$ by theorem 28.3.6 and thus $A\mathbf{J}(A) = (0)$ because $A\mathbf{J}(A)$ is an ideal of A. If $\mathbf{J}(A) = (0)$, A is semisimple and if $\mathbf{J}(A) = A$, then $A^2 = (0)$. Therefore, if A is a simple radical ring, then $D_r(A) = A$ and thus A has no maximal right ideal. □

Now we shall show the existence of a simple radical ring in the sense of Jacobson. The original construction of Sąsiada has been simplified by Sąsiada and Cohn [1].

PROPOSITION 32.3. Let R be the ring of all power series in non-commutating indeterminates x and y over a field K, I the set of those elements of R which have no inverse element in R; E the set of elements of R with constant term 1. Let

$$u = x - yx^2 y. \tag{1}$$

If $c = \sum_{i=1}^{n} a_i u b_i$ where b_1 possesses an inverse element, then there exist elements

$$a_1' \in R,\ b_1' \in E,\ b_i' \in I \quad (i \geqq 2)$$

such that

$$c = a_1' u b_1' + \sum_{i=2}^{n} a_i u b_i'.$$

PROOF. If $b_1=\beta b_1'$ with $\beta\in K$, $b_1'\in E$, and β_i is the constant term of b_i, then we have

$$c = a_1ub_1+\sum_{i=2}^{n} a_iub_i = \left(a_1\beta+\sum_{i=2}^{n}\beta_ia_i\right)ub_1'+\sum_{i=2}^{n} a_iu(b_i-\beta_1b_1'). \square$$

PROPOSITION 32.4. With notation as above, if

$$cx+dy = \sum_{i=1}^{n} a_iub_i \quad (b_1\in I), \tag{2}$$

then there exist elements $b_i^*\in R$ with $b_1^*\in I$ and with

$$cx+dy = cu(1+b_1^*)+\sum_{i=2}^{n} a_iub_i^*. \tag{3}$$

PROOF. By proposition 32.3 we can choose $b_1\in E$ and $b_i\in I$ $(i=2)$ in (2) without changing a_i $(i=2)$. If we write $b_1=1+b_1'x+b_1''y$, $b_i=b_i'x+b_i''y$ $(i\geqq 2)$, and compare the coefficients of x in (2), then we have

$$c = a_1+a_1ub_1'+\sum_{i=2}^{n} a_iub_i'.$$

By a formal expansion we have $(1+ub_1')^{-1}=1+ug$ with some $g\in R$, and hence it follows that

$$c(1+ug) = a_1+\sum_{i=2}^{n} a_iud_i$$

where $d_i\in R$. If we solve this equation for a_1 and substitute the solution into (2), we obtain

$$cx+dy = c(1+ug)ub_1-\sum_{i=2}^{n} a_iud_iub_1+\sum_{i=2}^{n} a_iub_i =$$

$$= cu(1+gu)b_1+\sum_{i=2}^{n} a_iu(b_i-d_iub_1). \square$$

PROPOSITION 32.5. If the product dy has the form

$$dy = \sum_{i=1}^{n} a_iub_i, \tag{4}$$

where n is chosen as small as possible, then $b_i\in I$ $(1\leqq i\leqq n)$.

PROOF. If $b_1\in I$, then by proposition 32.4 there exist elements $b_i^*\in R$ with $dy= = \sum_{i=2}^{n} a_iub_i^*$, which contradicts the minimality of n. $\square$

PROPOSITION 32.6. The ideal (u) of R generated by $u=x-yx^2y$ contains no monomial, and hence we have $x\notin(u)$, $y\notin(u)$.

PROOF. If (u) contains a monomial, then since $1 \in R$ there exist elements $a_i, b_i \in R$ with

$$c = \sum_{i=1}^{n} a_i u b_i. \tag{5}$$

Choose c such that n is as small as possible, and with this n choose c such that the degree δ of c is as small as possible. Then $\delta = 1$ since $(u) \neq R$.

Moreover, if $c = c_1 y$ with a monomial c_1 then $b_i \in I$ by proposition 32.5, and we obtain a representation of type (5) for c_1 by comparison of the coefficients of y. This contradicts the minimality of δ. Hence there exists a monomial c_1 with

$$c_1 x = \sum_{i=1}^{n} a_i u b_i. \tag{6}$$

Suppose that for $r < n$ we have found monomials $p_1, \ldots, p_r$ of positive degree, and elements $d_1, d_2, \ldots, d_n \in R$ with

$$c_1 p_1 \ldots p_r x = \sum_{i=1}^{r} c p_1 \ldots p_{i-1} u d_i + \sum_{j=r+1}^{n} a_j u d_j \tag{7}$$

such that

$$d_i \notin I \tag{8}$$

for an $i = 1, 2, \ldots$.

We shall show the validity of (7) and (8) with $r+1$ instead of r.

Now we distinguish two cases.

(a) If $d_j \in I$ for every $j > r$, then $d_k \notin I$ for some $k \leqq r$, and we write

$$d_k = \lambda_k + d_k' x + d_k'' y.$$

The comparison of the coefficients of x yields

$$c p_1 \ldots p_r = \sum_{i=1}^{r} c p_1 \ldots p_{i-1} \lambda_i + \sum_{i=1}^{r} c p_1 \ldots p_{i-1} u d_i' + \sum_{j=r+1}^{n} a_j u d_j.$$

If we carry over the first sum to the left-hand side in this equation, we have an equation of the type

$$c_1 p_1 \ldots p_{s-1}(1+f)\lambda_s = \sum_{i=1}^{r} c_1 p_1 \ldots p_{i-1} u d_i' + \sum_{j=r+1}^{n} a_j u d_j$$

where $f \in I$, and s is the least index with $\lambda_s \neq 0$. If we carry over the s-th term of the first sum to the left-hand side and multiply both sides of the equation by $(\lambda_s(1+f) - u d_s')^{-1}$ then we obtain a representation of type (5) for $c_1 p_1 \ldots p_{s-1}$ with fewer than n terms, which contradicts the minimality of n.

(b) If $d_j \notin I$ for some $j > r$, e.g. for $j = r+1$, then we apply proposition 32.4 to (7), and we have

$$c_1 p_1 \ldots p_r x = \sum_{i=1}^{r} c_1 p_1 \ldots p_{i-1} u d_i + c_1 p_1 \ldots p_r u(1 + d_{r+1}) + \sum_{j=r+2}^{n} a_j u d_j$$

with some elements $d_i \in R$ and $d_{r+1} \in I$. Hence it follows that

$$c_1 p_1 \dots p_r yx^2y = \sum_{i=1}^{r} c_1 p_1 \dots p_{i-1} ud_i + c_1 p_1 \dots p_r ud_{r+1} + \sum_{j=r+2}^{n} a_j ud_j .$$

By proposition 32.5 the elements d_j are contained in I, and by a comparison of the coefficients of y we obtain

$$c_1 p_1 \dots p_r yx^2 = \sum_{i=1}^{r} c_1 p_1 \dots p_{i-1} ud_i + c_1 p_1 \dots p_r ud_{r+1} + \sum_{j=r+2}^{n} a_j ud_j . \tag{9}$$

If all elements d of (9) are contained in I, then we have by comparison of the coefficients of x

$$c_1 p_1 \dots p_r yx = \sum_{i=1}^{r} c_1 p_1 \dots p_{i-1} ud_i + c_1 p_1 \dots p_r ud_r + \sum_{j=r+2}^{n} a_j ud_j . \tag{10}$$

If all elements d of (10) are still contained in I, then we repeat the above procedure. Applying proposition 32.5, we have

$$c_1 p_1 \dots p_r yx = \sum_{i=1}^{r} c_1 p_1 \dots p_{i-1} ud_i + c_1 p_1 \dots p_r ud_{r+1} + \sum_{j=r+2}^{n} a_j ud_j$$

(with a new choice of the element d at every step). If we carry over $cp_1 \dots p_r ud_{r+1}$ to the left-hand side and multiply both sides by $(1-ud_{r+1})^{-1}$ then we have a shorter representation of the type (5) for $c_1 p_1 \dots p_r$, which contradicts the minimality of n. Hence (10) is impossible and thus $d_i \in I$ in (9) for some i. Thus, the proof is completed by induction.

For $r=0$, (7) is just (6) and thus here $b_i \notin I$ for some i, because otherwise we could cancel x in (6), and hence obtain a representation for c_1 of the type (5) which contradicts the minimality of δ. Hence (7) and (8) are valid for $r=n$, but then the consideration of (a) yields again a contradiction. □

EXAMPLE 32.7 (Sąsiada [1]). Let R be the ring of all power series in non-commutating indeterminates x and y over a field K. Let I be the ideal of all power series with constant term 0. Then I is a radical ring in the sense of Jacobson.

By proposition 32.6, $x \notin (x-yx^2y)_R$ for the ideal generated by $x-yx^2y$ in R. Therefore we also have even $x \notin (x-yx^2y)_I$ for the ideal, generated by $x-yx^2y$ in I. By Zorn's lemma there exists an ideal M of I maximal with respect to the conditions $M \supseteq (x-yx^2y)_I$ and $x \notin M$. Write $S=I/M$, $\bar{x}=x+M$ and $\bar{y}=y+M$. Then S is a radical ring in the sense of Jacobson. The ideal T generated by $\bar{x}$ in S is minimal in S by the choice of M. Obviously we have $T^2=T$ since $\bar{x}=\bar{y}\cdot\bar{x}\cdot\cdot\bar{x}\cdot\bar{y}$. Hence T is a simple radical ring in the sense of Jacobson. For, if $U\neq(0)$

is an ideal of T, then $U^*=U+US+SU+SUS$ is an ideal of S with $(U^*)^3\leqq U$. We have $U^*=T$ because $U^*\geqq U\neq(0)$ and T is minimal. Hence, we have $U==T$ since $(U^*)^3=T^3=T\leqq U$. Therefore T is a simple ring. (This simple Jacobson radical ring contains divisors of zero.)

§ 33. THE DENSITY THEOREM FOR RIGHT PRIMITIVE RINGS

If A is a right primitive ring, then there exists a maximal right ideal R of A with $(R:A)=(0)$. Hence $Ax\leqq R$ implies $x=0$. Then A/R is a right A-module, and thus A can be homomorphically mapped into the full endomorphism ring E of the Abelian group $(A/R)^+$. The kernel of this homomorphism consists of those elements y of A for which $Ay\leqq R$ holds, and thus this kernel equals zero, i.e. the homomorphism is an isomorphism. Hence A is a subring of the full endomorphism ring E of the Abelian group $(A/R)^+$.

PROPOSITION 33.1 (Schur). If A is a right primitive ring, then in the full endomorphism ring E of the Abelian group $(A/R)^+$ the centralizer $Z=\{\alpha;\, \alpha\in E,\, \alpha r==r\alpha,\, r\in A\}$ of the subring A is a division ring S, and thus $M=(A/R)^+$ is a vector space over S.

PROOF. Since A/R is an irreducible right A-module and every element α of Z is an A-endomorphism, we have either $(A/R)\alpha=(0)$ or $(A/R)\alpha=A/R$. In the second case the kernel of α equals (0) by the irreducibility of A/R. Therefore α is an automorphism and hence it possesses an inverse α^{-1}. Hence Z is a division ring. □

PROPOSITION 33.2. If G is a subspace of finite dimension of the vector space A/R over the centralizer Z and v is an element of A/R with $v\notin G$, then there exists an element $a\in A$ with $Ga=(0)$ and $va\neq 0$.

PROOF. Assume that the statement holds for every subspace of dimension $n-1$, and we prove that it is also valid for every subspace of dimension n. Let $b_1, b_2, \dots$ $\dots, b_{n-1}, b_n$ be the basis elements of G and v an element of A/R with $v\notin G$.

Assume that $va=0$ also holds for every $a\in A$ with $Ga=(0)$. We shall deduce a contradiction.

Let B be the set of those elements $a\in A$, for which $b_1a=\dots=b_{n-1}a=0$. Then B is a right ideal of A. By the induction hypothesis there exists an element $c\in A$ with $b_1c=\dots=b_{n-1}c=0$ and $b_nc\neq 0$. Therefore $b_nB\neq(0)$.

Since A/R is a simple right A-module, we have $A/R=b_nB$, where b_n has the form $b_n=c_n+R$.

Consider the mapping $b_n b \to vb\,(b \in B)$. If $b_n b' = b_n b''$, then

$$b_n(b'-b'') = 0, \quad b'-b'' \in B,$$
$$b_1(b'-b'') = \ldots = b_{n-1}(b'-b'') = 0,$$

and thus $G(b'-b'') = (0)$. Hence $v(b'-b'') = 0$ and $vb' = vb''$ follows by our assumption. Thus the mapping $b_n b \to vb$ is uniquely determined. We shall show that the mapping $b_n b \to vb$ is contained in the centralizer. For, if $a \in A$, then we have $(b_n b)a = b_n(ba) \to vba$, since ba belongs to B. Moreover, $v(ba) = (vb)a$. Thus $b_n b \to vb$ commutes with every $a \in A$. Hence there exists an element $\alpha \in Z$ with $b_n b\alpha = vb$ for every $b \in B$. Since $b_n b\alpha = b_n \alpha b$ we have $(b_n \alpha - v)b = 0$ for every $b \in B$. If $b_n \alpha - v$ is not contained in the subspace generated by $b_1, b_2, \ldots, b_{n-1}$, then there exists an element $b_1 \in B$ which does not annihilate $b_n \alpha - v$, and this is a contradiction. Hence $b_n \alpha - v$ is contained in the subspace, generated by $b_1, b_2, \ldots, b_{n-1}$, and thus we have $v \in G$, which is a contradiction. Hence the statement holds also for n.

Since the statement can be proved by the above procedure for $n = 1$, proposition 33.2. is proved. □

A ring A is called a *dense ring of linear transformations* of a vector space V over a division ring S, if there exists an element $a \in A$ with $x_1 a = y_1$, $x_2 a = y_2, \ldots$ $\ldots, x_n a = y_n$ for every set of linearly independent elements $x_1, x_2, \ldots, x_n$ of V over S and for every set of elements $y_1, y_2, \ldots, y_n$ of V.

THEOREM 33.3 (Jacobson's density theorem). Every right primitive ring A is a dense ring of all linear transformations of a vector space over a division ring.

PROOF. Let $x_1, x_2, \ldots, x_n$ be a finite set of linearly independent elements of A/R over Z, and G_i the subspace of A/R generated by all x_j, $j \neq i$. Then $x_i \notin G_i$, and by proposition 33.2 there exist elements $t_i \in A$ with $x_i t_i = z_i \neq 0$ and $x_j t_i = 0$ for $j \neq i$.

By proposition 33.2 we have $z_i A \neq (0)$ for every z_i, and even $A/R = z_i A$, because A/R is irreducible. Hence for every i there exists $u_i \in A$ with $z_i u_i = y_i$. Now if $\beta = t_1 u_1 + \ldots + t_n u_n$, then we have $x_1 \beta = y_1$, $x_2 \beta = y_2, \ldots, x_n \beta = y_n$. □

From the density theorem one can immediately deduce

THEOREM 33.4 (Wedderburn–Artin structure theorem). A ring A is a semisimple ring (in the sense of Jacobson) with minimum condition on right ideals if and only if A is a direct sum of finitely many ideals S_i, each isomorphic to the ring of all linear transformations of a vector space of finite dimension over a division ring.

PROOF. First let A be a primitive ring with minimum condition on right ideals Then there exists an irreducible right A-module M, and M is a vector space over the centralizer Z. If M has an infinite dimension over the division ring Z, then there exist linearly independent elements $x_i \in M$ ($i=1, 2, 3, \ldots$). If I_k is the right annihilating right ideal of the subspace $\{x_1, \ldots, x_k\}$, then we have $I_1 \geqq I_2 \geqq I_3 = \ldots$. By the density theorem there exists $c_k \in I_k$ with $c_k \notin I_{k+1}$ and thus $I_1 > I_2 > I_3 > \ldots$ is a strictly descending chain of right ideals of A, which contradicts the minimum condition. Therefore M has finite dimension over Z, and thus A is isomorphic to the ring of all linear transformations of a vector space of finite dimension over a division ring. Hence A is isomorphic to a full matrix ring Z_n over a division ring Z.

In particular it follows that every primitive ideal P of a ring with minimum condition on right ideals is maximal in A because A/P is simple.

Now, let A be an arbitrary semisimple ring (in the sense of Jacobson) with minimum condition on right ideals. Then there exist primitive ideals P_α of A with $\bigcap_\alpha P_\alpha = (0)$, where each P_α is maximal in A. By the minimum condition, there exists a finite number of primitive ideals P_i with $\bigcap_i P_i = (0)$ ($i=1, 2, \ldots, n$) such that

$$S_i = \bigcap_{j \neq i} P_j = P_1 \cap \ldots \cap P_{i-1} \cap P_{i+1} \cap \ldots \cap P_n$$

differs from the zero-ideal (0). Hence we have

$$S_i \cap P_i = (0) \quad \text{and} \quad P_i + S_i = A,$$

because P_i is maximal in A. Hence $A = S_i \oplus P_i$. Now we put $T_r = P_1 \cap P_2 \cap \ldots \ldots \cap P_r$. We shall prove that

$$A = S_1 \oplus S_2 \oplus \ldots \oplus S_k \oplus T_k$$

by induction on k. The statement obviously holds for $k=1$. Assume it is already proved for $k=m$. By the second isomorphism theorem we have

$$T_m/(T_m \cap P_{m+j}) = (T_m + P_{m+j})/P_{m+j} = A/P_{m+j}$$

for $j=1, 2, \ldots, n-m$. Hence the ideal $T_m \cap P_{m+j}$ is maximal in T_m. Moreover, we have $\bigcap_{j=1}^{n-m} (T_m \cap P_{m+j}) = (0)$, but no proper subset of these ideals has the intersection (0).

In a similar way we obtain

$$T_m = T_{m+1} \oplus \bigcap_{j=2}^{n-m} (T_m \cap P_{m+j}) = T_{m+1} \oplus S_{m+1}.$$

Hence we also have

$$A = S_1 \oplus S_2 \oplus \ldots \oplus S_m \oplus S_{m+1} \oplus T_{m+1}$$

and thus the statement is valid for $m+1$. For $k=n$ we get

$$A = S_1 \oplus S_2 \oplus \ldots \oplus S_n,$$

where every ideal S_i is a simple ring, namely, every S_i is a full matrix ring over a division ring.

Conversely, $\mathbf{J}(A)=(0)$ for the Jacobson radical $\mathbf{J}(A)$ of a ring A, of the above type, and with minimum condition. For, the ideal $P_i=\sum_{j\neq i} S_j$ is primitive, and we obviously have $\bigcap_i P_i=(0)$. Consequently $\mathbf{J}(A)=(0)$. The minimum condition holds in $A=S_1\oplus S_2\oplus\ldots\oplus S_i$, because A, being a right A-module, has a composition series. Obviously, it is sufficient to prove the assertion for one simple summand S_i. There exist, e.g. in S_1 a finite number of maximal right ideals R_j with $\bigcap_j R_j=(0)$ $(j=1, 2, \ldots, m)$, but with $\bigcap_{i\neq j} R_i\neq(0)$ for every j. Then, by the second isomorphism theorem,

$$S_1 > R_1 > (R_1\cap R_2) >\ldots> (R_1\cap\ldots\cap R_m) = (0)$$

is a composition series for S_1. □

Finally, we mention the well-known

PROPOSITION 33.5 (Hopkins). The Jacobson radical $\mathbf{J}(A)$ is nilpotent in every ring A with minimum condition on right ideals.

PROOF. In the descending chain

$$\mathbf{J}(A) = J \geqq J^2 \geqq J^3 \geqq\ldots$$

there exists an exponent k with $N=J^k=J^{k+1}$ by the minimum condition. If $N\neq(0)$, then let $\mathfrak{M}$ be the set of the right ideals R of A with

$$R \leqq N \quad\text{and}\quad RN \neq (0).$$

$\mathfrak{M}$ is not empty since $N\in\mathfrak{M}$. Let I be a minimal right ideal of A in $\mathfrak{M}$. Then we have $IN\neq(0)$, and thus there exists an $i\in I$ with $iN\neq(0)$. Therefore we have $iN=I$ since $(iN)N\neq(0)$ and $iN=I$. Therefore, there exists an $n\in N$ with $in=i$ and an $m\in A$ with $n+m-nm=0$. Hence

$$i = i-i(n+m-nm) = (i-in)-(i-in)m = 0.$$

This contradicts the condition $I=iN\neq(0)$.

Therefore $J^k=N=(0)$. □

§ 34. RADICALS WHICH COINCIDE WITH THE JACOBSON RADICAL ON THE CLASS OF RINGS WITH MINIMUM CONDITION ON RIGHT IDEALS

For the subject-matter of this section, we refer the reader to Divinsky [5]. In the preceding section we have seen that the Jacobson radical is nilpotent in every ring with minimum condition on right ideals.

On the one hand, we wish to determine all radicals $\mathbf{R}$ for which the $\mathbf{R}$-radical rings A of $\mathbf{K}$ are nilpotent, i.e. $A \in \mathbf{K}$ and $A=\mathbf{R}(A)$ imply $A=\mathbf{J}(A)$. We say that a radical of this kind *weakly coincides* with $\mathbf{J}$.

On the other hand, we want to characterize those radicals $\mathbf{R}$, for which $A \in \mathbf{K}$ always implies $\mathbf{R}(A)=\mathbf{J}(A)$. We say that a radical of this type *strongly coincides* with $\mathbf{J}$.

These questions have been investigated by Kurosh [3] and more exactly by Divinsky [5].

By the Wedderburn–Artin structure theorem every ring A with minimum condition on right ideals and without nilpotent ideals is a direct sum of a finite number of simple rings $A_1, A_2, \ldots, A_n$, where A_i $(i=1, 2, \ldots, n)$ is a full matrix ring of type $n_i \times n_i$ over a division ring S_i.

Let $\mathbf{S}$ denote the lower radical determined by the class of all zero-rings of prime number order, $\mathbf{D}$ denote the lower radical, determined by the class of those nilpotent rings which are nil radicals of a ring of $\mathbf{K}$, and $\mathbf{T}$ denote the upper radical determined by the class of all full matrix rings over a division ring.

PROPOSITION 34.1. A radical $\mathbf{R}$ weakly coincides with the Jacobson radical $\mathbf{J}$ on the class $\mathbf{K}$ of all rings with minimum condition on right ideals if and only if $\mathbf{S} \leqq \mathbf{R} \leqq \mathbf{T}$. But there exists a ring A of $\mathbf{K}$ such that its $\mathbf{S}$-radical is smaller than its $\mathbf{J}$-radical.

PROOF. Let $\mathbf{R}$ be a radical with $\mathbf{S} \leqq \mathbf{R} \leqq \mathbf{T}$ and A an $\mathbf{R}$-radical ring with $A \in \mathbf{K}$. If A is not nilpotent, then by the Wedderburn–Artin structure theorem A can be homomorphically mapped onto a full matrix ring over a division ring. This is in contradiction to the fact that A is also a $\mathbf{T}$-radical ring since $\mathbf{S} \leqq \mathbf{R} \leqq \mathbf{T}$. Conversely, let A be an $\mathbf{R}$-radical ring with $A=\mathbf{J}(A)$ and $A \in \mathbf{K}$. Then by proposition 33.5 every homomorphic image B of A is nilpotent, and thus $B^n=(0)$, $B^{n-1} \neq (0)$ holds for some n. A minimal right ideal of B^{n-1} is a zero-ring of prime number order and hence A is an $\mathbf{S}$-radical ring. Furthermore, A is also a $\mathbf{T}$-radical ring. By the extremal properties of $\mathbf{S}$ and $\mathbf{T}$ we have $\mathbf{S} \leqq \mathbf{R} \leqq \mathbf{T}$.

Now let the ring A be the set of all sums $\alpha x + \beta e$ where α and β are rational numbers; let the multiplication be given by

$$x^2 = e^2 - e = xe - x = ex - x = 0,$$

and let equality and addition be defined componentwise. Since the element e is the unity element of A, every ideal of A is a subspace of the space A^+ of dimension 2 over the rational number field. $N=\{\alpha x\}$ is the only non-trivial ideal of A, and since A is commutative, $A \in \mathbf{K}$. Then N is the Jacobson radical of A, and we shall show that N is $\mathbf{S}$-semisimple. Let us assume that N contains a non-zero $\mathbf{S}$-ideal I. The ideal I is of a certain degree over $\mathbf{S}$. Let γ be the smallest ordinal number such that N contains an ideal I of degree γ over $\mathbf{S}$. Then γ is not a limit ordinal number, and thus $\gamma-1$ exists. Hence every homomorphic image of I and therefore I itself, contains a non-zero ideal I' of degree $\gamma-1$. Since $N^2=(0)$; I' is an ideal of N, which contradicts the minimality of γ if $\gamma \geqq 2$. Therefore $\gamma=1$. This is impossible because I is infinite, and a ring of degree 1 over **S** is of prime number order. □

REMARK 34.2. The ideal N in the above proof is **S**-semisimple, every homomorphic image N' of N is an **S**-radical ring. Moreover, every homomorphic image N' or N is of degree 2 over **S**. For, the additive group N^+ is of rank 1, and $kN=$ $=N$ for every natural number $k \neq 0$. Thus N' is a direct sum of zero-rings with Prüferian additive group $Z(p^\infty)$ for some prime numbers p. Every group $Z(p^\infty)$ contains a minimal subgroup which, as a zero-ring, is of degree 1 over **S**, and this shows that N' is of degree 2 over **S**.

We have the following result of Divinsky [5]:

THEOREM 34.3. A radical **R** strongly coincides with the Jacobson radical **J** on the class **K**, if $\mathbf{D} \leqq \mathbf{R} \leqq \mathbf{T}$.

PROOF. If **R** strongly coincides with **J** on K, then we have $\mathbf{D} \leqq \mathbf{R} \leqq \mathbf{T}$ by the definition of **D** and **T**.

Conversely, let $\mathbf{D} \leqq \mathbf{R} \leqq \mathbf{T}$ and A be a ring of **K**. We have in general $\mathbf{J} \leqq \mathbf{T}$. But by the Wedderburn–Artin structure theorem we have $\mathbf{J}(A)=\mathbf{T}(A)$. Moreover, $\mathbf{D} \leqq \mathbf{J}$ and we have $\mathbf{D}(A)=\mathbf{J}(A)$ by the definition of **D** and by theorem 33.4. Hence $\mathbf{D}(A)=\mathbf{J}(A)=\mathbf{T}(A)$. Therefore $\mathbf{R}(A)=\mathbf{J}(A)$ since $\mathbf{D} \leqq \mathbf{R} \leqq \mathbf{T}$. □

THEOREM 34.4. **D** is strictly smaller than the Baer lower nil radical **B**.

PROOF. We shall show that the zero-ring $Z(\infty)$, whose additive group is the infinite cyclic group, is both **D**-semisimple and a **B**-radical ring. If $Z(\infty)$ contains a non-zero **D**-ideal I, then $Z(\infty)$ is itself a **D**-ring since $I \cong Z(\infty)$. Hence $Z(\infty)$ is of a certain degree α over the class **K** of those nilpotent rings which are nil radicals of a ring with minimum condition on right ideals. Let α be chosen as small as possible. Then $\alpha-1$ exists, and every homomorphic image H of $Z(\infty)$ contains a non-zero ideal I of degree $\alpha-1$ over **K**. If $H=Z(\infty)$ this certainly implies $\alpha=1$ by the minimality of α and because $I \cong Z(\infty)$. Therefore there exists a ring A with minimum condition on right ideals and with the nilpotent radical $\mathbf{J}(A)$ such that $\mathbf{J}(A)$ can be homomorphically mapped onto $Z(\infty)$. Therefore, we take $J=\mathbf{J}(A)$, $J/K \cong Z(\infty)$ where K is an ideal of J. Since the nil radical of A/J^2 is J/J^2, we may assume that $J^2=0$ since $(A/J^2)/(J/J^2) \cong A/J$. Since $J/K=Z(\infty)$ every element of J has the form $mz+k$ where $k \in K$, m is an integer, and $z \in J$ is such that its image is a fixed generator of $Z(\infty)$. Thus, $m=0$ if $mz \in K$.

If $A=\mathbf{J}(A)$, then the minimum condition on right ideals holds also in $Z(\infty)$

since $A/K \cong Z(\infty)$. This is a contradiction. Therefore $A \neq \mathbf{J}(A)$, and thus A has idempotent elements e_α, as is well known (e.g. Jacobson [2], propositions 3.8.3, 3.8.4). Let e_0 be an idempotent element, whose right annihilator is minimal in the set of those right ideals R_{e_α} which are right annihilators of the idempotent elements e_α of A. We shall show that $R_{e_0} = \{y;\, y \in A,\, e_0 y = 0\}$ is nilpotent. For, if R_{e_0} is not nilpotent, then there exists a non-zero idempotent element e' in R_{e_0} with $e_0 e' = 0$. Let $e = e_0 + e' - e' e_0$. We have $e \neq 0$ since $(e')^2 = e' \neq 0$ and $e_0 e' = 0$. Moreover, it is easy to see that $e^2 = e$. $ez = 0$ obviously implies $e_0 z = 0$ since $e_0 e = e_0$. Therefore we have $R_e \leqq R_{e_0}$. Now $e' \in R_{e_0}$, $e' \notin R_e$ because $e' \in R_e$ implies $0 = ee' = e_0 e' + e' - e' e_0 e' = e'$. Thus we have $R_e \subset R_{e_0}$ which contradicts the minimality of R_{e_0}.

Hence R_{e_0} is nilpotent. Then we have the Pierce decomposition $A = e_0 A + (1 - e_0) A$ with $(1 - e_0) A \leqq R_{e_0} \leqq J$. Since the elements of J with $J^2 = (0)$ have the form $mz + k$, we have $z(1 - e_0) A = (0)$.

Since $ze_0 \in J$ there exist an integer m and an element $k \in K$ with $ze_0 = mz + k$. Assume that $m \neq 1$. Then the cyclic additive group $\{2^n(z - ze_0)\}$ is a right ideal of A since $(z - ze_0)A = (0)$. If

$$2^n(z - ze_0) = 2^n(z - mz - k) = 0$$

for a natural number n, then we have $2^n(1 - m)z \in K$, which is impossible since $m \neq 1$. Hence the right ideals $\{2^n(z - ze_0)\}$ differ from zero and by the minimum condition on right ideals there exists an n such that

$$\{2^n(z - ze_0)\} = \{2^{n+1}(z - ze_0)\}.$$

Hence the existence of an integer s such that

$$2^n(z - ze_0) = s2^{n+1}(z - ze_0)$$

follows. Since

$$2^n(z - mz - k) = s2^{n+1}(z - mz - k)$$

we have

$$2^n(1 - 2s)(1 - m)z = (2^n - 2^{n+1}s)k \in K$$

which is impossible if $m \neq 1$. Hence $m = 1$ and thus $ze_0 = z + k$. If $2^n zA = (0)$ for an exponent n, then we have the contradiction $2^n z = -2^n k \in K$ since $2^n ze_0 = 2^n(z + k) = 0$. Thus $2^n zA$ differs from zero. By the minimum condition on right ideals there exists an integer n with $2^n zA = 2^{n+1} zA$, and thus there exists an element $a \in A$ with

$$2^n ze_0 = 2^{n+1} za.$$

Since $za \in J$ there exist an integer l and an element $k_1 \in K$ with $za = lz + k_1$, whence we have the contradiction

$$2^n(1 - 2l)z = 2^{n+1}k_1 - 2^n k \in K$$

because $2^n(z + k) = 2^{n+1}(lz + k_1)$. Hence the nil radical $\mathbf{J}$ of a ring A with minimum condition on right ideals cannot be homomorphically mapped onto $Z(\infty)$, and thus we have $\mathbf{D} < \mathbf{B}$. □

Szász [3] has also discussed the question, when a general radical on the class of rings with minimum condition on principal right ideals (MHR-rings) coincides with the Jacobson radical. In order to study this, two further concrete radicals $\mathbf{R}_1$ and $\mathbf{R}_2$ have been introduced:

$\mathbf{R}_1$ is the upper radical determined by the class of all simple MHR-rings.

$\mathbf{R}_2$ is the lower radical determined by the class of those rings which are upper nil radicals of MHR-rings.

Obviously, $\mathbf{R}_1 < \mathbf{R}_2$ holds. But we do not want to enter here into the details of $\mathbf{R}_1$ and $\mathbf{R}_2$. It would be interesting to study $\mathbf{R}_1$ and $\mathbf{R}_2$, and their relations to other concrete radicals (solution in Gardner [10]).

PROBLEM 45. Find a necessary and sufficient condition for all non-zero elements of a simple Jacobson radical ring to have an infinite number of distinct quasi-inner automorphic images! (An isomorphism is said to be *quasi-inner,* if it is of the form

$$\varphi_a \colon x \to (1-a)x(1-a)^{-1}.)$$

PROBLEM 46. Does there exist an ordered simple Jacobson radical ring? (Examine them.)

PROBLEM 47. What is the Jacobson radical of a polynomial ring $A[\lambda]$ over an arbitrary ring A?

PROBLEM 48. Can every quasi-modular right ideal of a ring be embedded into a quasi-modular maximal right ideal?

PROBLEM 49. By an *Ω-ring* we understand a ring A with a quasi-modular maximal right ideal R, which is not modular in A. (Szász [17] has proved the existence of such rings.) Does there exist an Ω-ring which is even an MHR-ring?

PROBLEM 50. Does there exist an Ω-ring which, as a ring, can be generated by the right ideal R, mentioned in problem 49, and by a finite number of elements $a_1, a_2, \ldots, a_n \in A$?

PROBLEM 51. Does there exist an Ω-ring which possesses a quasi-modular, maximal, but non-modular right ideal R and a finite number of elements $a_1, a_2, \ldots, a_n \in A$ such that for every $x \in A$ one of the relations $a_i x \notin R$ holds? (The same problem for $n=2$.)

PROBLEM 52. Let $\mathfrak{m}$ denote the cardinality of the set of the modular maximal right ideals and $\mathfrak{n}$ denote the cardinality of the set of the quasi-modular, maximal, but non-modular right ideals in an Ω-ring. Does each of the three possibilities $\mathfrak{m}>\mathfrak{n}$, $\mathfrak{m}=\mathfrak{n}$ and $\mathfrak{m}<\mathfrak{n}$ casually occur for certain Ω-rings?

PROBLEM 53. Must $\mathfrak{m}=\mathfrak{n}$, if a ring is the discrete direct sum of $\mathfrak{m}$ idempotent minimal right ideals, and at the same time, the discrete direct sum of $\mathfrak{n}$ idempotent minimal left ideals, where $\mathfrak{m}$ and $\mathfrak{n}$ are infinite cardinalities?

PROBLEM 54. Let A be a ring in which the intersection of any two modular right ideals is again modular in A. Do the sets Q_a and Q_b of the form $Q_a = \{a+x-ax;\ x\in A\}$ always have a non-empty intersection? Must A contain a left unity element (see Szász [19])?

PROBLEM 55. Let **K** be the class of all subdirectly irreducible rings, whose Jacobson radical is (0). Examine the upper radical determined by the class **K**.

PROBLEM 56. Consider radicals and find structure theorems for rings, in which every left ideal contained in a principal right ideal, has a right unity element (or a left unity element). (These rings are generalizations of the discrete direct sums of division rings, see also Szász [2].)

PROBLEM 57 (Kertész). Let A be a ring, M a right A-module, $D(M)$ the Frattini submodule of M and take $K(M)=\{m;\ m\in M,\ mA\leqq D(M)\}$. Determine all rings A, for which $K(M)\cdot A=(0)$ for every right A-module M. (By Szász [3] A is a semisimple ring with minimum condition on right ideals if and only if A is an MHR-ring with unity element satisfying the condition $K(M)A=(0)$ for every right A-module M.)

CHAPTER V

THE BROWN–McCOY RADICAL

In this chapter we introduce the Brown–McCoy radical by means of special modules. It is shown that both the Brown–McCoy radical and the Jacobson radical can be considered as (F, Ω_1, Ω)-groups. The Brown–McCoy radical can be regarded also as a (k, l, m, n)-radical for some natural numbers k, l, m and n. In particular, we examine the Brown–McCoy radical of a full matrix ring. Following the important investigations of Suliński, we discuss the upper radicals determined by a class of simple rings with unity element, and the completely non-special Brown–McCoy semisimple rings. We shall also discuss a classification of the Brown–McCoy semisimple rings using topological methods and transfinite ordinal numbers. In connection with a problem of Szász, the Brown–McCoy radical rings with a module-theoretic property will be discussed. Finally, we mention certain open problems.

§ 35. THE DEFINITION OF THE BROWN–McCOY RADICAL

For the subject-matter of this section we refer the reader to Andrunakievich and Ryabukhin [1].

Let $\mathbf{M}_A$ be the class of all right A-modules M over the ring A with the property that, for every $m \in M$ and for every ideal B of A with $B \nsubseteq (0:M)_A$ there exists an element $b \in B$ with $mb=m$. If $M \in \mathbf{M}_A$ then M is called a *Brown–McCoy module.*

PROPOSITION 35.1. $\mathbf{M} = \cup \mathbf{M}_A$ is a special class of modules.

PROOF. First we shall show that every module M of $\mathbf{M}_A$ is a prime module. If B is an ideal of A with $B \nsubseteq (0:M)_A$ and m is an element of M with $mB=(0)$, then there exists an element $b \in B$ with $mb=m$ and hence $m=mb\ mB=0$. Thus $m=0$.

If $M \in \mathbf{M}_A$ and B is an ideal of A with $MB \neq (0)$, then we can show that $M \in \mathbf{M}_B$ as follows. There exists an element $b \in B$ with $mab=ma$ for every $m \in M$ and every $a \in A$. Hence we have $ab-a \in (0:M)_A$ and thus $A=B+(0:M)_A$. But hence every A-module is also a B-module and vice versa, and thus $\mathbf{M}_B = \mathbf{M}_A$.

If $M \in \mathbf{M}_B$ and if B is an ideal of the ring A, then we can prove that $MB \in \mathbf{M}_A$ in the following way. We have $(0:MB)_A = (0:M)_B : A$. Now, if C is an ideal of A with

$C \not\subseteq (0:MB)_A$, then we have $MBC \neq (0)$ and $BC \subseteq B$. Since BC is an ideal of B with $BC \not\subseteq (0:M)_B$, there exists $b^* \in BC$ with $mb^* = m$ for every $m \in M$. Hence we also have $mbb^* = mb$ with $b^* \in BC \subseteq C$ for every $m \in M$ and every $b \in B$. Therefore we obtain $MB \in \mathbf{M}_A$.

If $M \in \mathbf{M}_{\bar{A}}$ where $\bar{A} = A/P$, then we have $M \in \mathbf{M}_A$ with $P \subseteq (0:M)_A$. Conversely, if $M \in \mathbf{M}_A$ and $P \subseteq (0:M)_A$, then we also have $M \in \mathbf{M}_{\bar{A}}$ with $\bar{A} = A/P$. Thus, $M = \bigcup \mathbf{M}_A$ is a special class of modules, indeed. □

The intersection $G = \bigcap_{M \in \mathbf{M}_A} (0:M)_A$ is called *the Brown–McCoy radical* of the ring A. If $\mathbf{M}_A$ is empty, then A is called *a radical ring in the sense of Brown and McCoy*.

THEOREM 35.2. $G = \bigcap_{M \in \mathbf{M}_A} (0:M)_A$ is a radical in the sense of Amitsur and Kurosh and G is a special radical. Furthermore, $A/(0:M)_A$ is a simple ring with unity element, and thus A/G is a subdirect sum of simple rings with unity element.

PROOF. The first statement of the theorem follows from section 15. Now, if $M \in \mathbf{M}_A$ and B is an ideal of A satisfying the condition $B \not\subseteq (0:M)_A$, then there exists $b \in B$ with $mab = ma$ for every $m \in M$ and every $a \in A$. Hence we have $a - ab \in (0:M)_A$ and $A = B + (0:M)_A$. Therefore $(0:M)_A$ is a maximal ideal of A, and the simple ring $A/(0:M)_A$ contains a right unity element $b + (0:M)_A$. If $b + (0:M)_A$ is not the unity element of $A/(0:M)_A$, then $(1-b)A + A(1-b) + (0:M)_A = A$ and also $(1-b)A + (0:M)_A = A$ because $A(1-b) \subseteq (0:M)_A$. But hence we obtain the contradiction $A^2 = A(1-b)A + A(0:M)_A \subseteq (0:M)_A$. Thus $b + (0:M)_A$ is the unity element of $A/(0:M)_A$. □

CONSEQUENCE 35.3. G is the intersection of those ideals I of the ring A for which A/I is a simple ring with unity element. $\mathbf{G}(I) = I \cap \mathbf{G}(A)$ for every ideal I of the ring A.

PROOF. The first statement follows at once from theorem 35.2. The second statement also follows from theorem 35.2 and from the fact that every special radical is supernilpotent and thus hereditary. □

§ 36. THE BROWN–McCOY RADICAL AS AN (F, Ω)-SUBGROUP AND AS AN (F, Ω_1, Ω)-SUBGROUP

Now we introduce the notion of the Brown–McCoy radical using a general consideration due to Brown and McCoy [4]. This is a generalization of the original definition of Brown and McCoy [1]. This method makes possible a common discussion of the Jacobson radical, the Brown–McCoy radical, and certain other radicals in rings, non-associative rings, groups, etc.

Let Ω be a group, written additively, which nevertheless need not be commutative in general. Let Ω be a set of endomorphisms of $\mathfrak{G}$ containing all inner automorphisms of $\mathfrak{G}$. The Ω-subgroup generated by the element $a \in \mathfrak{G}$ will be denoted by (a). Assume that there exists a unique mapping $\mathbf{F}$ which associates to every element a of $\mathfrak{G}$ a subgroup $\mathbf{F}(a)$, and the following conditions are fulfilled:

(P_1) $\mathbf{F}(a+b) \leqq \mathbf{F}(a)+(b)$.
(P_2) $b \in \mathbf{F}(a)$ implies $\mathbf{F}(a+b) \leqq \mathbf{F}(a)$.

Then $\mathfrak{G}$ is said to be *an* (**F**, Ω)-*group.*

It must be remarked that in general $\mathbf{F}(a)$ need not be an Ω-subgroup. But if $\mathbf{F}(a)$ is an Ω-subgroup, then axiom (P_1) implies axiom (P_2).

An element $a \in \mathfrak{G}$ is called **F**-regular if $a \in \mathbf{F}(a)$ holds. A subset of $\mathfrak{G}$ is called **F**-*regular,* if all its elements are **F**-regular. (a) is a normal subgroup of the $(\mathbf{F}, \Omega)$-group $\mathfrak{G}$, because Ω contains all inner automorphisms of $\mathfrak{G}$. Similarly $(a)+(b)$ is a normal subgroup of $\mathfrak{G}$, and $(a+b) \leqq (a)+(b)$.

Now we have the important general

THEOREM 36.1. If $\mathfrak{G}$ is an (**F**, Ω)-group, then the set $N = \{a; a \in \mathfrak{G}, (a) \text{ is } \mathbf{F}\text{-regular}\}$ is an **F**-regular Ω-subgroup of $\mathfrak{G}$ containing all **F**-regular Ω-subgroups of $\mathfrak{G}$.

PROOF. If z and w are arbitrary elements of N, then we shall show that $(z-w)$ is an **F**-regular Ω-subgroup. For, if $a \in (z-w)$, then $a = u-v$ holds with suitable $u \in (z)$ and $v \in (w)$ since $(z-v) \leqq (z)+(v)$. Here u is an **F**-regular element since $u \in (z)$ and $z \in N$. Thus we have $u = a+v \in \mathbf{F}(a+v) \leqq \mathbf{F}(a)+(v)$ by axiom (P_1). Hence there exist elements $b \in \mathbf{F}(a)$ and $v_1 \in (v)$ with

$$a+v = -b+v_1,$$

whence we have $a = -b+v_1-v = v_2-b$ with a suitable element v_2 of the normal subgroup (v). Hence we obtain $a+b = v^2 \in (v) \leqq (w)$. Hence it follows that $a+b \in \mathbf{F}(a+b)$ since $w \in N$. Also by axiom (P_2) $a+b \in \mathbf{F}(a+b)$ since $b \in \mathbf{F}(a)$. Since $\mathbf{F}(a)$ is a subgroup and $b \in \mathbf{F}(a)$, it follows that $a \in \mathbf{F}(a)$. Thus the arbitrary element $a = u-v \in (z-w)$; therefore the Ω-subgroup $(z-w)$ is **F**-regular, and hence $z-w \in N$.

If $w \in \Omega$ and $z \in N$, then we have $zw \in (z)$, $(zw) \leqq (z)$ and thus $zw \in N$. Hence N is an Ω-subgroup of G.

Now, if H is an **F**-regular Ω-subgroup of $\mathfrak{G}$, and b is an element of H, then we have $(b) \leqq H$. Therefore (b) is **F**-regular, and thus we have $b \in N$, therefore $H \leqq N$. □

N is called the *largest* **F**-*regular* Ω-*subgroup of the* (**F**, Ω)-*group* $\mathfrak{G}$.

If K is an Ω-subgroup of $\mathfrak{G}$, then the factor group $\mathfrak{G}/K$ is an Ω-group under the natural definition $(a+K)w = aw+K$. Conversely, every Ω-homomorphic image

$\mathfrak{G}^*$ of $\mathfrak{G}$ is Ω-isomorphic to a group $\mathfrak{G}/K$, where K is an Ω-subgroup, the kernel of the Ω-homomorphism. The following theorem gives a natural procedure to make $\mathfrak{G}^*$ also an $(\mathbf{F}, \Omega)$-group:

THEOREM 36.2. If $a \to a^*$ is an Ω-homomorphism of an $(\mathbf{F}, \Omega)$-group $\mathfrak{G}$ onto an Ω-group $\mathfrak{G}^*$, and if we take $(\mathbf{F}(a))^* = \mathbf{F}(a^*)$, then $\mathfrak{G}^*$ also becomes an $(\mathbf{F}, \Omega)$-group.

PROOF. First we show that $a^* = b^*$ implies $\mathbf{F}(a^*) = \mathbf{F}(b^*)$. For, if K is the kernel of the Ω-homomorphism of $\mathfrak{G}$ onto $\mathfrak{G}^*$, then $a = b + k$ with $k \in K$ since $a^* = b^*$. By axiom (P_1) we have $\mathbf{F}(a) = \mathbf{F}(b+k) \leqq \mathbf{F}(b) + (k)$, and thus $\mathbf{F}(a^*) \leqq F(b^*)$. In a similar way we also have $\mathbf{F}(b^*) \leqq \mathbf{F}(a^*)$, and thus $\mathbf{F}(a^*) = \mathbf{F}(b^*)$. Now

$$\mathbf{F}(a^*+b^*) = \mathbf{F}((a+b)^*) = (\mathbf{F}(a+b))^* \leqq (\mathbf{F}(a)+(b))^* = \mathbf{F}(a^*)+(b^*).$$

Therefore, axiom (P_1) holds in $\mathfrak{G}^*$.

Now, if b^* $\mathbf{F}(a^*) = (\mathbf{F}(a)^*)$, then $b - c \in K$ with a suitable element $c \in \mathbf{F}(a)$. Hence $b^* \in \mathbf{F}(a^*)$ implies

$$\mathbf{F}(a^*+b^*) = \mathbf{F}(a^*+c^*) = (\mathbf{F}(a+c))^* \leqq \mathbf{F}((a))^* = \mathbf{F}(a^*),$$

and thus axiom (P_2) holds in $\mathfrak{G}^*$. Hence $\mathfrak{G}^*$ is an $(\mathbf{F}, \Omega)$-group. □

By theorems 36.2 and 36.1, $\mathbf{N}(\mathfrak{G}^*)$ has a meaning for every Ω-homomorphic image $\mathfrak{G}^*$ of an $(\mathbf{F}, \Omega)$-group $\mathfrak{G}$.

THEOREM 36.3. If $\mathfrak{G}$ is an $(\mathbf{F}, \Omega)$-group and $\mathbf{N}(\mathfrak{G})$ is the largest $\mathbf{F}$-regular Ω-subgroup of $\mathfrak{G}$, then $\mathbf{N}(\mathfrak{G}/\mathbf{N}(\mathfrak{G})) = (0)$.

PROOF. If a^* denotes the coset $a + \mathbf{N}(\mathfrak{G})$, $b^* \in \mathbf{N}(\mathfrak{G}/\mathbf{N}(\mathfrak{G}))$ and $a \in (b)$, then $a^* \in (b^*)$. Since $a^* \in \mathbf{F}(a)^* = (\mathbf{F}(a))^*$ there exists an element $c \in \mathbf{F}(a)$ with $a + c \in \mathbf{N}(\mathfrak{G})$. Hence we have $a \neq c \in \mathbf{F}(a+c) = F(a)$ by axiom (P_2), and since $c \in \mathbf{F}(a)$ we have $a \in \mathbf{F}(a)$, therefore $b \in \mathbf{N}(\mathfrak{G})$ and thus $b^* = 0$. □

Now we consider a result which is in a certain sense the converse of theorem 36.2:

THEOREM 36.4. If $a \to \mathbf{F}(a)$ is a mapping which associates to every element a of an Ω-group $\mathfrak{G}$ a uniquely determined Ω-subgroup $\mathbf{F}(a)$ of $\mathfrak{G}$ such that in every Ω-homomorphic image $\mathfrak{G}^*$ of $\mathfrak{G}$, $a^* = b^*$ implies $(\mathbf{F}(a))^* = (\mathbf{F}(b))^*$, then $\mathfrak{G}$ is an $(\mathbf{F}, \Omega)$-group.

PROOF. If a and b are arbitrary elements of $\mathfrak{G}$, we take $\mathfrak{G}^* = \mathfrak{G}/(b)$. Then $(a+b)^* = a^*$ implies $(\mathbf{F}(a+b))^* = (\mathbf{F}(a))^*$ by our assumption and thus $\mathbf{F}(a+b) \leqq \leqq \mathbf{F}(a) + (b)$ which is precisely axiom (P_1). Since $\mathbf{F}(a)$ is now an Ω-subgroup of $\mathfrak{G}$ by assumption, axiom (P_1) implies axiom (P_2). □

Now we give the definition of an $(\mathbf{F}, \Omega_1, \Omega)$-group and discuss some related notions which have also been introduced by Brown and McCoy [4].

Let $\mathfrak{G}$ be a not necessarily commutative group, written additively. Let $\mathfrak{J}$ be the group of all inner automorphisms of $\mathfrak{G}$, and if Ω_1, and Ω are fixed sets of endomorphisms of $\mathfrak{G}$ satisfying the condition $\mathfrak{J} \leqq \Omega_1 \leqq \Omega$. Let (a) denote the Ω-subgroup of $\mathfrak{G}$ generated by the element $a \in \mathfrak{G}$, and let $(a)_1$ denote the Ω_1-subgroup of $\mathfrak{G}$ generated by a. Then obviously $(a)_1 \leqq (a)$.

We assume that there exists a mapping **F** which associates to every element $a \in \mathfrak{G}$ an Ω_1-subgroup $\mathbf{F}(a)$ such that for every $a, b \in \mathfrak{G}$ the condition

$$\text{(P)} \qquad \mathbf{F}(a+b) \leqq \mathbf{F}(a) + (b)_1$$

holds.

Then $\mathfrak{G}$ is said to be an (**F**, Ω_1, Ω)-*group*. Since $\mathbf{F}(a)$ is now an Ω_1-subgroup, by assumption, $b \in \mathbf{F}(a)$ implies $(b)_1 \leqq \mathbf{F}(a)$, and thus we infer the axioms (P_1) and (P_2) from (P). Hence every (**F**, Ω_1, Ω)-group is also obviously an (**F**, Ω)-group. Hence theorems 36.1, 36.2, 36.3 and 36.4 are also valid for the (**F**, Ω_1, Ω)-groups.

An Ω_1-subgroup I of the (**F**, Ω_1, Ω)-group $\mathfrak{G}$ is called *modular*, if there exists an element $e \in \mathfrak{G}$ with $e \notin I$ such that $\mathbf{F}(e) \leqq I$. A modular Ω_1-subgroup is called *large* if the above element e is contained in every Ω_1-subgroup K with $K > I$. By Zorn's lemma every modular Ω_1-subgroup can be embedded into a large modular Ω_1-subgroup.

If I is an arbitrary subgroup of the (**F**, Ω_1, Ω)-group $\mathfrak{G}$ then let I' be the largest Ω-subgroup of $\mathfrak{G}$ contained in I, so that $I' = \{a;\ a \in \mathfrak{G},\ (a) = I\}$.

We have the following

THEOREM 36.5. If $\mathfrak{G}$ is an (**F**, Ω_1, Ω)-group, then the set $N = \{a;\ a \in \mathfrak{G},\ (a)$ is **F**-regular$\}$ coincides with the intersection D of those Ω-subgroups M', for which M is a large modular Ω_1-subgroup of $\mathfrak{G}$.

PROOF. If $a \notin N$, then there exists an element $b \in (a)$ with $b \notin \mathbf{F}(b)$. Then $\mathbf{F}(b)$ is a modular Ω_1-subgroup which by Zorn's lemma can be embedded into a large modular Ω_1-subgroup M with $b \notin M$. Now $b \in (a)$ implies $a \notin M'$ and thus $a \notin D$. Hence we have $D \leqq N$.

Conversely, if $a \notin D$, then there exists a large modular Ω_1-subgroup M with $a \notin M'$. Hence there exists an element $b \in (a)$ with $b \notin M$. Since M is a large modular Ω_1-subgroup, there exists an element $e \in \mathfrak{G}$ with $e \notin M$ and $\mathbf{F}(e) \leqq M$ such that every Ω_1-subgroup which is properly larger than M contains the element e. Therefore $e \in (M, b)_1$, i.e. e has a representation $e = m + c$ with $m \in M$ and $c \in (b)_1$. If $a \in N$, then since $c \in (b)_1 \leqq (b)_1 \leqq (a)$ and $a \in N$ we have $c \in \mathbf{F}(c)$ and thus

$$e - m \in F(e-m) \leqq \mathbf{F}(e) + (m)_1 \leqq M.$$

But then we have the contradiction $e \in M$. Therefore $a \notin D$ implies $a \in N$. Thus $N \leqq D$ Hence we have $N = D$. □

An $(\mathbf{F}, \Omega_1, \Omega)$-group $\mathfrak{G}$ is called *primitive* if $\mathfrak{G}$ contains a large modular Ω_1-subgroup M with $M'=(0)$. If $\mathfrak{G}$ is primitive then $N(\mathfrak{G})=(0)$ by theorem 36.5.

PROPOSITION 36.6. If M is a large modular Ω_1-subgroup of the $(\mathbf{F}, \Omega_1, \Omega)$-group $\mathfrak{G}$, then the factor group $\mathfrak{G}/M'$ is primitive.

PROOF. We denote the natural Ω-homomorphic images in $\mathfrak{G}^*=\mathfrak{G}/M$ by an asterisk. First we shall show that M^* is a large modular Ω_1-subgroup of $\mathfrak{G}^*$. We take $e\in\mathfrak{G}$ with $e\notin M$ and $\mathbf{F}(e)\leqq M$ such that every Ω_1-subgroup, properly containing M, also contains the element e. Then $\mathbf{F}(e^*)=(\mathbf{F}(e))^*\leqq M^*$. If $e^*\in M$, then we have $e\in M+M'$, which is a contradiction. Therefore $e^*\notin M^*$. Hence M^* is modular in $\mathfrak{G}^*$. Now, let M_1^* be an Ω_1-subgroup properly containing M^*. Then M_1^* is the image of an Ω_1-subgroup M_1 of $\mathfrak{G}$ properly containing M. Consequently, $e\in M$ and thus $e^*\in M_1^*$. Hence M^* is a large modular Ω_1-subgroup.

Now, if $a^*\in(M^*)'$, then we have $(a)^*=(a^*)\leqq M$ and thus $(a)\leqq M+M'=M$. Therefore $a\in M'$ and hence $a^*=0$. Thus $\mathfrak{G}/M'$ is primitive. □

We have the following

THEOREM 36.7. In an $(\mathbf{F}, \Omega_1, \Omega)$-group $\mathfrak{G}$ we have $N=(0)$ if and only if $\mathfrak{G}$ is a subdirect sum of primitive $(\mathbf{F}, \Omega_1, \Omega)$-groups.

PROOF. If $N=(0)$ we have $\bigcap_\alpha M_\alpha'=(0)$ by theorem 36.6, where M_α runs over all large modular Ω_1-subgroups of $\mathfrak{G}$. Hence $\mathfrak{G}$ is isomorphic to a subdirect sum of the groups $\mathfrak{G}/M_\alpha'$. But every factor group $\mathfrak{G}/M_\alpha'$ is a primitive $(\mathbf{F}, \Omega_1, \Omega)$-group by proposition 36.6.

Conversely, if $\mathfrak{G}$ is Ω-isomorphic to a subdirect sum of all primitive $(\mathbf{F}, \Omega_1, \Omega)$-groups, then $\mathfrak{G}$ contains a set of Ω-subgroups) K_α with $\mathfrak{G}\cong\mathfrak{G}/K_\alpha$ and $\bigcap_\alpha K_\alpha=(0)$. Now $N(\mathfrak{G})$ is mapped onto $N(\mathfrak{G}_\alpha)$ by the natural homomorphism $\mathfrak{G}\to\mathfrak{G}/K$. But $N(\mathfrak{G}_\alpha)=(0)$, because $\mathfrak{G}_\alpha$ is primitive. Hence $N\leqq K_\alpha$. Therefore

$$N\leqq\bigcap K_\alpha$$

and thus $N=(0)$. □

Now if $\mathfrak{G}$ is the additive group of a ring A, Ω is the set of all right multiplications and all left multiplications together with the identity automorphism, $\Omega_1=\Omega$, and if $\mathbf{F}(a)$ is an ideal of A with $(\mathbf{F}(a))^*=\mathbf{F}(a^*)$ for every ring homomorphism $a\to a^*$, then $\mathfrak{G}$ is an $(\mathbf{F}, \Omega_1, \Omega)$-group. The corresponding radical N is called the **F**-*radical* of the ring A (cf. theorem 36.5). A large modular Ω_1-subgroup is an ideal $M=M'$, for which there exists an element $a\in A$ with $a\notin M$ and $\mathbf{F}(a)\leqq M$ such that every ideal M_1 of A, properly containing M, also contains e. Hence the **F**-primitive ring A/M is subdirectly irreducible with $\bar{a}=a+M\neq\bar{0}$ and $\mathbf{F}(\bar{a})=(\bar{0})$.

In particular, we take $\mathbf{F}(a)=(1-a)A+A(1-a)A$, so that $\mathbf{F}(a)$ is the ideal generated by $(1-a)A$ in A. Since the axioms (P_1), (P_2) and (P) are fulfilled, the ring A is an $(\mathbf{F}, \Omega_1, \Omega)$-group with the above $\Omega_1=\Omega$. Now if A is **F**-primitive then there exists an $a\neq 0$ in A with $\mathbf{F}(a)=(0)$, i.e. with $(1-a)A+A(1-a)A=(0)$. Then A is subdirectly irreducible, and it has a left unity element, even a unity element. Therefore A is a simple ring with unity element. Hence the **F**-radical N of A coincides with the Brown–McCoy radical **G** of A.

Now, if $\mathfrak{G}$ is again the additive group of a ring A, Ω_1 is the set of all right multiplications together with the identity automorphism, Ω is the union of Ω_1 and all left multiplications, and $\mathbf{F}(a)$ is the right ideal $(1-a)A$, then an Ω_1-subgroup is a right ideal of A, and an Ω-group is an ideal of A. A large modular Ω_1-subgroup is a modular maximal right ideal in the sense of Chapter IV. A is **F**-primitive precisely when A contains a modular maximal right ideal R, for which $R'=(0)$, i.e. when (0) is the largest (two-sided) ideal of A contained in R. Hence an **F**-primitive ring is precisely a primitive ring in the sense of Jacobson. Therefore the corresponding **F**-radical N coincides now with the Jacobson radical.

§ 37. THE BROWN–McCOY RADICAL AS A (k, l, m, n)-RADICAL

In this section we shall explicitly define a radical, the (k, l, m, n)-radical, for every quadruple of non-negative integers, k, l, m, n (see Szász [4]). The (k, l, m, n)-radical has e.g. the following properties:

(1) The Jacobson radical is contained both in the $(k, 0, 1, 1)$-radical and in the $(0, l, 1, 1)$-radical for every $k\geqq 0$, $l\geqq 0$.

(2) If the elements of the ring A form a von Neumann regular semigroup with respect to the operation $x\circ y=x+y-xy$ (i.e. for every $a\in A$ there exists $b\in A$ with $a=a\circ b\circ a$), then A is both a $(k, 0, 1, 1)$-radical ring and a $(0, l, 1, 1)$-radical ring for every $k\geqq 0$, $l\geqq 0$.

(3) Every (k, l, m, n)-semisimple ring with minimum condition for (two-sided) principal ideals is two-sided completely reducible.

(4) The Brown–McCoy radical **G** coincides with a (k, l, m, n)-radical for some k, l, m, n.

We start by giving some definitions and notation. Let k, l, m, n be non-negative integers and write $a\circ b=a+b-ab$, $a^{(0)}=0$, $a^{(1)}=a$, $a^{(k+1)}=a^{(k)}\circ a$. Let a be a fixed element. We denote the sum of all (two-sided principal ideals of the form $(k\cdot a^{(l)}-a^{(m)}\circ x\circ a^{(n)})$, where x runs over all elements of the ring, by $(k, l, m, n)(a)$. An element $a\in A$ is called *(k, l, m, n)-regular*, if $a\in(k, l, m, n)(a)$. Since $(k, l, m, n)(a)$ is an ideal of A, this (k, l, m, n)-regularity is an **F**-regularity in the sense of the preceding section (see Brown and McCoy [1]). In this way an **F**-radical N can be determined. This radical is denoted by $(k, l, m, n)A$. Every (k, l, m, n)-semisimple ring is a subdirect sum of (k, l, m, n)-primitive rings. Here a ring A is called *(k, l, m, n)-primitive* if A is subdirectly irreducible and (k, l, m, n)-semisimple. Moreover, a ring A is (k, l, m, n)-primitive if and only if A is subdirectly irreducible, and if in the minimal ideal M of A there exists a non-zero element d with $k \cdot d^{(l)} = d^{(m)} \circ x \circ d^{(n)}$ for every $x \in A$. This element d is the *generalization of the unity element of a ring.*

We have the following

THEOREM 37.1. In every ring A the Brown–McCoy radical $\mathbf{G}$ coincides with the radicals $(1, 1, 1, 1)A$, $(1, 1, 1, 0)A$, $(1, 1, 0, 1)A$ and $(1, 2, 1, 1)A$.

PROOF. It is sufficient to show that every (k, l, m, n)-primitive ring is a simple ring with unity element for the k, l, m, n, mentioned in the theorem.

If P is a (k, l, m, n)-primitive ring, then P is subdirectly irreducible and in the minimal ideal M of P there exists an element d with $(k, l, m, n)(d) = (0)$ and $d \neq 0$. Then we have $k \cdot d^{(l)} = d^{(m)} \circ x \circ d^{(n)}$ for every $x \in P$. In particular, if $x = 0$, then we have $k \cdot d^{(l)} = d^{(m)} \circ d^{(n)} = d^{(m+n)}$. Hence $d^{(m)} \circ x \circ d^{(n)} = d^{(m+n)}$ for every $x \in P$, and thus $x = d^{(m)}x + xd^{(n)} - d^{(m)}xd^{(n)} \in M$ for every $x \in P$. Therefore we have $P = M$, and thus P has only the trivial ideals (0) and P.

Moreover we also have $(1-d^{(m)})P(1-d^{(n)}) = (0)$, $d = d \cdot d^{(m+n)}$ and $d = kd \cdot d^{(l)}$. In all four cases, mentioned in the theorem, $k=1$. Hence $d = d \cdot d^{(l)}$ and $d^{(l)} = d^{(m+n)}$.

If $m = n = l = 1$, then we have $d^2 = d \neq 0$, and thus P is not a zero-ring. Since P is simple (and not a zero-ring), P does not contain a non-zero nilpotent right ideal. Now $R = (1-d)P$ is a nilpotent right ideal since $(1-d)P(1-d) = (0)$. Thus $R = (0)$. Consequently, d is a left unity element of P, which is also the unity element of P, because $L = P(1-d) + P(1-d)P$ is nilpotent. Hence $L = (0)$.

If $k = l = m = 1$ and $n = 0$, then we have

$$(k, l, m, n)(a) = \sum_{a \in A} (a - a \circ x \circ a^{(0)}) = \sum_{x \in A} (x - ax) = (1-a)A + A(1-a)A.$$

Hence $(1, 1, 1, 0)A = \mathbf{G}(A)$ by the original definition of the Brown–McCoy radical.

The case $k = l = n = 1$ and $m = 0$ is similar to the preceding case and thus we have $(1, 1, 0, 1)A = \mathbf{G}(A)$.

Now, if $l = 2$ and $k = m = n = 1$, then we have $d = 2d^2 + d^3 = 0$ since $d = d \cdot d^{(2)}$. Then P is not a zero-ring because $d = 2d^2 - d^3 \neq 0$. Since P is simple, P has no non-zero nilpotent one-sided ideals. Now $R = (1-d)P$ is a nilpotent right ideal and $L = P(1-d)$ is a nilpotent left ideal of P since $(1-d)P(1-d) = 0$. Therefore, $(1-d)P = P(1-d) = (0)$. Hence we have $(d-d^2)P = (1-d)dP = 0$ and $P(d-d^2) = (0)$. Since P contains no non-zero one-sided annihilators, we have $d^2 = d$, and

d is the unity element of P since $(1-d)P=P(1-d)=(0)$. Therefore $(1, 2, 1, 1)A=$ $=\mathbf{G}(A)$. □

REMARKS. I. Every ring is a (k, l, m, n)-radical ring for the $(0, 0, 0, 0)$-, $(k, 0, 0, 1)$-, $(0, l, 0, 1,)$-, $(k, 0, 1, 0)$-, $(0, l, 1, 0)$-, $(2, 1, 1, 0)$-, $(2, 1, 0, 1)$-, and $(2, 1, 1, 1)$-radicals.

II. Every $(k, 0, 1, 1)$-primitive ring P and every $(0, l, 1, 1)$-primitive ring P is a simple ring with unity element and with the property $2P=P$. Moreover, every $(3, 1, 1, 1)$-primitive ring P, every $(3, 1, 1, 0)$-primitive ring and every $(3, 1, 0, 1)$-primitive ring is a simple ring with unity element and with the property $2P=(0)$. Hence a non-zero ring cannot be e.g. a $(0, l, 1, 1)$-primitive ring and a $(3, 1, 0, 1)$-primitive ring at the same time.

III. If $\mathbf{W}(a)=A(1-a)A(1-a)A$, then $\mathbf{W}(a)\leqq(1, 2, 1, 1)(a)$, and the $\mathbf{W}$-primitive rings are the following:

1. The simple rings with unity element.
2. The zero-rings with the additive group

$$Z(p^k) \quad (k = 1, 2, \ldots, \text{ or } k = \infty).$$

3. The rings P with $P^3=(0)$, $P^2 \neq (0)$,

$$(P^2)^+ \cong Z(p^k) \quad (k = 1, 2, \ldots, \text{ or } k = \infty),$$

in which $PM=MP$ $(=0)$ holds for the minimal ideal M of P. Such a ring is e.g. $P=\{a_1, a_2, a_3, \ldots, b_1, b_2, b_3, \ldots\}$ with the relations

$$a_i^2-b_i = pa_1 = b_j-pb_{j+1} = a_ia_j = a_i^3 = 0$$

$$(i \neq j;\ i, j = 1, 2, 3, \ldots).$$

§ 38. THE BROWN–McCOY RADICAL OF A FULL MATRIX RING

We need some preliminaries in order to determine the Brown–McCoy radical of a full matrix ring A_n over a ring A. For the subject-matter of this section we refer the reader to Brown and McCoy [2].

We take

$$\mathbf{H}(a) = A(1-a)A = \left\{ \sum_{i=1}^{m} (x_iy_i - x_iay_i) \right\}$$

for every element $a\in A$. Then $\mathbf{H}(a)$ is an ideal of A such that $\mathbf{H}(a)\leqq(1-a)A+$ $+A(1-a)A\leqq\mathbf{G}(a)$.

PROPOSITION 38.1. If $A(a)A=\mathbf{H}(a)$, then $\mathbf{H}(a)=A^2$.

PROOF. $\mathbf{H}(a) \leqq A^2$ is obvious. Therefore if $A(a)A \leqq \mathbf{H}(a)$ we have $\mathbf{H}(a) = A^2$. □

PROPOSITION 38.2. If I is an ideal of the ring A, $a \in I \cap \mathbf{G}(a)$ and $\mathbf{H}'(a) = I(1-a)I$, then we have $\mathbf{H}'(a) = I^2$.

PROOF. Since $a \in I \cap (1-a)A + A(1-a)A)$ we have $IaI \leqq I(1-a) + I(1-a) \leqq \mathbf{H}'(a)$ and thus $\mathbf{H}'(a) = I^2$ by proposition 38.1. □

THEOREM 38.3. For the Brown–McCoy radical $\mathbf{G}(A)$ the following conditions are equivalent:

(1) $x \in \mathbf{G}(A)$.
(2) $y \in \mathbf{G}(y)$ for every $y \in AxA$.
(3) $y \in \mathbf{H}(y)$ for every $y \in AxA$.

PROOF. (2) follows from (1) since $AxA \leqq (x) \leqq \mathbf{G}(A)$. (3) follows from (2), because if $y \in AxA$ we have $\mathbf{H}(y) = A^2$ for $I = A$ by proposition 38.2. We have $y \in \mathbf{H}(y)$ since $AxA \leqq A^2$ and thus y is $\mathbf{H}$-regular. (1) follows from (3). For, if $y \in (x)$, then $y^3 \in AxA$, and thus we have $y^3 \in \mathbf{H}(y^3) \leqq \mathbf{G}(y^3)$. If we take $s = -\sum_{i=1}^{n-1} a^i$, then $a^n = a - (as - s)$, and thus $a^n \in \mathbf{G}(a^n)$ implies $a \in \mathbf{G}(a)$ by the following

PROPOSITION 38.4. If $a - c \in \mathbf{G}(a-c)$ and $c \in \mathbf{G}(a)$ then $a \in \mathbf{G}(a)$ for $\mathbf{G}(x) = (1-x)A + A(1-x)A$.

PROOF. Since $(a-c) \in (1-a+c)A + A(1-a+c)A$ we have $a \in (1-a)A + A(1-a)A + c + cA + Ac$, and since $c \in \mathbf{G}(a)$ it follows that $a \in \mathbf{G}(a)$. □

This also proves theorem 38.3. □

PROPOSITION 38.5. If $x, y \in A$, $M = (a_{ij}) \in A_n$, e_{ij} is a matrix unit and $(x)^{ij} = xe_{ij}$, $(y)^{ij} = ye_{ij}$, then

$$(x)^{ip} M (y)^{qj} = (xa_{pq}y)^{ij}.$$

PROOF is clear. □

PROPOSITION 38.6. If $M = (a_{ij}) \in A_n$, and if every a_{ij} is $\mathbf{H}$-regular in A (i.e. if $a_{ij} \in \mathbf{H}(a_{ij})$ for every a_{ij}), then M itself is $\mathbf{H}$-regular in A_n.

PROOF. Since $a_{ij} \in \mathbf{H}(a_{ij})$ we have $Aa_{ij}A \leqq \mathbf{H}(a_{ij})$, and thus $\mathbf{H}(a_{ij}) = A^2$ by proposition 38.1. Hence, we have e.g. $a_{ij} \in A^2 = \mathbf{H}(a_{11})$. Therefore, there exist natural numbers m_{ij} and elements $x_{ijk}, y_{ijk} \in A$ $(k = 1, 2, \ldots)$ such that

$$a_{ij} = \sum_{k=1}^{m_{ij}} (x_{ijk} a_{11} y_{ijk} - x_{ijk} y_{ijk}).$$

Hence by proposition 38.5 it follows that

$$M = \sum_{i,j=1}^{n} \left(\sum_{k=1}^{m_{ij}} (x_{ijk})^{i1} M (y_{ijk})^{ij} - (x_{ijk})^{i1} (y_{ijk})^{1j} \right).$$

Therefore $M \in \mathbf{H}(M)$. □

THEOREM 38.7. $\mathbf{G}(A_n) = (\mathbf{G}(A))_n$.

PROOF. Let $N = \mathbf{G}(A_n)$. We shall first show that $(\mathbf{G}(A))_n \subseteq N$. Take

$$B = (b_{ij}) \in (\mathbf{G}(A))_n, \quad X = (x_{ij}) \in A_n, \quad Y = (y_{ij}) \in A_n.$$

Then $X(b_{ij})^{ij} Y = Z \in (Ab_{ij}A)_n$ where $Z = (z_{ij})$, $z_{pq} = x_{pi} b_{ij} y_{jq}$. Hence we obviously have

$$A_n (b_{ij})^{ij} A_n \subseteq (Ab_{ij}A)_n.$$

By proposition 38.3 $Ab_{ij}A$ is H-regular in A, because $b_{ij} \in \mathbf{G}(A)$. Hence $A_n(b_{ij})^{ij}A_n$ is $\mathbf{H}$-regular in A_n by proposition 38.6. By theorem 38.3 $(b_{ij})^{ij} \in N$ follows, and thus $B = \sum (b_{ij}) \in N$. Therefore $(\mathbf{G}(A))_n = N$. Conversely, we shall prove that $N \subseteq (\mathbf{G}(A))_n$. If $M = (a_{ij}) \in N$ and $x, y \in A$ we have $(xa_{pq}y)^{ij} \in N$ by proposition 38.5. Let $b = a_{pq}$ be an arbitrary fixed element of the matrix $M = (a_{ij})$ and take $s \in AbA$. Then N contains the scalar matrix $S = (s)$. Consequently $S \in \mathbf{G}(S)$ in A_n. There exist $X, X_k, Y_k \in A_n$, such that

$$S = SX - X + \sum_k (X_k S Y_k - X_k Y_k).$$

By comparison of the elements in the (1,1)-th place we obtain $s \in \mathbf{G}(s)$ in A. Hence every element s of AbA is G-regular, whence we have $b \in \mathbf{G}(A)$, $M \in (\mathbf{G}(A))_n$ and $N \subseteq (\mathbf{G}(A))_n$. □

§ 39. THE UPPER RADICAL DETERMINED BY A CLASS OF SIMPLE RINGS WITH UNITY ELEMENT

For the subject-matter of this section we refer the reader to Suliński [1, 4].

In Chapter IV we have seen that the Jacobson radical is smaller than the upper radical determined by the class of all primitive rings. However, the Brown–McCoy radical coincides with the upper radical determined by the class of all simple and semisimple rings in the sense of Brown and McCoy.

Namely, we have the following

THEOREM 39.1. The Brown–McCoy radical **G** coincides with the upper radical **U**, determined by the class of all simple rings with unity element.

PROOF. Since every simple ring with unity element is **G**-semisimple, we obviously have $\mathbf{G} \leqq \mathbf{U}$ by the definition of the upper radical **U**. Now, if A is a ring which is not a **G**-radical ring, then $A/\mathbf{G}(A)$ is a subdirect sum of simple rings with unity element and thus A can be homomorphically mapped onto a non-zero simple ring with unity element. Hence A is not a **U**-radical ring. Therefore, if A is a **U**-radical ring, then A is also a **G**-radical ring. Hence $\mathbf{U} \leqq \mathbf{G}$ and since $\mathbf{G} \leqq \mathbf{U}$ we have $\mathbf{G} = \mathbf{U}$. □

The following result of Suliński shows that the Brown–McCoy radical is the smallest hereditary radical with the property that every semisimple ring is a subdirect sum of simple rings.

THEOREM 39.2. Let **U** be the upper radical determined by a class **K** of simple rings. Then every ring in **K** contains a unity element if and only if both of the following conditions are fulfilled:

(1) every **U**-semisimple ring is a subdirect sum of rings belonging to the class **K**;

(2) the radical **U** is hereditary.

PROOF. First we assume that every simple ring in **K** has a unity element, and we show the validity of condition (1) and (2). Let A be a **U**-semisimple ring. Consider the class $\mathfrak{M}$ of those ideals M_α of A, for which $A/M_\alpha \in \mathbf{K}$. A certainly possesses such ideals M_α because A is not a **U**-radical ring. In particular, if A is simple, then condition (1) trivially holds. Now we take $D = \bigcap_{M_\alpha \in \mathfrak{M}} M_\alpha$. If $D=0$ then condition (1) holds by the definition of $\mathfrak{M}$ and that of the subdirect sum.

But if D differs from zero, then the ideal D is again **U**-semisimple, and thus D can be homomorphically mapped onto a simple ring T with unity element and with $T \in \mathbf{K}$. Let $f \in D$ be an element mapped by this homomorphism $h: D \to T$ onto e. Therefore $hf = e$.

Now we define a further homomorphism k by

$$k:\ x \to k(x) = h(fxf).$$

Since

$$k(x+y) = h(f(x+y)f) = h(fxf) + h(fyf) = k(x) + k(y)$$

and

$$k(xy) = h(fxyf) = h(fx)\cdot h(yf) = h(fx)\cdot h(f)\cdot h(f)\cdot h(yf) =$$

$$= h(fxf)\cdot h(fyf) = k(x)\cdot k(y),$$

k is in fact a ring-homomorphism of A onto T, which coincides with h on the ideal D. In particular, $k(f)=e\in T$, and thus f is not contained in the kernel of the homomorphism k. But since $T\in K$ the kernel of k is an ideal $M_n\in\mathfrak{M}$. Hence we have $k(f)=0$ because $f\in D=\cap M_\alpha = M_\alpha$. This yields the contradiction $e=0$. Therefore condition (1) holds.

Now let A be a **U**-radical ring and assume that there exists an ideal M of A which is not a **U**-radical ring. Then A cannot be homomorphically mapped onto a simple ring of **K**, but there exists a homomorphism $h: M\to T$ onto a simple ring T of **K**.

Since T has a unity element e, there exists an element f with $h(f)=e$. As we have seen above, the mapping $k(x)=h(fxf)$ is an extension of h, which homomorphically maps the ring A onto T. This is a contradiction. Therefore if M is not a **U**-radical ring, then A is not a **U**-radical ring, and thus **U** is hereditary. Hence condition (2) holds.

Conversely, now we assume that conditions (1) and (2) are fulfilled for the upper radical, determined by the class **K** of simple rings. We shall show that every ring in **K** has a unity element. Assume that a ring of A of the class **K** has no unity element. Then we can embed A into a ring A' with unity element. Let A' consist of all pairs (n, a) where n is a natural number and $a\in A$. The equality and the addition of the pairs is defined componentwise and the multiplication by

$$(n_1, a_1)(n_2, a_2) = (n_1 n_2, n_1 a_2 + n_2 a_1 + a_1 a_2).$$

A' is a ring with the unity element $(1, 0)$. In A' the pairs $(0, a)$ form an ideal which is isomorphic to the ring A. Since A is simple, this ideal which can be identified with A is minimal in A'. Moreover, A'/A is isomorphic to the ring I of the rational integers.

We shall show that A is contained in the intersection D of all maximal ideals M of A'. For, if $A\not\subseteq D$, then there exists a maximal ideal M of A' with $A\not\subseteq M$. Then we have $A\cap M=(0)$ by the minimality of A in A', and we have $A+M=A'$ by the maximality of M in A. Therefore, we obtain the ring direct decomposition $A'=A\oplus M$. Hence A is an endomorphic image of the ring A' with unity element, and thus A also has a unity element. But this fact contradicts the assumption and hence $A\subseteq D$. Consequently A' is not a subdirect sum of simple rings and by condition (1), A' is not a **U**-semisimple ring. If $\mathbf{U}(A')$ is the **U**-radical of A', then by condition (1), $A'/\mathbf{U}(A')$ is a subdirect sum of rings of **K**, and thus $A'/\mathbf{U}(A')$ contains a set of maximal ideals M_α of A', whose intersection coincides with $\mathbf{U}(A')$. As above, $A\subseteq D$, and since $D\subseteq\cap M_\alpha=\mathbf{U}(A')$ we have $A=\mathbf{U}(A')$. Since the radical **U** is hereditary by condition (2), A is a **U**-radical ring which contradicts the assumption $A\in\mathbf{K}$. Hence every ring A in **K** has a unity element. □

§ 40. SPECIAL AND COMPLETELY NON-SPECIAL SEMISIMPLE RINGS IN THE SENSE OF BROWN AND McCOY

In this section we discuss some results of Suliński [4, 5] concerning semisimple rings in the sense of Brown and McCoy. In the set of all modular ideals of a semisimple ring we define a topology and we topologically formulate various algebraic properties of the ring and its ideals.

By means of the minimal set of a ring, the special part and the completely non-special part of an arbitrary semisimple ring are introduced.

In this section radical will always mean the Brown–McCoy radical. Let A be a semisimple ring. The set of all modular maximal ideals M_α of A will be denoted by $\mathfrak{M}$. For $\mathfrak{N} \leqq \mathfrak{M}$ define

$$I(\mathfrak{N}) = \bigcap_{M_\alpha \in \mathfrak{N}} M_\alpha .$$

Let $\overline{\mathfrak{N}}$ be the set of all M_α with $M_\alpha \supseteqq I(\mathfrak{N})$. Then $I(\mathfrak{N}) = I(\overline{\mathfrak{N}})$. We shall show that the closure operator so defined enables us to introduce a topology. $\mathfrak{M}$ is called *the structure space* of A. On the other hand, $\mathfrak{M}$ is a subspace of the Jacobson structure space (cf. Suliński [5]).

For $M \in \mathfrak{M}$ we denote the factor ring A/M by A_M, and the natural homomorphism of A onto A_M by h_M. Let $C(\mathfrak{N})$ be the complete direct sum of the rings A_M with $M \in \mathfrak{N} \leqq \mathfrak{M}$. Every element $x \in C(\mathfrak{N})$ can be considered as a function $x(M)$, where $M \in \mathfrak{N}$ and $x(M) \in A_M$.

The semisimple ring A is *subdirectly embedded* into $C(\mathfrak{N})$, if there exists a subring $S \leqq C(\mathfrak{N})$ such that A and S are isomorphic, and there exists an element $s \in S$ with $s(M) = a$ for every $M \in \mathfrak{N}$ and for every $a \in A$.

We have the following

THEOREM 40.1. Let A be a semisimple ring and $\mathfrak{M}$ be its structure space. A can be subdirectly embedded into $C(\mathfrak{N})$ if and only if $I(\mathfrak{N}) = (0)$ (i.e. $\overline{\mathfrak{N}} = \mathfrak{M}$).

PROOF. We shall show first that if $I(\mathfrak{N}) = (0)$ then A can be subdirectly embedded into $C(\mathfrak{N})$. Define the function $x(M)$ by

$$x(M) = h_M(x) \in A_M \quad (M \in \mathfrak{N})$$

for every $x \in A$. Then $x \to x(M)$ is a homomorphism of A onto a subring S of $C(\mathfrak{N})$. If there exists a non-zero $x \in A$ with $x(M) = 0$ for every $M \in \mathfrak{N}$, then $x \in M$ for every ideal $M \in \mathfrak{N}$. But this is impossible since $I(\mathfrak{N}) = (0)$. Hence $x \to x(M)$ is an isomorphism of A onto S. Since A is mapped by h_M onto A_M, the ring A is subdirectly embedded into $C(\mathfrak{N})$.

Conversely, let A be subdirectly embedded into $C(\mathfrak{N})$. Then every element $x \in I(\mathfrak{N})$ is mapped onto $0 \in C(\mathfrak{N})$. Hence x itself equals 0. Therefore $I(\mathfrak{N})=(0)$. □

On the basis of the isomorphism of A and S we shall identify the element $x \in A$ with the function $x(M)$.

PROPOSITION 40.2. $x \in I(\mathfrak{N})$ for $\mathfrak{N} \leqq \mathfrak{M}$ if and only if $x(M)=0$ for every $M \in \mathfrak{N}$.

PROOF is trivial by theorem 40.1. □

An ideal K of the semisimple ring A is called *representable* if there exists a set $\mathfrak{N} \leqq \mathfrak{M}$ with $K=I(\mathfrak{N})$.

PROPOSITION 40.3. A representable ideal $K=I(\mathfrak{N})$ of the semisimple ring A can be subdirectly embedded into $C(\mathfrak{M} \setminus \overline{\mathfrak{N}})$. In particular, every ideal $M \in \mathfrak{M}$ can be subdirectly embedded into $C(\mathfrak{M} \setminus \{M\})$.

PROOF. If $M \in \mathfrak{M} \setminus \overline{\mathfrak{N}}$ is, then $M+I(\mathfrak{N})=A$, and since

$$K/(K \cap M) \cong (K+M/M) = \big(I(N)+M\big)/M = A/M = A_M$$

$K \cap M$ is a modular maximal ideal of K. Hence the intersection of all ideals $K \cap M$ with $M \in \mathfrak{M} \setminus \overline{\mathfrak{N}}$ is just the ideal $K \cap I(\mathfrak{M} \setminus \overline{\mathfrak{N}}) = I(\overline{\mathfrak{N}}) \cap I(\mathfrak{M} \setminus \overline{\mathfrak{N}}) = (0)$. Hence K can be subdirectly embedded into $C(\mathfrak{M} \setminus \overline{\mathfrak{N}})$ by theorem 40.1. The second statement follows from the first. □

THEOREM 40.4. If A is a semisimple ring and $\mathfrak{M}$ is its structure space, then the following conditions hold:

I. $\overline{\mathfrak{N}_1 \cup \mathfrak{N}_2} = \overline{\mathfrak{N}}_1 \cup \overline{\mathfrak{N}}_2$ for every $\mathfrak{N}_1, \mathfrak{N}_2 \leqq \mathfrak{M}$;

II. $\overline{\{M\}} = \{M\}$ for every $M \in \mathfrak{M}$;

III. $\overline{\emptyset} = \emptyset$ for the empty set $\emptyset$;

IV. $\overline{\overline{\mathfrak{N}}} = \overline{\mathfrak{N}}$ for every $\mathfrak{N} \leqq \mathfrak{M}$.

Thus, M is a topological T_1-space.

PROOF. It is clear that $\mathfrak{M}$ satisfies conditions II, III and IV. Also obviously $\overline{\mathfrak{N}}_1 = \overline{\mathfrak{N}_1 \cup \mathfrak{N}_2}$, $\overline{\mathfrak{N}}_2 \leqq \overline{\mathfrak{N}_1 \cup \mathfrak{N}_2}$ and thus $\overline{\mathfrak{N}}_1 \cup \overline{\mathfrak{N}}_2 = \overline{\mathfrak{N}_1 \cup \mathfrak{N}_2}$. Now take $M \in \overline{\mathfrak{N}_1 \cup \mathfrak{N}_2}$. Then $M \geqq I(\mathfrak{N}_1 \cup \mathfrak{N}_2) = I(\mathfrak{N}_1) \cap I(\mathfrak{N}_2)$. Therefore $x(M)=0$ (for every $x \in$ $\in I(\mathfrak{N}_1) \cap I\mathfrak{N}_2)$ by proposition 40.2. Suppose that $M \notin \overline{\mathfrak{N}}_1$ and $M \notin \overline{\mathfrak{N}}_2$. Therefore $M \in \mathfrak{M} \setminus \overline{\mathfrak{N}}_1$ and $M \in \mathfrak{M} \setminus \overline{\mathfrak{N}}_2$. If we apply proposition 40.3 to $I(\mathfrak{N}_1)$ and $I(\mathfrak{N}_2)$, we obtain $y \in I(\mathfrak{N}_1)$ and $z \in I(\mathfrak{N}_2)$ with $y(M)=z(M)=1_M$, where 1_M denotes the unity element of the simple ring $A_M=A/M$. Then we have $x(M)=y(M)\cdot z(M)=1_M \neq 0$, and $x=yz \in I(\mathfrak{N}_1) \cap I(\mathfrak{N}_2)$. This contradiction shows $\overline{\mathfrak{N}_1 \cup \mathfrak{N}_2} = \overline{\mathfrak{N}}_1 \cup \overline{\mathfrak{N}}_2$ and thus also the validity of condition I for $\mathfrak{M}$. □

We recall that the *annihilator* K^* of the ideal K in the ring A consists of the elements $x\in A$ with $xK=Kx=(0)$.

PROPOSITION 40.5. $(I(\mathfrak{N}))^*=I(\mathfrak{M}\setminus\overline{\mathfrak{N}})$ holds for every $\mathfrak{N}\subseteq\mathfrak{M}$. In particular we have $M^*=I(\mathfrak{M}\setminus\{M\})$.

PROOF. Assume that $x\in(I(\mathfrak{N}))^*$ but $x\notin I(\mathfrak{M}\setminus\overline{\mathfrak{N}})$. Then there exists an ideal $M\in\mathfrak{M}\setminus\overline{\mathfrak{N}}$ with $x(M)\neq 0$ by proposition 40.2. Moreover, there exists $y\in I(\mathfrak{N})$ with $y(M)=1_M\in A_M$ by proposition 40.3. Hence we have $xy\neq 0$ which is impossible since $x\in(I(\mathfrak{N}))^*$. Thus, $(I(\mathfrak{N}))^*\subseteq I(\mathfrak{M}\setminus\overline{\mathfrak{N}})$. Conversely, if $x\in I(\mathfrak{M}\setminus\overline{\mathfrak{N}})$, then we have $xy\in I(\mathfrak{M}\setminus\mathfrak{N})\cap I(\overline{\mathfrak{N}})=(0)$ for every $y\in I(\mathfrak{N})$. Therefore $x\in(I(\mathfrak{N}))^*$. □

For $\mathfrak{N}\subseteq\mathfrak{M}$, $\overline{\mathfrak{N}}$ is called a *closed domain* if $\overline{\mathfrak{N}}=\overline{\mathrm{Int}\,\overline{\mathfrak{N}}}$, where $\mathrm{Int}\,\overline{\mathfrak{N}}=$ $=\mathfrak{M}\setminus\overline{\mathfrak{M}\setminus\overline{\mathfrak{N}}}$.

We have the following

THEOREM 40.6. Let A be a semisimple ring, $\mathfrak{M}$ its structure space, and take $\mathfrak{N}\subseteq\mathfrak{M}$. Then $(I(\mathfrak{N}))^{**}=I(\mathfrak{N})$ if and only if $\overline{\mathfrak{N}}$ is a closed domain.

PROOF. By proposition 40.5 we have

$$(I(\mathfrak{N}))^{**} = I(\mathfrak{M}\setminus(\overline{\mathfrak{M}\setminus\overline{\mathfrak{N}}})) = I(\mathrm{Int}\ \overline{\mathfrak{N}}) = I(\overline{\mathrm{Int}\ \overline{\mathfrak{N}}}).$$

Hence $\overline{\mathrm{Int}\,\overline{\mathfrak{N}}}=\overline{\mathfrak{N}}$ implies

$$(I(\mathfrak{N}))^{**} = I(\overline{\mathrm{Int}\ \overline{\mathfrak{N}}}) = I(\overline{\mathfrak{N}}) = I(\mathfrak{N}),$$

and conversely,

$$I(\mathfrak{N}) = (I(\mathfrak{N}))^{**}$$

(i.e. $I(\overline{\mathfrak{N}})=I(\overline{\mathrm{Int}\,\overline{\mathfrak{N}}})$) implies $\overline{\mathfrak{N}}=\overline{\mathrm{Int}\,\overline{\mathfrak{N}}}$. □

If every ideal K of a ring A is representable, then A is called *strongly semisimple* (Andrunakievich [10, 11, 12]).

THEOREM 40.7. If A is a strongly semisimple ring, $\mathfrak{M}$ its structure space,

$$\mathfrak{N} \subseteq \mathfrak{M} \quad \text{and} \quad \mathrm{Fr}\ \mathfrak{N} = \overline{\mathfrak{N}}\cap\overline{\mathfrak{M}\setminus\mathfrak{N}},$$

then

$$I(\mathrm{Fr}\ \overline{\mathfrak{N}}) = I(\overline{\mathfrak{N}})\oplus(I(\mathfrak{N}))^*.$$

In particular, the strongly semisimple ring A is the ring direct sum of the ideal $K=I(\mathfrak{N})$ and the ideal K^* if and only if $\overline{\mathfrak{N}}$ is an open set in $\mathfrak{M}$, i.e. if $\mathrm{Fr}\,\overline{\mathfrak{N}}=\emptyset$.

PROOF. Consider $L=I(\mathfrak{N})+(I(\mathfrak{N}))^*$. There exists an $\mathfrak{N}_1\subseteq\mathfrak{M}$ with $L=I(\mathfrak{N}_1)$. If $M\in\overline{\mathfrak{N}}_1$, we have $M\supseteq I(\mathfrak{N}_1)$ and $M\supseteq(I(\mathfrak{N}))^*=I(\mathfrak{M}\setminus\overline{\mathfrak{N}})$. Hence $M\in\overline{\mathfrak{N}}\cap\overline{\mathfrak{M}\setminus\overline{\mathfrak{N}}}=$ $=\mathrm{Fr}\,(\overline{\mathfrak{N}}$. Conversely, if $M\in\mathrm{Fr}\,(\overline{\mathfrak{N}})$, then we have $M\in\overline{N}\cap\overline{M\setminus\overline{N}}$ and thus $M\in\mathfrak{N}$ and $M\supseteq I(\mathfrak{M}\setminus\overline{\mathfrak{N}})$. Hence $M=L$. Therefore $M\in\overline{\mathfrak{N}}_1$. □

PROPOSITION 40.8. If $\mathfrak{N}\subseteq\mathfrak{M}$ and $M\in\mathfrak{M}\setminus\overline{\mathfrak{N}}$, then $M^*=M\cap(I(\mathfrak{N}))^*\cap I(\mathfrak{N})$.

PROOF. By proposition 40.5 we have

$$(M\cap I(\mathfrak{N}))^* = (I(\mathfrak{N}\cup\{M\}))^* = I(M\setminus\overline{N\cup\{M\}}).$$

Another application of proposition 40.5 yields

$$(M\cap I(\mathfrak{N}))^*\cap I(\mathfrak{N}) = I(\mathfrak{M}\setminus\overline{\mathfrak{N}\cup\{M\}})\cap I(\overline{\mathfrak{N}}) = I(\mathfrak{M}\setminus\{M\}) = M^*,$$

because

$$\mathfrak{M}\setminus\{M\} = (\mathfrak{M}\setminus\overline{\mathfrak{N}\cup\{M\}})\cup\overline{N}. \square$$

PROPOSITION 40.9. If $M\in\mathfrak{M}$, then either $M^*=(0)$ or $M^*=A_M\cong A/M$.

PROOF. We associate to each $x\in M^*$ the element $x(M)\in A_M$. This gives an isomorphism of M^* onto a subring S of A_M. But S is an ideal of A_M, because there exist $x\in M^*$ and $y\in A$ with $x(M)=s$ and $y(M)=a_M$ for every $s\in S$ and $a_M\in\in A_M$. Therefore, $xy\in M^*$ implies $xy(M)=x(M)\cdot y(M)=s\cdot a_M\in S$. Since A_M is a simple ring, we have $S=(0)$ or $S=A_M$. Thus $M^*=(0)$ or $M^*=A/M$. $\square$

PROPOSITION 40.10. If the ideal K of the semisimple ring A is a simple ring with unity element, then there exists an $M\in\mathfrak{M}$ with $K=M^*\cong A_M$.

PROOF. Since the ideal K has a unity element, we have the ring direct decomposition $A=K\oplus K^*$. By the minimality of K in A, $K^*=M\in\mathfrak{M}$, and we obtain $M^*=K\cong A_M=A/M$ by proposition 40.9. $\square$

By the *minimal set* of the semisimple ring A we mean the intersection $\mathfrak{D}$ of those sets $\mathfrak{N}\subseteq\mathfrak{M}$, for which $I(\mathfrak{N})=(0)$.

PROPOSITION 40.11. A modular maximal ideal M belongs to the minimal set $\mathfrak{D}$ if and only if $M^*\neq(0)$.

PROOF. Let $M\in\mathfrak{D}$. If $M^*=(0)$, then we have $I(\mathfrak{M}\setminus\{M\})=(0)$. Therefore $\mathfrak{M}\setminus\{M\}\supseteq\mathfrak{D}$, and thus we have the contradiction $M\notin\mathfrak{D}$. Conversely, if $M\notin\mathfrak{D}$, then there exists an $\mathfrak{N}\subseteq\mathfrak{M}$ with $M\notin\mathfrak{N}$ and $I(\mathfrak{N})=(0)$. Hence $\mathfrak{N}\subseteq\mathfrak{M}\setminus\{M\}$, and thus $M^*=I(\mathfrak{M}\setminus\{M\})\subseteq I(\mathfrak{N})=(0)$. $\square$

If $I(\mathfrak{D})=(0)$ for a semisimple ring A, then A is called *special*. By a *completely non-special ring* we mean a semisimple ring A with $I(\mathfrak{D})=A$ (i.e. with $\mathfrak{D}=\emptyset$).

THEOREM 40.12. Let A be a semisimple ring, and K a representable ideal. The ring K is special if and only if $K\cap I(\mathfrak{D})=(0)$. Moreover, the ring K is completely non-special if and only if $K\subseteq I(\mathfrak{D})$.

PROOF. Take $\mathfrak{N}\subseteq\mathfrak{M}$ and $K=I(\mathfrak{N})$. Then $K\cap M$ is a modular maximal ideal of K for every $M\in\mathfrak{M}\setminus\overline{\mathfrak{N}}$. Furthermore, let $\mathfrak{D}_K$ be the minimal set of K. Then

$K\cap M\in\mathfrak{D}$ holds if and only if $M\in\mathfrak{D}\cap(\mathfrak{M}\cap\overline{\mathfrak{N}})$. If $I_K(\mathfrak{D}_K)$ is the intersection of all ideals $K\cap M\in\mathfrak{D}_K$, then

$$I_K(\mathfrak{D}_K) = K\cap I(\mathfrak{D}\cap(\mathfrak{M}\setminus\overline{\mathfrak{N}})) = I(\overline{\mathfrak{N}})\cap I(\mathfrak{D}\cap(\mathfrak{M}\setminus\overline{\mathfrak{N}})) =$$
$$= I(\overline{\mathfrak{N}}\cup(\mathfrak{D}\cap(\mathfrak{M}\setminus\overline{\mathfrak{N}}))) = I(\overline{\mathfrak{N}}\cup\mathfrak{D}) = K\cup I(\mathfrak{D})$$

because we have

$$\mathfrak{N}\cap(\mathfrak{D}\cap(\mathfrak{M}\setminus\overline{\mathfrak{N}})) = \overline{\mathfrak{N}}\cup\mathfrak{D}.$$

The remaining parts of the proof follow from the definitions of special rings and completely non-special rings. □

THEOREM 40.13 (McCoy [2]). A semisimple ring A is special if and only if every non-zero ideal K of A contains such a simple ideal L of the ring K and L has a unity element.

PROOF. If A is non-special, then $I(\mathfrak{D})\neq(0)$ for the minimal set $\mathfrak{D}$. But if every non-zero ideal K of A has a simple ideal with unity element, then the ring $I(\mathfrak{D})$ also has a simple ideal L with unity element. Then L is a ring direct summand of $I(\mathfrak{D})$ and thus L is an ideal of A, because $\bar{L}\leqq I(\mathfrak{D})$ and $\bar{L}=L\oplus(\bar{L}\cap I_1)$ where $\bar{L}=L+AL+LA+ALA$ and $I(\mathfrak{D})=L\oplus I_1$. Therefore, there exists an $M\in\mathfrak{M}$ with $M^*=I(\mathfrak{M}\setminus\{M\})=L\leqq I(\mathfrak{D})$ by propositions 40.5 and 40.10. But since $M^*= =L\neq(0)$, $M\in\mathfrak{D}$ by proposition 40.11. Hence $M\supseteqq I(\mathfrak{D})\supseteqq L=M^*$ holds and thus we obtain $L^2=(0)$ which contradicts the existence of unity element in L. Therefore, if every non-zero ideal K of A has a simple ideal with unity element, then A is special.

Conversely, if A is a special semisimple ring, and K is an arbitrary non-zero ideal of A, then $I(\mathfrak{D})=(0)$, and by theorem 40.1, A can be subdirectly embedded into the complete direct sum $C(\mathfrak{D})$. Then there exists an $M\in\mathfrak{D}$ with $x(M)\neq0$ for every $x\in K$. Moreover, we have $M^*\cong A_M$ for every $M\in\mathfrak{D}$, by propositions 40.9, and 40.11. Let y be an element of $M^*\leqq A$ with $y(M)=1_M\in A_M$. Then we have $M^*\cap K\neq(0)$ because $xy\neq0$ and $xy\in K\cap M^*$. Since M^* is simple, it follows that $K\geqq M^*\cong A_M$. □

THEOREM 40.14. A semisimple ring A is completely non-special if and only if A contains no simple ideal with unity element.

PROOF. Let A be a completely non-special semisimple ring. If A has a simple ideal L with unity element, then there exists an $M\in\mathfrak{M}$ with $L=M^*\cong A_M$ by proposition 40.10. Then $M\in\mathfrak{D}$ by proposition 40.11. This contradicts the assumption $\mathfrak{D}=\emptyset$. Conversely, if A is not a completely non-special semisimple ring, then $\mathfrak{D}\neq\emptyset$, and there exists an $M\in\mathfrak{M}$ with $M\in\mathfrak{D}$. Then M^* differs from zero by proposition 40.11, and M^* is a simple ring with unity element by proposition 40.9. □

PROPOSITION 40.15. A semisimple ring is completely non-special if and only if the annihilator of every modular maximal ideal is zero.

PROOF is clear by theorem 40.14 and proposition 40.9. □

REMARK 40.16. The rational number field is a special semisimple ring, and the ring of the rational integers is a completely non-special semisimple ring.

PROPOSITION 40.17. $I(\mathrm{Int}\,\overline{\mathfrak{D}}\cup(\mathfrak{M}\setminus\overline{\mathfrak{D}}))=(0)$.

PROOF. By the additivity of the closure operator we have

$$\overline{\mathrm{Int}\cup\overline{\mathfrak{D}}(\mathfrak{M}\setminus\overline{\mathfrak{D}})} = \overline{\mathfrak{M}\setminus(\mathfrak{M}\setminus\overline{\mathfrak{D}})}\cup\overline{\mathfrak{M}\setminus\overline{\mathfrak{D}}} \geqq (\mathfrak{M}\setminus(\overline{\mathfrak{M}\setminus\overline{\mathfrak{D}}}))\cup\overline{\mathfrak{M}\setminus\overline{\mathfrak{D}}} = \mathfrak{M}.\square$$

PROPOSITION 40.18. $\mathrm{Int}\,\overline{\mathfrak{D}}\geqq\mathfrak{D}$.

PROOF. By proposition 40.17 we have $\mathrm{Int}\,\overline{\mathfrak{D}}\cup(\mathfrak{M}\setminus\overline{\mathfrak{D}})=\mathfrak{D}$. Moreover, $(\mathfrak{M}\setminus\overline{\mathfrak{D}})\cap\mathfrak{D}\leqq(\mathfrak{M}\setminus\overline{\mathfrak{D}})\cap\overline{\mathfrak{D}}=\emptyset$ and thus it follows that $\mathrm{Int}\,\overline{\mathfrak{D}}\geqq\mathfrak{D}$. □

THEOREM 40.19. The closure $\overline{\mathfrak{D}}$ of the minimal set $\mathfrak{D}$ of a semisimple ring A is a closed domain in the structure space $\mathfrak{M}$ of A.

PROOF. By proposition 40.18 we have $\overline{\mathrm{Int}\,\overline{\mathfrak{D}}}\geqq\overline{\mathfrak{D}}$. Conversely $\mathrm{Int}\,\overline{\mathfrak{D}}\leqq\overline{\mathfrak{D}}$ implies $\overline{\mathrm{Int}\,\overline{\mathfrak{D}}}\leqq\overline{\mathfrak{D}}$. Therefore $\overline{\mathrm{Int}\,\overline{\mathfrak{D}}}=\overline{\mathfrak{D}}$. □

If A is a semisimple ring with structure space $\mathfrak{M}$ and minimal set $\mathfrak{D}$, then by the *special part* of A we mean the ideal $S(A)=I(\mathfrak{M}\setminus\overline{\mathfrak{D}})$ and by the *completely non-special part* of A we mean the ideal $N(A)=I(\mathfrak{D})$.

PROPOSITION 40.20. $(N(A))^*=S(A)$, $(S(A))^*=N(A)$ and thus $(N(A))^{**}=$ $=N(A)$ and $(S(A))^{**}=S(A)$.

PROOF. By theorems 40.6 and 40.19 we have $(I(\mathfrak{D}))^{**}=I(\mathfrak{D})$ and thus $(N(A))^{**}=$ $=N(A)$. Moreover $S(A)=I(\mathfrak{M}\setminus\overline{\mathfrak{D}})=(I(\mathfrak{D}))^*=(N(A))^*$. Since

$$(S(A))^* = (I(\mathfrak{M}\setminus\overline{\mathfrak{D}}))^* = I(\mathfrak{M}\setminus(\overline{\mathfrak{M}\setminus\overline{\mathfrak{D}}})) = I(\mathrm{Int}\;\overline{\mathfrak{D}}) = I(\overline{\mathrm{Int}\;\overline{\mathfrak{D}}}) = I(\overline{\mathfrak{D}})$$

we have $(S(A))^{**}=(I(\overline{\mathfrak{D}}))^*=I(\mathfrak{M}\setminus\overline{\mathfrak{D}})=S(A)$. □

THEOREM 40.21. The special part $S(A)$ of a semisimple ring A is a special ring, and the completely non-special part $N(A)$ of A is a completely non-special ring.

PROOF.

$$S(A)\cap I(\mathfrak{D}) = I(\mathfrak{M}\setminus\overline{\mathfrak{D}})\cap I(D) = (0),$$

and

$$N(A)\cap I(\mathfrak{D}) = I(\mathfrak{D})\cap I(\mathfrak{D}) = I(\mathfrak{D}) = N(A).$$

An application of theorem 40.12 completes the proof. □

THEOREM 40.22. If a representable ideal K of the semisimple ring A is a special ring, then $K \leqq S(A)$. If a representable ideal K of the semisimple ring A is a completely non-special ring, then $K \leqq N(A)$.

PROOF. If $K=I(\mathfrak{N})$ with $\mathfrak{N} \leqq \mathfrak{M}$ is a special ring, then $K \cap I(\mathfrak{D})=(0)$ by theorem 40.12. Therefore, we have $xy \in K \cap I(\mathfrak{D})=(0)$ for every $x \in K$ and $y \in I(\mathfrak{D})$. Thus $K \leqq (I(\mathfrak{D}))^* = I(\mathfrak{M} \setminus \overline{\mathfrak{D}}) = S(A)$. If $K=I(\mathfrak{N})$ with $\mathfrak{N} \leqq \mathfrak{M}$ is a completely non-special ring, then we have $K \cap I(\mathfrak{D})=K$ by theorem 40.12. Hence it follows that $K \leqq I(\mathfrak{D}) = N(A)$. □

THEOREM 40.23. A strongly semisimple ring A is the ring direct sum of its special part and its completely non-special part if and only if the closure $\overline{\mathfrak{D}}$ of the minimal set $\mathfrak{D}$ of A is an open set of the structure space $\mathfrak{M}$, i.e. if $\mathrm{Fr}\, \overline{\mathfrak{D}} = \emptyset$.

PROOF follows from theorem 40.7 and proposition 40.20. □

§ 41. A CLASSIFICATION OF THE SEMISIMPLE RINGS IN THE SENSE OF BROWN AND McCOY

In this section we develop some results of Suliński [5]. We discuss the factor ring of a semisimple ring with respect to its special part and its completely non-special part. An ordinal number which is called the type of the ring, is associated to every semisimple ring.

If A is a semisimple ring in the sense of Brown and McCoy, and $\mathfrak{M}$ its structure space, i.e. the set of all modular maximal ideals M of A, then we can embed A subdirectly into the complete direct sum $C(\mathfrak{M})$ of the simple rings $A_M = A/M$ with unity element. Hence every element $x \in A$ can be considered as a function $x(M)$ with $M \in \mathfrak{M}$ and $x(M) \in A/M$. For every $\mathfrak{N} \subseteq \mathfrak{M}$ we define a mapping $h_{\mathfrak{N}}$ of A by $h_{\mathfrak{N}}(x) = \bar{x}$ where

$$\bar{x}(M) = \begin{cases} x(M) & \text{if} \quad M \in \mathfrak{N}, \\ 0 & \text{if} \quad M \in \mathfrak{M} \setminus \mathfrak{N}. \end{cases}$$

$h_{\mathfrak{N}}$ obviously maps the ring A homomorphically onto a subring A' of $C(\mathfrak{N})$ and the kernel of the homomorphism $h_{\mathfrak{N}}$ is the ideal $I(\mathfrak{N})$ of A.

PROPOSITION 41.1. The ring $\bar{A} = AI/(\mathfrak{N})$ can be subdirectly embedded into the complete direct sum $C(\mathfrak{N})$.

PROOF. The kernel of $h_{\mathfrak{N}}$ is $I(\mathfrak{N})$, and the subring A' of $C(\mathfrak{N})$ is isomorphic to $\bar{A} = A/I(\mathfrak{N})$ with $h_{\mathfrak{N}}(A) = A'$. Hence $\bar{A}$ is subdirectly embedded into $C(\mathfrak{N})$. □

PROPOSITION 41.2. Let $\mathfrak{N} \leqq \mathfrak{M}$, $\mathfrak{N}_1 \leqq \overline{\mathfrak{N}}$, and let $I_{\mathfrak{N}}(\mathfrak{N}_1)$ be the intersection of those ideals $M/I(\mathfrak{N})$ of $A/I(\mathfrak{N})$ for which $M \in \mathfrak{N}_1$. Then $I_{\mathfrak{N}}(\mathfrak{N}_1) = I(\mathfrak{N}_1)/I(\mathfrak{N})$.

PROOF. We take $h_{\mathfrak{N}}(x)=\bar{x}\in I_{\mathfrak{N}}(\mathfrak{N}_1)$. Then we have $\bar{x}(M)=0$ for every $M\in\mathfrak{N}_1\leqq\overline{\mathfrak{N}}$, because $\bar{A}$ is subdirectly embedded into $C(\mathfrak{N})$. Hence the inverse image x of $\bar{x}$ is contained in $I(\mathfrak{N}_1)$. Therefore, $\bar{x}\in I(\mathfrak{N}_1)/I(\mathfrak{N})$. Conversely, we have $x\in I(\mathfrak{N}_1)$ and $\bar{x}(M)=0$ for every $M\in\mathfrak{N}_1$ and $\bar{x}\in I(\mathfrak{N}_1)/I(\mathfrak{N})$. Hence $x\in I_{\mathfrak{N}}(\mathfrak{N}_1)$. □

Let $D(\mathfrak{M})$ denote the discrete direct sum of the rings $A_M=A/M$ $(M\in\mathfrak{M})$. If the rings A_1 and A_2 are subdirectly embedded into the complete direct sum $C(\mathfrak{M})$, we say that A_1 and A_2 are *in a similar situation* in $C(\mathfrak{M})$ if $A_1\cap D(\mathfrak{M})=A_2\cap D(\mathfrak{M})$. Since the completely non-special semisimple rings have no simple ideal with unity element as we have seen, $D(\mathfrak{M})\cap A=(0)$ for every completely non-special ring subdirectly embedded into $C(\mathfrak{M})$ and thus two completely non-special rings are always embedded into $C(\mathfrak{M})$ in a similar situation.

PROPOSITION 41.3. If the special (completely non-special) semisimple rings A_1 and A_2 are subdirectly embedded into $C(\mathfrak{M})$ in a similar situation, and if $A_1\leqq A\leqq A_2$, then A is again a special (completely non-special) ring, and it is subdirectly embedded into $C(\mathfrak{M})$ in a similar situation as A_1 and A_2.

PROOF. $A_1\cap D(\mathfrak{M})\leqq A\cap D(\mathfrak{M})\leqq A_2\cap D(\mathfrak{M})$, and since $A_1\cap D(\mathfrak{M})=A_2\cap D(\mathfrak{M})$ we have

$$A\cap D(\mathfrak{M}) = A_1\cap D(\mathfrak{M}) = A_2\cap D(\mathfrak{M}).$$

Let $\mathfrak{D}$, $\mathfrak{D}_1$ and $\mathfrak{D}_2$ be the minimal sets of A, A_1 and A_2, respectively. Then $M\in\mathfrak{D}$ is equivalent to $M^*\neq(0)$ where M^* denotes the annihilator of M in A, as in the preceding section. Therefore, since $A\cap D(\mathfrak{M})=A_1\cap D(\mathfrak{M})=A_2\cap D(\mathfrak{M})$, $M\in\mathfrak{D}$ if and only if there exist modular maximal ideals M_i of A_i $(i=1, 2)$ satisfying $M_1\in\mathfrak{D}_1$ and $M_2\in\mathfrak{D}_2$, $M_1\leqq M\leqq M_2$ and $A_1/M_1\cong A/M\cong A_2/M_2$. If A_1 and A_2 are special rings, then we have $I_1(\mathfrak{D}_1)\leqq I(\mathfrak{D})\leqq I(\mathfrak{D}_2)=(0)$. Consequently $I(\mathfrak{D})=$ $=(0)$. But if A_1 and A_2 are completely non-special, then we obviously have $D=\emptyset$ since $\mathfrak{D}_1=\mathfrak{D}_2=\emptyset$. □

THEOREM 41.4. Let A be a semisimple ring, $\mathfrak{M}$ its structure space, $\mathfrak{D}$ the minimal set of A. Let $S(A)$ be the special part and $N(A)$ the completely non-special part of A. Finally, let $\operatorname{Int}\overline{\mathfrak{D}}=\mathfrak{M}\setminus(\overline{\mathfrak{M}\setminus\overline{\mathfrak{D}}})$. Then $A/N(A)$ is a special ring and it can be subdirectly embedded into $C(\operatorname{Int}\overline{\mathfrak{D}})$ in a similar situation as $S(A)$. Moreover, $A/S(A)$ is a completely non-special ring and it can be subdirectly embedded into $C(\mathfrak{M}\setminus\overline{\mathfrak{D}})$.

PROOF. By proposition 41.1, $A=A/N(A)$ and $A''=A/S(A)$ can be subdirectly embedded into $C(\operatorname{Int}\overline{\mathfrak{D}})$, and $C(\mathfrak{M}\setminus\overline{\mathfrak{D}})$, respectively. This is because $N(A)=$ $I(\overline{\mathfrak{D}})=I(\operatorname{Int}\overline{\mathfrak{D}})$ and $S(A)=I(M\setminus\overline{\mathfrak{D}})$, and $S(A)$ and $N(A)$ can be subdirectly embedded into $C(\operatorname{Int}\overline{\mathfrak{D}})$ and $C(\mathfrak{M}\setminus\overline{\mathfrak{D}})$, respectively, by proposition 40.3.

We now show that $A'=A/N(A)$ and $S(A)$ have similar situation at their subdirect embedding into $C(\operatorname{Int}\bar{\mathfrak{D}})$. For, if $M^*=(0)$, then $M\in\mathfrak{D}\subseteq\operatorname{Int}\bar{\mathfrak{D}}$ and thus $(M/N(A))^*\neq(0)$. Conversely, assume that $(M/N(A))^*\neq(0)$ for $M\supseteq N(A)=I(\bar{\mathfrak{D}})$. We have $(M/N(A))^*=(M/I(\bar{\mathfrak{D}}))^*=I_{\bar{\mathfrak{D}}}(\bar{\mathfrak{D}}\setminus\{M\})$ by proposition 41.2 and by the results of the preceding section. If $\bar{x}\in I_{\bar{\mathfrak{D}}}(\bar{\mathfrak{D}}\setminus\{M\})$ with $\bar{x}\neq 0$, then $x\in I(\bar{\mathfrak{D}}\setminus\{M\})$ where x is the inverse image of $\bar{x}$. If $M\notin\operatorname{Int}\bar{\mathfrak{D}}$, then we have $\bar{\mathfrak{D}}\setminus\{M\}\supseteq\operatorname{Int}\bar{\mathfrak{D}}$, and thus

$$x\in I(\bar{\mathfrak{D}}\setminus\{M\})\subseteq I(\operatorname{Int}\bar{\mathfrak{D}})=I(\bar{\mathfrak{D}})=N(A),$$

which contradicts the condition $\bar{x}\neq 0$. Hence $M\in\operatorname{Int}\bar{\mathfrak{D}}$. There exists an element $y\in S(A)$ with $y(M)=1_M\in A_M$, because the ring $S(A)=I(\mathfrak{M}\setminus\bar{\mathfrak{D}})$ is subdirectly embedded into $C(\operatorname{Int}\bar{\mathfrak{D}})$. Hence $xy(M)=\bar{x}\cdot\bar{1}\neq 0$, and since

$$xy\in I(\bar{\mathfrak{D}}\setminus\{M\})\cap I(\mathfrak{M}\setminus\bar{\mathfrak{D}})=I(\mathfrak{M}\setminus\{M\})=M^*$$

we have $M^*\neq(0)$.

By the preceding $\mathfrak{D}$ is also the minimal set of $A'=A/N(A)$. Hence $I_{\operatorname{Int}\bar{\mathfrak{D}}}(\mathfrak{D})=$ $=I(\mathfrak{D})/I(\operatorname{Int}\bar{\mathfrak{D}})=I(\bar{\mathfrak{D}})/I(\bar{\mathfrak{D}})=(0)$ by proposition 41.2, and thus A' is a special ring.

We shall now show that $A''=A/S(A)$ is completely non-special. We shall deduce a contradiction to the assumption $(M/S(A))^*\neq(0)$ for some $M\in\mathfrak{M}\setminus\bar{\mathfrak{D}}$. If x is an inverse image of an element $\bar{x}\in(M/S(A))^*=I_{\mathfrak{M}\setminus\bar{\mathfrak{D}}}(\mathfrak{M}\setminus\bar{\mathfrak{D}}\setminus\{M\})$ with $\bar{x}\neq 0$, then $x\in I(\mathfrak{M}\setminus\bar{\mathfrak{D}}\setminus\{M\})$. Since $N(A)$ is subdirectly embedded into $C(\mathfrak{M}\setminus\bar{\mathfrak{D}})$, there exists an element $y\in N(A)=I(\mathfrak{D})$ with $y(M)=1_M\in A_M$ and hence $xy\neq 0$. Now we have $xy\in I((\mathfrak{M}\setminus\mathfrak{D})\setminus\{M\})\cap I(\bar{\mathfrak{D}})=I(\mathfrak{M}\setminus\{M\})=M^*$. This yields the contradiction $M^*\neq(0)$, because M^* is a simple ideal with unity element and A is a completely non-special ring. □

Let A be a semisimple ring, $\mathfrak{M}$ its structure space, and $\mathfrak{D}$ the minimal set of A. Let $\mathfrak{M}'$ denote the *derived set* of $\mathfrak{M}$, i.e. the set of M, for which $M\in\overline{\mathfrak{M}\setminus\{M\}}$. Hence $\mathfrak{M}'$ consists precisely of the ideals of M with $M^*=(0)$. Then $\mathfrak{D}=\mathfrak{M}\setminus\mathfrak{M}'$.

Now we define the sets $\mathfrak{M}_\gamma$ and $\mathfrak{D}_\gamma$ for every ordinal number γ in the following way: take $\mathfrak{M}_0=\mathfrak{M}$, $\mathfrak{M}_1=\operatorname{Fr}\bar{\mathfrak{D}}_1$, $\mathfrak{D}_1=\mathfrak{D}$, where $\operatorname{Fr}\bar{\mathfrak{D}}_1=\bar{\mathfrak{D}}_1\cap\overline{\mathfrak{M}\setminus\bar{\mathfrak{D}}_1}$. Assume that $\mathfrak{D}_\beta$ and $\mathfrak{M}_\beta$ are defined for every ordinal number $\beta<\alpha$. If the ordinal number $\alpha-1$ exists, then we take $\mathfrak{D}_\alpha=\mathfrak{M}_{\alpha-1}\setminus\mathfrak{M}'_{\alpha-1}$, where $\mathfrak{M}'_{\alpha-1}$ denotes the derived set of $\mathfrak{M}_{\alpha-1}$, i.e. the set of $M\in\mathfrak{M}$, for which $M\in\overline{\mathfrak{M}_{\alpha-1}\setminus\{M\}}$. Moreover, we take $\mathfrak{M}_\alpha=\bar{\mathfrak{D}}_\alpha\cap\overline{\mathfrak{M}_{\alpha-1}\setminus\bar{\mathfrak{D}}_\alpha}$. But if α is a limit ordinal number, take $\mathfrak{M}_\alpha=\bigcap_{\beta<\alpha}\mathfrak{M}_\beta$ and $\mathfrak{D}_\alpha=\mathfrak{M}_\alpha\setminus\mathfrak{M}'_\alpha$, where $\mathfrak{M}'_\alpha$ again denotes the derived set of $\mathfrak{M}_\alpha$.

Thus we have two descending chains of subsets of $\mathfrak{M}$:

$$\mathfrak{M}_0\supseteq\mathfrak{M}_1\supseteq\mathfrak{M}_2\supseteq\dots$$

and

$$\bar{\mathfrak{D}}_1\supseteq\bar{\mathfrak{D}}_2\supseteq\bar{\mathfrak{D}}_3\supseteq\dots.$$

There exists an ordinal number τ with $\mathfrak{M}_\tau=\mathfrak{M}_{\tau+1}$. Then $\mathfrak{M}_{\tau+1}=\overline{\mathfrak{D}}_{\tau+1}\cap\overline{\mathfrak{M}\setminus\overline{\mathfrak{D}}_{\tau+1}}$. Hence we have $\overline{\mathfrak{D}}_{\tau-1}=\mathfrak{M}$ and thus $\mathfrak{M}_\tau=\emptyset$. □

The least ordinal number τ, for which $\mathfrak{M}_\tau=\emptyset$, is called the *type of the semisimple ring A*.

We remark that every set $\mathfrak{M}_\alpha$ $(\alpha<\tau)$ is closed in the structure space $\mathfrak{M}$ of A.

Proposition 41.5. Let A be a semisimple ring of the type τ. Then $\mathfrak{M}_\alpha$ is the structure space, and $\mathfrak{D}_{\alpha+1}$ is the minimal set of $A_\alpha=A/I(\mathfrak{M})$ for every $\alpha<\tau$. Moreover, $A'_\alpha=A_\alpha/N(A_\alpha)$ is a special ring and it can be subdirectly embedded into $C(\operatorname{Int}\overline{\mathfrak{D}}_{\alpha+1})$ in a similar situation as $S(A_n)$, where $\operatorname{Int}\overline{\mathfrak{D}}_{\alpha+1}=\mathfrak{M}_\alpha\setminus\overline{\mathfrak{M}_\alpha\setminus\overline{\mathfrak{D}}_{\alpha+1}}$ is valid. The ring $A''=A_\alpha/S(A_\alpha)$ is completely non-special and it can be subdirectly embedded into $C(\mathfrak{M}\setminus\overline{\mathfrak{D}}_{\alpha+1})$.

Proof follows from proposition 41.1 and theorem 41.4. □

Proposition 41.6. If A is a strongly semisimple ring, and $\mathfrak{M}$ is its structure space, then $I(\mathfrak{N}_1)+I(\mathfrak{N}_2)=I(\overline{\mathfrak{N}}_1\cap\overline{\mathfrak{N}}_2)$ for all $\mathfrak{N}_1, \mathfrak{N}_2\leqq\mathfrak{M}$.

Proof. Since A is strongly semisimple, there exists an $\mathfrak{N}\leqq\mathfrak{M}$ with $I(\mathfrak{N}_1)+$ $+I(\mathfrak{N}_2)=I(\mathfrak{N})$. If $M\in\overline{\mathfrak{N}}$, then $\mathfrak{M}\geqq I(\mathfrak{N})=I(\mathfrak{N}_1)+I(\mathfrak{N}_2)$. Therefore $M\geqq I(\mathfrak{N}_1)$ and $M\geqq I(\mathfrak{N}_2)$. Consequently $M\in\overline{\mathfrak{N}_1\cap\mathfrak{N}_2}$. Conversely, if $M\in\overline{\mathfrak{N}}_1\cap\overline{\mathfrak{N}}_2$, then we have $M\geqq I(\mathfrak{N}_1)$ and $M\geqq I(\mathfrak{N}_2)$. Hence $M\supseteqq I(\mathfrak{N}_1)+I(\mathfrak{N}_2)=I(\mathfrak{N})$ and thus $M\in\overline{\mathfrak{N}}$. □

Theorem 41.7. If $\mathfrak{D}$ is the minimal set of a strongly semisimple ring A and if we take $\mathrm{Fr}\,\overline{\mathfrak{D}}=\mathfrak{D}\cap\overline{\mathfrak{M}\setminus\overline{\mathfrak{D}}}$, then $I(\mathrm{Fr}\,\overline{\mathfrak{D}})=S(A)\oplus N(A)$. If the strongly semisimple ring A is of type 1, then

$$A = S(A)\oplus N(A).$$

Proof. $S(A)=I(\mathfrak{M}\setminus\overline{\mathfrak{D}})$ and $N(A)=I(\overline{\mathfrak{D}})$ holds. By proposition 41.6 we have $S(A)+N(A)=I((\mathfrak{M}\setminus\overline{\mathfrak{D}})\cap\overline{\mathfrak{D}})=I(\mathrm{Fr}\,\overline{\mathfrak{D}})$. Since $S(A)\cap N(A)=(I(\mathfrak{M}\setminus\overline{\mathfrak{D}})\cap I(\overline{\mathfrak{D}})=$ $=I(\mathfrak{M})=(0)$, the sum $S(A)+N(A)$ is a ring direct sum. □

Proposition 41.8. Every ideal K of a strongly semisimple ring A is again strongly semisimple.

Proof. There exists an $\mathfrak{N}\leqq\mathfrak{M}$ with $K=I(\mathfrak{N})$. Hence A/K is a subdirect sum of the simple rings A/M $(M\in\mathfrak{N})$ with unity element. Therefore, every homomorphic image A/K of A is semisimple. Since K/K^2 is a nilpotent ideal in the semisimple ring A/K^2, we have $K=K^2$. Hence every ideal K of A is idempotent. Now, if L is an ideal of the ring K, and $\bar{L}=L+LA+AL+ALA$ is the ideal generated by L in A, then since $\bar{L}\leqq K$, we have the relation

$$\bar{L} = (\bar{L})^3 \leqq K(L+LA+AL+ALA)K \leqq L \leqq \bar{L}.$$

Hence $L=\bar{L}$. Therefore, L is an ideal of A (cf. Andrunakievich [6]). Hence there exist $\mathfrak{N}_1 \leqq \mathfrak{M}$ with $L=I(\mathfrak{N}_1)$ and $\mathfrak{N}_2 \leqq \mathfrak{M}$ with $\mathfrak{N}_1 \geqq \mathfrak{N}_2$ such that $K=I(\mathfrak{N}_2)$. Since $K=I(\mathfrak{N}_2)$ can be subdirectly embedded into $C(\mathfrak{M}\setminus\overline{\mathfrak{N}}_1)$; $K\cap M$ is a maximal ideal of K for $M\in\mathfrak{M}\setminus\overline{\mathfrak{N}}_2$. But the intersection of all ideals $K\cap M$ with $M\in\mathfrak{N}_1\cap(\mathfrak{M}\setminus\overline{\mathfrak{N}}_2)$ is

$$K\cap I(\overline{\mathfrak{N}}_1\cap(\mathfrak{M}\setminus\mathfrak{N}_2)) = I(\mathfrak{N}_2)\cap I(\overline{\mathfrak{N}}_1\cap(\mathfrak{M}\setminus\overline{\mathfrak{N}}_2)) =$$
$$= I(\mathfrak{N}_2\cup(\mathfrak{N}_1\cap(\mathfrak{M}\setminus\mathfrak{N}_2))) = I(\mathfrak{N}_1) = L,$$

and thus every ideal L of the ring I is representable. □

A semisimple ring A is called *decomposable* if A is the ring direct sum of its special and completely non-special part.

THEOREM 41.9. If A is a strongly semisimple ring of type τ, then the ring $C_\alpha=I(\mathfrak{M}_{\alpha+1})/I(\mathfrak{M}_\alpha)$ is decomposable and strongly semisimple for every $\alpha<\tau$.

PROOF. We have seen that $\mathfrak{M}_\alpha$ is the structure space and $\mathfrak{D}_{\alpha+1}$ is the minimal set of $A_\alpha=/I(\mathfrak{M}_\alpha)$. Hence $C_\alpha=I(\mathfrak{M}_{\alpha+1})/I(\mathfrak{M}_\alpha)$ holds for $\mathfrak{M}_{\alpha+1}=\mathfrak{D}_{\alpha+1}\cap\overline{\mathfrak{M}_\alpha\setminus\mathfrak{D}_{\alpha+1}}$. Since $A_\alpha=A/I(\mathfrak{M}_\alpha)$ is also strongly semisimple, we have $C_\alpha=I_{\mathfrak{M}\alpha}(\mathfrak{M}_{\alpha+1})=I(\mathfrak{M}_{\alpha+1})/$ $/I(\mathfrak{M}_\alpha)=S(A_\alpha)\oplus N(A_\alpha)$ by theorem 41.7. Since $S(A)$ is a special ring and $N(A_\alpha)$ is a completely non-special ring, $S(A_\alpha)\leqq S(C_\alpha)$ and $N(A_\alpha)\leqq N(C_\alpha)$, and thus we also have $C_\alpha=S(C_\alpha)\oplus N(C_\alpha)$.

C_α is strongly semisimple because C_α is an ideal of the strongly semisimple ring $A_\alpha=A/I(\mathfrak{M}_\alpha)$.

If A is a strongly semisimple ring of the type τ, then the sequence of the decomposable and semisimple rings

$$C_0,\, C_1,\, C_2,\, \ldots,\, C_\alpha,\, \ldots \quad (\alpha<\tau) \qquad (*)$$

is called the *composition sequence* of the ring A, where

$$C_\alpha = I(\mathfrak{M}_{\alpha+1})/I(\mathfrak{M}_\alpha). \square$$

We have the following

THEOREM 41.10. Let A be a strongly semisimple ring of type τ and let $(*)$ be the composition sequence of A.

Then the following hold:

I. Every ring C_α is decomposable and strongly semisimple.

II. For every $\alpha<\tau$, there exists an ascending chain of special and strongly semisimple rings

$$S(C_\alpha) = S_{\alpha+1}(C_\alpha) \subset \ldots \subset S_\beta(C_\alpha) \subset \ldots \subset S_\tau(C_\alpha) = A' = A_\alpha/N(A_\alpha)$$

such that $S_\beta(C_\alpha)$ is an ideal of $S_{\beta+1}(C_\alpha)$ and $S_{\beta+1}(C_\alpha)/S_\beta(C_\alpha)$ is isomorphic to C_β $(\alpha<\beta<\tau)$.

III. For every $\alpha<\tau$, there exists an ascending chain of completely non-special and strongly semisimple rings

$$N(C_\alpha) = N_{\alpha+1}(C_\alpha) \subset \ldots \subset N_\beta(C_\alpha) \subset \ldots \subset N_\tau(C_\alpha) = A''_\alpha = A_\alpha/S(A_\alpha)$$

such that $N_\beta(C_\alpha)$ is an ideal of $N_{\beta+1}(C_\alpha)$ and $S_{\beta+1}(C_\alpha)/S_\beta(C_\alpha)$ is isomorphic to $C_\beta\,(\alpha<\beta<\tau)$.

IV. Every ring $S_\beta(C_\alpha)$ $(\alpha<\beta<\tau)$ is subdirectly embedded into $C(\operatorname{Int}\overline{\mathfrak{D}}_{\alpha+1})$ in a similar situation as $S(C_\alpha)$ and A'_α. Moreover, every ring $N_\beta(C_\alpha)$ $(\alpha<\beta<\tau)$ is subdirectly embedded into $C(\mathfrak{M}\setminus\overline{\mathfrak{D}}_{\alpha+1})$.

Proof. The validity of I is precisely the content of theorem 41.9.

Now we take $K_{\beta\alpha}=I(\mathfrak{M}_\beta)/I(\mathfrak{M}_\alpha)$ for $\alpha<\beta$. By proposition 41.2, $K_{\beta\alpha}=I_{\mathfrak{M}\alpha}(\mathfrak{M}_\beta)$ for $A_\alpha=A/I(\mathfrak{M}_\alpha)$. In particular, we have $K_{\alpha+1,\alpha}=C_\alpha$ and $K_{\tau\alpha}=I_{\mathfrak{M}\alpha}(\emptyset)=A_\alpha$. The ideals $K_{\beta\alpha}$ form an ascending chain for a fixed α. For, $\mathfrak{M}_{\beta_1}>\mathfrak{M}_{\beta_2}$ for $\alpha<\beta_1<\beta_2<\tau$, and thus $I_{\mathfrak{M}_\alpha}(\mathfrak{M}_{\beta_1})<I_{\mathfrak{M}_\alpha}(\mathfrak{M}_{\beta_2})$. In particular, we obtain $K_{\beta\alpha}=I_{\mathfrak{M}_\alpha}(\mathfrak{M}_\beta)\geqq I_{\mathfrak{M}_\alpha}(\mathfrak{M}_{\alpha+1})=$ $=C_\alpha=S(C_\alpha)\oplus N(C_\alpha)$, and hence we can form the factor rings $S_\beta(C_\alpha)=K_{\beta\alpha}/N(C_\alpha)$ and $N_\beta(C_\alpha)=K_{\beta\alpha}/S(C)_\alpha$.

$$S_{\alpha+1}(C_\alpha) = (S(C_\alpha)\oplus N(C_\alpha))/N(C) \cong S(C_\alpha)$$

and $S_\tau(C_\alpha)=A_\alpha/N(C_\alpha)=A'_\alpha$. Analogously we have

$$N_{\alpha+1}(C_\alpha) \cong (S(C_\alpha)\oplus N(C_\alpha))/S(C_\alpha) \cong N(C_\alpha)$$

and $N_\tau(C_\alpha)\cong A_\alpha/S(C_\alpha)=A''_\alpha$.

Furthermore, we have

$$S_{\beta+1}(C_\alpha)/S_\beta(C_\alpha) \cong (K_{\beta+1,\alpha}/N(C_\alpha))/(K_{\beta\alpha}/N(C_\alpha)) \cong$$
$$\cong K_{\beta+1,\alpha}/K_{\beta\alpha} \cong I(\mathfrak{M}_{\beta+1})/I(\mathfrak{M}_\beta) = C_\beta,$$

and, similarly, also $N_{\beta+1}(C_\alpha)/N_\beta(C_\alpha)\cong C_\beta$.

Since $S(A_\alpha)=S(C_\alpha)\leqq S_\beta(C'_\alpha)=A'_\alpha$ and $N(A_\alpha)=N(C_\alpha)\leqq N_\beta(C_\alpha)\leqq A''_\alpha$ we have by theorem 41.4 and propositions 41.3 and 41.8, that every ring $S_\beta(C_\alpha)$ is special, strongly semisimple and subdirectly embedded into $C(\operatorname{Int}\overline{\mathfrak{D}}_{\alpha+1})$ in a similar situation as $S(A_\alpha)$ and $\bar{A}'_\alpha$. Moreover, every ring $N_\beta(C_\alpha)$ is completely non-special, strongly semisimple and subdirectly embedded into the complete direct sum $C(\mathfrak{M}_\alpha\setminus\overline{\mathfrak{D}}_{\alpha+1})$. □

Remark 41.11. We mention the following result of Suliński without proof: For every sequence (∗) of rings C_α satisfying conditions I, II, III and IV of theorem 41.10, there exists a strongly semisimple ring A of type τ such that the sequence (∗) coincides with the composition sequence of the ring A. Hence the theory of Suliński, discussed in sections 40 and 41, appears to play a role in ring theory similar to that of the classical Prüfer–Ulm–Zippin theory of countable primary Abelian groups (cf. Kurosh [1], and Fuchs [1]).

§ 42. BROWN–McCOY RADICAL RINGS WITH A MODULE-THEORETICAL PROPERTY

For the subject-matter of this section we refer the reader to the papers [15, 21] of Szász.

As is well known, a ring A is a *Brown–McCoy radical ring* if and only if A cannot be homomorphically mapped onto a non-zero simple ring with unity element.

Kertész has raised the following question [1] (p. 97, problem 1):

Does a ring A necessarily possess a unity element if the maximal trivial submodule of an arbitrary A-module is always a direct summand?

By a *trivial A-submodule* M_0 of the A-module M here we mean a submodule satisfying the condition $M_0A=(0)$.

In this connection Szász [15] has studied six classes of rings, the E_i-rings (for $i=0, 1, 2, 3, 4, 5$). The E_2-rings and E_5-rings have been discussed in detail by Szász. He has also given some criteria for the existence of the unity element in a ring by means of the E_2-rings and the E_5-rings.

A ring A is called an *E_2-ring,* if the maximal trivial A-submodule M_0 of every right A-module M is a direct summand of M.

A ring A is called an *E_5-ring,* if no homomorphic image of A has a non-zero left annihilator.

Every ring with unity element is an E_2-ring, we have this from the Pierce decomposition of the A-module M. Rings with unity element are also E_5-rings, but not Brown–McCoy radical rings. Later we shall show that every E_2-ring is also an E_5-ring but not a Jacobson radical ring. Szász has raised the following problem in [15]:

Does there exist a non-zero E_2-ring, which is a radical ring in the sense of Brown and McCoy?

The aim of this section is to examine this problem. The existence of these rings is not proved; however some of their properties are deduced under the assumption of their existence.

We need some preliminaries.

THEOREM 42.1 (cf. Szász [15]). (1) *Every homomorphic image A' of an E_5-ring A is again an E_5-ring.* (2) *$L \leqq LA$ for every left ideal L and $a \in aA+AaA$ for every element $a \in A$, of an E_5-ring A.*

PROOF. (1) Since every homomorphic image A'' of A' is also a homomorphic image of the E_5-ring A, the ring A'' has no non-zero left annihilator and thus A' is an E_5-ring. (2) The product LA is an ideal of A for every left ideal L, and since $l+LA$ is contained in the left annihilator of the E_5-ring A/LA for every $l \in L$, we have $l \in LA$ for every $l \in L$ and thus $L \leqq LA$. If, in particular, L is the principal left ideal $(a)_l$ generated by a in A, then $(a)_l \leqq (a)_l A$ by the preceding. Therefore $a \in aA + + AaA$. □

From the definition of E_2-rings it follows almost trivially that every homomorphic image A' of an E_2-ring A is also an E_2-ring, since every A'-module can be considered as an A-module. With the methods used in the proof of theorem 2.3.2 of Szász [15] the following result of Michler can be proved (communication in a letter):

PROPOSITION 42.2 (Michler). *$a \in aA$ for every element a of an E_2-ring A.*

PROOF. If $a \notin aA$ holds an element $a \in A$ of an E_2-ring, then by Zorn's lemma there exists a maximal right ideal R of A such that $a \notin R$ and $R \geqq aA$. Then the right A-module A/R is subdirectly irreducible, and thus A/R has a unique minimal right A-module $M_1/R = ((a)_r + R)/R$. Since $aA \leqq R$, M_1/R is contained in the maximal trivial A-submodule M_0/R of A/R. Since A is an E_2-ring, $A/R = M_0/R \oplus M_2/R$ for some M_2. Since A/R is subdirectly irreducible, we obtain $A = M_0$, and thus $M_0 A \leqq A^2 \leqq R$. But by theorem 42.1.2 we have $A = A^2$ (with $L = A$). This contradiction proves $a \in aA$ for every $a \in A$. □

THEOREM 42.3 (cf. Szász [15]). *(1) Every E_2-ring is an E_5-ring. (2) A non-zero E_2-ring A, cannot be a Jacobson radical ring.*

PROOF. (1) If l is a left annihilator of the E_2-ring A, then we have $l \in lA = (0)$ by theorem 42.2. Consequently $l = 0$. Hence A has no non-zero left annihilator, and since the homomorphic images of A are also E_2-rings, A is an E_5-ring. (2) By theorem 42.2 $a \in aA$ for every $a \in A$, and thus A is not a Jacobson radical ring if A is an E_2-ring. For, if we have $a = ab$ and $b + c - bc = 0$, then we obtain

$$a = a - a(b + c - bc) = (a - ab) - (a - ab)c = 0$$

for every $a \in A$. Consequently $A = (0)$. □

A non-zero element a of a ring A is said to be a *left multiplicator* if there exists a rational integer n such that $ax = nx$ for every $x \in A$.

THEOREM 42.4 (Szász [15]). *A ring A has a unity element if and only if A is an E_2-ring containing a left multiplicator.*

PROOF. If e is the unity element of a ring A, and M is a right A-module, then we have $M = M(1-e) \oplus Me$ with $M(1-e)A = (0)$, $MeA \leqq MA \leqq M$, $Me \geqq MeA$. Hence, $M(1-e)$ and Me are in fact right A-modules, and $M(1-e)$ is the maximal trivial A-submodule of M. Moreover, e is a left multiplicator of A.

Conversely, let A be an E_2-ring containing a left multiplicator $b \neq 0$. Then there exists a rational integer n with $(b+n)A=(0)$. Let us consider the canonical ring extension A_1 of A with unity element, if A has no unity element. Then the pair (b, n) is contained in the maximal trivial A-submodule A_0 of the right A-module A_1. Since A is left annihilator free by theorem 42.3, for every n there exists at most one $b \in A$ (eventually none) with $(b, n) \in A_0$. Now, let $(a, m) \in A_0$ be such that the absolute value $|m|$ is minimal in the set of $|n|$ with $(b, n) \in A_0$. Since A is left annihilator free, $|m| \neq 0$, and we may obviously assume $m > 0$. Now, if $n = mq + r$ with $0 \leqq r < m$ for an n from an arbitrary pair $(b, n) \in A_0$, then, we have $(b-qa, r) \in A_0$ since $(b-qa)x = bx - qax = -nx + qmx = -rx$ for every $x \in A$, which contradicts the minimality of m if $r \neq 0$. Therefore $r=0$. Hence $(b-qa)A=(0)$ and since A is left annihilator free by theorem 42.3, we have $b=qa$. Therefore $(b, n) = q(a, m)$. Accordingly, the additive group A_0^+ is cyclic and obviously infinite. Since A is an E_2-ring, there exists an A-submodule A_2 of A_1 with $A_1 = A_0 \oplus A_2$. Hence

$$(0, 1) = k(a, m) + (-ka, 1-km)$$

with $k(a, m) \in A_0$ and $(-ka, 1-km) \in A_2$. Since $A_0 \neq (0)$ we have $k \neq 0$. Also since

$$(x, 0) = (0, 1)x = (-ka, 1-km)x \in A_2$$

for every $x \in A$, we have $(0, 1-km) \in A_2$ because $(-ka, 0) \in A_2$. Since $m(0, 1-km) \in$ $\in A_2$ it follows that

$$(1-km)(a, m) = ((1-km)a, 0) + (0, m(1-km)) \in A_0 \cap A_2.$$

Therefore $1-km=0$, because (a, m) generates the infinite cyclic group A_0^+. Since $m>0$ we have $k=m=1$, and hence $e=-a$ is a left unity element of the ring A, because $ax+x=0$ for every $x \in A$. Furthermore, since every element of $A(1-e)$ is a left annihilator of A, and A is left annihilator free, e is a two-sided unity element of A. □

REMARK. For further criteria, for the existence of the unity element in a ring, formulated by means of E_2-rings, or E_5-rings, we refer to the paper [15] of Szász. $a \in aA$ obviously holds for every $a \in A$ in a ring A with a right unity element. Moreover, a simple Jacobson radical ring is an E_5-ring, in which $a \in aA$ only for $a=0$. The ring A of all linear transformations of finite rank of a vector space of infinite dimension over a division ring is a non-zero Brown–McCoy radical ring which is also regular in the sense of von Neumann. Hence $a \in aA$ for every element a of A.

Now we shall demonstrate some properties of non-zero Brown–McCoy radical rings which are also E_2-rings. The existence of such rings has been recently disproven by E. Kiss. We have the following

THEOREM 42.5 (Szász [21]). If there exists a non-zero Brown–McCoy radical ring A which is also an E_2-ring, then A has the following properties:

(1) A can be chosen a primitive ring (a left primitive ring, respectively), consequently a prime ring.

(2) Every element a of A is a left divisor of zero in A; if A is in addition a prime ring, then $axa=0$ with $xa\neq 0$, and thus every non-zero left ideal has a non-zero nilpotent element.

(3) $A(1-a)A=A$ for every $a\in A$.

(4) For every non-zero element $a\in A$ there exists an element $b\in A$ with $ab\notin Aa$, and hence A has a trivial centre; moreover, $Aa\neq A$ for every $a\in A$.

(5) $a\in(a+n)A+A(a+n)A$ for every $a\in A$ and for every rational integer n.

(6) There exist maximal left ideals in A, and $LA=A$ for every maximal left ideal L of A.

(7) The maximum condition does not hold for the left ideals of the form $L_a=\{x;\ x\in A,\ xa=x\}$ for every $a\in A$, and thus A is not a left Noetherian ring.

REMARK. According to property (2), for every $a\in A$ the left ideal Aa is sufficiently large in the sense that Aa contains a right divisor of zero $b=ba\neq 0$ with $ab=0$. But, according to property (4), the left ideal Aa is also sufficiently small in the sense that Aa does not contain all products ab.

PROOF. (1) If A is a Brown–McCoy radical ring and also an E_2-ring, then every homomorphic image inherits both these properties. Moreover, by theorem 42.3, A is not a Jacobson radical ring and thus $A\neq \mathbf{J}(A)$. Then the factor ring $A/\mathbf{J}(A)$ is a subdirect sum of primitive rings $S_\alpha=A/P_\alpha$ (respectively of left primitive rings $T_\beta=A/P'_\beta$). The rings S_α and T_β are prime rings, Brown–McCoy radical rings and also E_2-rings. Hence A can be chosen to be an S_α, or a T_β respectively.

(2) If an element $a\in A$ is not a left divisor of zero in A, then we get a contradiction as follows. By theorem 42.2, there exists an element $b\in A$ with $a=ab$. Then $a(1-b)A=(0)$. Since a is not a left divisor of zero in A, we have $(1-b)A=(0)$, and thus b is a left unity element of A. If M is a maximal ideal of A such that $b\notin M$, then A/M is a simple ring with unity element because $(A(1-b))^2=(0)$. This contradicts the definition of Brown–McCoy radical rings. Hence every element of A is a left divisor of zero.

If, in particular, A is a prime ring, then we have $cAd\neq(0)$, for every $c\neq 0$, $d\neq 0$ ($c, d\in A$). Consequently, we have $cA\cap Ad\neq(0)$ for arbitrary non-zero elements c and d of A. Moreover, by property (1), for every non-zero element $a\in A$ there exists a non-zero element $b\in A$ with $ab=0$. Therefore by the preceding there exist elements $x\in A$, $y\in A$ with $0\neq xa=by\in bA\cap Aa$. Therefore, we have $axa=0$ and $(xa)^2=0$.

(3) By theorem 42.1.2, we have $A(1-a)\subseteq A(1-a)A$ for every element $a\in A$, and thus $A/A(1-a)A$ is a ring with a right unity element $a+A(1-a)A$. If $a\notin A(1-a)A$, and M is an ideal maximal with respect to $a\notin M$ and $M\supseteq A(1-a)A$

(one such exists by Zorn's lemma), then A/M is a simple ring with unity element because $((1-a)A)^2 \leq M$, which contradicts the definition of Brown–McCoy radical rings. Therefore $a \in A(1-a)A$. Since $A(1-a) \leq A(1-a)A$ it follows that $A(1-a)A = A$.

(4) If there exists an element $a \in A$ with $ab \in Aa$ for every $b \in A$, then we have $aA = Aa$. By theorem 42.2, there exists an element $c \in A$ with $a = ac$, and by property (3) we have $A = A(1-c)A$. Consequently,

$$aA = a(A(1-c)A) \leq Aa(1-c)A = A \cdot 0 \cdot A = (0)$$

and thus $a=0$ by theorem 42.3.1. Therefore, if $a \neq 0$ we certainly have $aA \nsubseteq Aa$, and hence A has trivial centre. Moreover, $Aa \neq A$ holds for every $a \in A$.

(5) If there exists a non-zero element a in A and a rational integer n with

$$a \notin B = (a+n)A + A(a+n)A,$$

then $a+B$ is a left multiplicator in the factor ring A/B. Since A/B is also an E_2-ring, theorem 42.4 implies that A/B has a unity element and thus A/B has a maximal ideal M/B. But then A/M is a simple ring with unity element which is a contradiction. Therefore $a \in (a+n)A + A(a+n)A$.

(6) If A has no proper maximal left ideal, then $A = \Phi_l$ for the Frattini left ideal Φ_l of A. Since $\mathbf{J}(A) = \{x;\ x \in A,\ xA = \Phi_l\}$ we have $\mathbf{J}(A) = A$ for the Jacobson radical $\mathbf{J}(A)$ of A. This is impossible by theorem 42.3.2. Therefore A has a maximal left ideal L, for which $L \leq LA$, by theorem 42.1.2. If $L = LA$, L is a two-sided ideal, and then A/L has only trivial left ideals. Thus, in this case A/L is either a division ring or a zero-ring. But the first subcase contradicts the definition of Brown–McCoy radical rings and the second subcase contradicts the fact that $A^2 = A$ by theorem 42.1.2. Therefore $LA = A$ for every maximal left ideal L of A.

(7) By property (3), $A(1-b)A = A$ and thus $A(1-b) \neq (0)$ by theorem 42.3. Consequently, the left ideal

$$L_b = \{x;\ x \in A,\ xb = x\}$$

is a proper left ideal of A for every $b \in A$. Hence for every element $b \in A$, there exists an element $a \in A$ such that $a(1-b) \neq 0$. By theorem 42.2 there exists an element $c \in A$ with $a(1-b) = a(1-b)c$. Consequently with $a(1-b \circ c) = 0$ and $a \in L_{b \circ c}$, where $b \circ c = b + c - bc$. Since $L_b \leq L_{b \circ c}$ and $a \in L_b$, we have $L_b < L_{b \circ c}$ for some $c \in A$. Thus the maximum condition does not hold for left ideals of the form L_b. □

PROBLEM 58.† Does there exist a Brown–McCoy radical ring A, for which the maximal trivial A-submodule M_0 of every right A-module M is a direct summand of M? In addition can $aA = A$ for an element $a \in A$ occur?

PROBLEM 59. Let $\mathfrak{m}$ be an arbitrary cardinal number, A a ring, M a right A-module and K an A-submodule of M. Then K is called $\mathfrak{m}$-*homoperfect* in M, if the following four conditions are fulfilled:

(a) $MA+K=M$.

(b) M/K is completely reducible with the A-rank$<\mathfrak{m}$.

(c) M/K does not have a non-trivial A-submodule invariant under all A-endomorphisms of M/K.

(d) If φ is an A-homomorphism of a right A-module M/L onto M/K such that the conditions (a), (b) and (c) are fulfilled for L, then φ is an isomorphism.

Let $\mathbf{R}_{\mathfrak{m}}(M)$ be the intersection of all $\mathfrak{m}$-homoperfect A-submodules K of M.
Find the necessary and sufficient conditions for the equality $\mathbf{R}_{\aleph_0}(A)=\mathbf{G}(A)$, where A is considered as a right A-module and $\mathbf{G}(A)$ is the Brown–McCoy radical.

PROBLEM 60. Find the necessary and sufficient conditions for the equality $\mathbf{R}_2(A)=\mathbf{J}(A)$, where A is a ring, considered as a right A-module and $\mathbf{J}(A)$ is the Jacobson radical of A. (For $\mathbf{R}_{\mathfrak{m}}(M)$ see problem 59. Besides, in every right A-module M, $\mathbf{R}_2(M)$ coincides with $\mathbf{K}(M)$, cf. problem 57.)

PROBLEM 61. Investigate those rings, in which every subring is a Brown–McCoy radical ring.

PROBLEM 62. Call the $(\mathbf{F}, \Omega_1, \Omega)$-radical determined in the sense of section 36 by the ideals $\mathbf{F}(a)=(1-a)a\cdot I[a]$ *the polynomial radical of the ring* A, where $I[a]$ denotes the ring of polynomials in the indeterminate a. Denote this radical by $\mathbf{P}(A)$. Then obviously $\mathbf{U}(A)\leqq\mathbf{P}(A)\leqq\mathbf{J}(A)$ for the upper nil radical $\mathbf{U}(A)$ and the Jacobson radical $\mathbf{J}(A)$ of an arbitrary ring A. Is every polynomial radical ring with maximum condition for right ideals nilpotent?

CHAPTER VI

FURTHER CONCRETE RADICALS AND ZEROID-PSEUDORADICALS

In this chapter we shall show by means of $(\mathbf{F}, \Omega_1, \Omega)$-radicals that in a ring the maximal von Neumann regular ideal is a radical. As a two-sided analogue, we consider the maximal biregular ideal of a ring. In particular, the maximal von Neumann regular ideal, and the maximal biregular ideal of a full matrix ring will be discussed. In a ring A, whose factor ring $A/\mathbf{J}(A)$ with respect to the Jacobson radical $\mathbf{J}(A)$ is von Neumann regular, we deduce a criterion for the maximal von Neumann regular ideal $\mathbf{M}(A)$ of A to be (0). The Fuchsian zeroid-pseudoradical and its lattice-theoretic generalizations (due to Szász) will be treated in detail. For the definition of the *zeroid element* see Clifford–Miller [1]. A zeroid-pseudoradical is not in general a radical in the sense of Amitsur and Kurosh, but it is always an intersection of prime ideals. We first deal with lattice ordered groupoids. For ring-theoretic applications of the results we use the notions of blocks, weak blocks etc.

§ 43. THE MAXIMAL VON NEUMANN REGULAR IDEAL AS A RADICAL

For the subject-matter of this section we refer the reader to Brown and McCoy [3].

We remark that the *regularity of a ring A* in the sense of von Neumann is equivalent to the following condition: $a \in aAa$ for every element $a \in A$. An ideal-theoretic characterization of regular rings is due to Kovács [1]. A ring A is called *strongly regular* if for every $a \in A$ there exists an $x \in A$ with $a = a^2x$. Every strongly regular ring is von Neumann regular. For strongly regular rings we refer the reader e.g. to Szász [32], cf. also Lajos–Szász [1, 4].

Now, if A is an arbitrary ring, $\mathbf{F}(a) = aAa$, and Ω is the set of all right and left multiplications together with the identity automorphism of A^+, then the following statements hold:

(P_1) $\mathbf{F}(a+b) \subseteqq \mathbf{F}(a) + (b)$

(P_2) $b \in \mathbf{F}(a)$ implies $\mathbf{F}(a+b) \leqq \mathbf{F}(a)$.

Hence A is an $(\mathbf{F}, \Omega)$-group, in which the ideals are the Ω-subgroups (cf. § 36). Consequently, by the general results of section 36, the maximal von Neumann regular ideal $\mathbf{M}(A)$ of A exists, and $\mathbf{M}(A/\mathbf{M}(A))=(0)$. The von Neumann regular rings form a radical class.

THEOREM 43.1 The maximal von Neumann regular ideal $\mathbf{M}(A)$ of the ring A exists and is a hereditary radical.

PROOF. By the above, it is sufficient to show that $\mathbf{M}(A)$ is hereditary. But this follows from the fact that every ideal I of a von Neumann regular ring A is von Neumann regular. For, if $i \in I$, then there exists $a \in A$ with $i=iai$. Hence $(ai)^2=ai$ and thus $i=i(ai)^2=i(aia)i$ with $aia \in I$. □

PROPOSITION 43.2 If A is a von Neumann regular ring, then the full matrix ring A_n is again von Neumann regular.

PROOF. First consider the case $n=2$. For $a \in A$ we denote by a' an element $a' \in A$ such that $a=aa'a$. Now let

$$M_1 = \begin{pmatrix} a & b \\ c & d \end{pmatrix}$$

be an arbitrary element of the full matrix ring A_2. If we take

$$M_2 = \begin{pmatrix} 0 & 0 \\ b' & 0 \end{pmatrix}$$

and

$$M_3 = M_1 - M_1 M_2 M_1,$$

then it follows by a straightforward calculation that M_3 has the form

$$M_3 = \begin{pmatrix} g & 0 \\ h & i \end{pmatrix}.$$

Now, if

$$M_4 = \begin{pmatrix} g' & 0 \\ 0 & i \end{pmatrix}$$

and

$$M_5 = M_3 - M_3 M_4 M_3,$$

then M_5 has the form

$$M_5 = \begin{pmatrix} 0 & 0 \\ k & 0 \end{pmatrix}.$$

If finally

$$M_6 = \begin{pmatrix} 0 & k' \\ 0 & 0 \end{pmatrix},$$

then we have

$$M_5 - M_5 M_6 M_5 = 0.$$

A simple calculation shows that, in a ring A,

$$a - aya = (a - aya)z(a - aya)$$

implies $a = axa$ with $x = z - zay - yaz + yazay + y$.

Hence, since $M_5 = M_5 M_6 M_5$ and $M_5 = M_3 - M_3 M_4 M_3$, it follows that

$$M_3 = M_3 M_7 M_3$$

for some matrix $M_7 \in A_2$. Accordingly, because $M_3 = M_1 - M_1 M_2 M_1$ we also have

$$M_1 = M_1 M_8 M_1$$

for some matrix $M_8 \in A_2$. Thus proposition 43.2 is proved for $n=2$.

Because $(A_2)_2 \cong A_4$, A_4 and, similarly A_{2^k} are von Neumann regular for every positive integer k. For an arbitrary natural number n we choose k such that $2^k \geqq n$. If $M \in A_n$, then let $\overline{M}$ denote the matrix of A_{2^k} with M in the upper left corner, and zeros elsewhere. Since A_{2^k} is von Neumann regular, there exists a matrix $\overline{X} \in A_{2^k}$ with

$$M = \overline{M} \cdot \overline{X} \cdot \overline{M}.$$

Hence $M = MXM$. □

THEOREM 43.3. The maximal von Neumann regular ideal $\mathbf{M}(A_n)$ of the full matrix ring over a ring A is $(\mathbf{M}(A))_n$.

PROOF. $(\mathbf{M}(A))_n \leqq \mathbf{M}(A_n)$ by proposition 43.2. Conversely, if M is a matrix of $\mathbf{M}(A_n)$ and m_{ij} is an element of $\mathbf{M}(m_{ij} \in A)$, then the principal ideal (M) is von Neumann regular in A_n, and thus there exists a matrix $Y \in A_n$ with $M = MYM = MYMYM$. Hence we have

$$m_{ij} = \sum_{p,q} t_{pq} a_{pq} s_{pq}$$

with suitable elements $t_{pq}, s_{pq} \in A$. It can be immediately shown that in (M) there exists a matrix with $t_{pq} a_{pq} s_{pq}$ in the (1, 1)-th place and 0's elsewhere (cf. Brown and McCoy [3], lemma 5). Now if m is an arbitrary element of the principal ideal (m_{ij}) of A, then there exists an element $\hat{M}$ of (M) with m in the (1, 1)-th place and 0's elsewhere. Moreover, we have $\hat{M} = \hat{M} N \hat{M}$ with a matrix $N \in A_n$, because (M) is von Neumann regular. But this implies $m = mn_{11}m$, and thus m is regular. Consequently, $m_{ij} \in \mathbf{M}(A)$. Therefore $M \in (\mathbf{M}(A))_n$ and thus $\mathbf{M}(A_n) \leqq (\mathbf{M}(A))_n$. Hence, we have $\mathbf{M}(A_n) = (\mathbf{M}(A))_n$. □

THEOREM 43.4. If $\mathbf{M}(A)$ is the maximal von Neumann regular ideal, and $\mathbf{J}(A)$ is the Jacobson radical of a ring A, then the following conditions hold:

$$\mathbf{M}(A)\cap\mathbf{J}(A) = (0), \quad \mathbf{J}(A) \subseteq (\mathbf{M}(A))^*, \quad \mathbf{M}(A) \subseteq (\mathbf{J}(A))^*,$$

and

$$\mathbf{M}(A)\cap(\mathbf{M}(A))^* = (0),$$

where $*$ denotes the annihilator. Moreover, $\mathbf{J}(A)$ is the Jacobson radical of $(\mathbf{M}(A))^*$ and $\mathbf{M}(A)$ is the maximal von Neumann regular ideal of $(\mathbf{J}(A))^*$.

PROOF. Since $\mathbf{J}(A)$ contains no non-zero idempotent element, and $a=axa$ implies $(ax)^2=ax$ holds, we have $\mathbf{J}(A)\cap\mathbf{M}(A)=(0)$. Hence $\mathbf{J}(A)\cdot\mathbf{M}(A)=\mathbf{M}(A)\cdot\mathbf{J}(A)=(0)$ holds and thus $\mathbf{J}(A)\subseteq(\mathbf{M}(A))^*$ and $\mathbf{M}(A)\subseteq(\mathbf{J}(A))^*$. If $a\in\mathbf{M}(A)\cap(\mathbf{M}(A))^*$, then there exists $x\in A$ with $a=axa$. Because of $xa\in(\mathbf{M}(A))^*$ we have $a=a(xa)=0$. Now $\mathbf{J}((\mathbf{M}(A))^*)=\mathbf{J}(A)\cap(\mathbf{M}(A))^*$ and $\mathbf{J}(A)\subseteq(\mathbf{M}(A))^*$ imply $\mathbf{J}(A)=\mathbf{J}((\mathbf{M}(A))^*)$. Analogously we have $\mathbf{M}(A)=\mathbf{M}((\mathbf{J}(A))^*)$. □

THEOREM 43.5. Let A be a ring with Jacobson radical $\mathbf{J}(A)$, such that $A/\mathbf{J}(A)$ is von Neumann regular. Then $\mathbf{M}(A)=(0)$ if and only if the annihilator $(\mathbf{J}(A))^*$ of $\mathbf{J}(A)$ is contained in $\mathbf{J}(A)$.

PROOF. If $(\mathbf{J}(A))^*\subseteq\mathbf{J}(A)$, then, by theorem 43.4, we have $\mathbf{M}(A)\subseteq(\mathbf{J}(A))^*$, and $\mathbf{M}(A)\subseteq(\mathbf{J}(A))^*\subseteq\mathbf{J}(A)$ implies $\mathbf{M}(A)=(0)$.

Conversely, suppose $\mathbf{M}(A)=(0)$ and let $A/\mathbf{J}(A)$ be von Neumann regular. We shall first show that $\mathbf{J}(A)\cap((\mathbf{J}(A))^*)^2=(0)$ by induction. Assume that there exists an element $j\in\mathbf{J}(A)$ with

$$j = \sum_{i=1}^{n} a_i b_i \quad (a_i,\ b_i\in(\mathbf{J}(A))^*).$$

Since $A/\mathbf{J}(A)$ is von Neumann regular, there exist elements $x_i\in A$ with $a_i-a_ix_ia_i=j_i\in\mathbf{J}(A)$. Hence, because $b_i\in(\mathbf{J}(A))^*$, we obtain the following equation:

$$j = \sum_{i=1}^{n} (a_ix_ia_i+j_i)b_i = \sum_{i=1}^{n} a_ix_ia_ib_i. \tag{1}$$

If $n=1$, then we have $j=a_1x_1a_1b_1=a_1x_1j=0$, because $j=a_1b_1$ and $a_1\in(\mathbf{J}(A))^*$. If $n\neq1$, then we have

$$a_nb_n = j-\sum_{i=1}^{n-1} a_ib_i.$$

Hence (1) implies

$$j = \sum_{i=1}^{n-1} a_ix_ia_ib_i+a_nx_n\left(j-\sum_{i=1}^{n-1} a_ib_i\right) = \sum_{i=1}^{n-1} (a_ix_i-a_nx_n)a_ib_i.$$

By the induction assumption

$$j = \sum_{i=1}^{n-1} c_i d_i \quad \text{and} \quad c_i,\, d_i \in (\mathbf{J}(A))^*$$

implies $j=0$, and this can be applied here because $(a_i x_i - a_n x_n)a_i \in (\mathbf{J}(A))^*$ and $b_i \in (\mathbf{J}(A))^*$. Thus $\mathbf{J}(A) \cap ((\mathbf{J}(A))^*)^2 = (0)$. Now, if $a \in ((\mathbf{J}(A))^*)^2$, then there exists an $x \in A$ with $a - axa \in \mathbf{J}(A)$, because $A/\mathbf{J}(A)$ is von Neumann regular. But $a - axa \in ((\mathbf{J}(A))^*)^2$ and $\mathbf{J}(A) \cap ((\mathbf{J}(A))^*)^2 = (0)$. Hence we have $a = axa$ and thus $((\mathbf{J}(A))^*)^2 \subseteq \mathbf{M}(A)$. By the assumption $\mathbf{M}(A) = 0$ and so $((\mathbf{J}(A))^*)^2 = (0)$. Thus $(\mathbf{J}(A))^* \subseteq \mathbf{J}(A)$, because $\mathbf{J}(A)$ contains every nilpotent ideal. □

Remark 43.6. The assumption in theorem 43.5 that $A/\mathbf{J}(A)$ is von Neumann regular, is fulfilled if A is e.g. a right Artinian ring, or an MHR-ring (a ring with minimum condition for principal right ideals) or a right linearly compact ring.

§ 44. THE MAXIMAL BIREGULAR IDEAL OF A RING

For the subject-matter of this section we refer the reader to Andrunakievich [6].

As is well known, a ring A is called *biregular* (Arens and Kaplansky [1]), if every principal ideal can be generated by a central idempotent. Biregular rings are a two-sided analogue of regular rings in the sense of von Neumann. From the definition it follows that every principal ideal of a biregular ring is a ring direct summand. It is easy to prove that the principal ideals of a biregular ring with unity element form a Boolean algebra.

Proposition 44.1. The centre Z of a biregular ring A is biregular.

Proof. Let (a) denote the principal ideal generated by $a \in Z$ in Z. Then there exists an idempotent element $e = ma + ax \in Z$, where m is a rational integer, $x \in A$, and $ae = ea$, since A is biregular and $a \in Z$. Since

$$e = e^2 = m^2 a^2 + 2max + a^2 x^2$$

we have $e = ay = ya$ with an $y \in A$. Hence we have $e = y^2 a^2 = (y^2 a)a$. Now we shall show that $y^2 a \in Z$ from which it will follow that $e \in (a)$ because $a \in Z$. Since $e \in Z$, $et = te$ for every $t \in A$, and thus $(ya)t = t(ya)$. Therefore, we have

$$(y^2 a)t = y(ya)t = yt \cdot ya = y(ty)a = ya(ty) = (tya)y = t \cdot y^2 a.$$

Hence $y^2 a \in Z$.

An element a of an arbitrary ring A is called *biregular* if the principal ideal generated by a in A has a unity element. An ideal B of an arbitrary ring A is said to be *biregular*, if B consists of only biregular elements. In this case B is a *biregular ring*.

PROPOSITION 44.2. The sum of two biregular ideals is also a biregular ideal.

PROOF. If B_1 and B_2 are two biregular ideals, and b is an arbitrary element of B_1+B_2, then $b=b_1+b_2$ where $b_1 \in B_1$ and $b_2 \in B_2$. Since the principal ideals (b_1) and (b_2) have unity elements e_1 and e_2, respectively, we have

$$(b) = (b_1+b_2) \leqq (b_1)+(b_2) = (e_1)+(e_2).$$

Now $e_{12}=e_1+e_2-e_1e_2$ is the unity element of $(e_1)+(e_2)$. Hence

$$(b) = (b)[(e_1)+(e_2)] = (be_1)+(be_2).$$

Since $be_1 \in B_1$ and $be_2 \in B_2$ the principal ideals (be_1) and (be_2) have unity elements, and thus (b) also has a unity element. Hence b is biregular. □

Since the union of a chain of biregular ideals is trivially biregular, there exists a unique maximal biregular ideal. The maximal biregular ideal of a ring A will be denoted by $\mathbf{B}(A)$.

Obviously every homomorphic image of a biregular ring is also biregular. But it is still an open question whether $\mathbf{B}(A/\mathbf{B}(A))=(0)$. We write $a \cdot (a)=\mathbf{F}(a)$ for an element a of a ring A. Then $a \in A$ is said to be *weakly regular* if $a \in \mathbf{F}(a)$. An ideal S of A is called *weakly regular* if S consists only of weakly regular elements (cf. § 14). By § 36, the maximal weakly regular ideal $\mathbf{S}(A)$ of A is a radical, and $\mathbf{B}(A) \leqq \mathbf{S}(A)$. Moreover, we have $\mathbf{B}(A/\mathbf{S}(A))=(0)$.

Every regular ring in the sense of von Neumann is weakly regular. Also every biregular ring is weakly regular. For, if A is biregular, then $(a)=(e)$ for every $a \in A$, where e is a central idempotent element such that $a=ae$.

PROPOSITION 44.3. Every ideal K of a biregular ideal B of a ring A is an ideal of A. If B is a biregular ideal and C is an arbitrary ideal of a ring A, then $B \cap C=BC=CB$.

PROOF. Take $a \in A$, $k \in K$. Then $aK \in B$. Since B is biregular, there exists a unity element e in aK. Hence $eaK=aK$. Since $ea \in B$ we have $aK=eaK \in BK \leqq K$ and thus $AK \leqq K$. Similarly we have $KA \leqq K$.

Clearly, $BC \leqq B \cap C$. Since $B \cap C$ is also biregular if B is biregular, we have $B \cap C=(B \cap C)^2 \leqq BC \leqq B \cap C$. Hence $BC=B \cap C=BC$. □

PROPOSITION 44.4. If $\mathbf{B}(A)$ is the maximal biregular ideal and $\mathbf{G}(A)$ the Brown–McCoy radical of a ring A, and I^* denotes the annihilator of an ideal I, then we have:

(1) $$\mathbf{B}(A) \cap \mathbf{G}(A) = (0);$$

(2) $$\mathbf{G}(A) \leqq (\mathbf{B}(A))^*, \quad \mathbf{B}(A) \leqq (\mathbf{G}(A))^*;$$

(3) $$\mathbf{B}(A) \cap (\mathbf{B}(A))^* = (0);$$

(4) $$\mathbf{G}(A) = \mathbf{G}((\mathbf{B}(A))^*), \quad \mathbf{B}(A) = \mathbf{B}((\mathbf{G}(A))^*).$$

PROOF. Since every principal ideal of $\mathbf{B}(A)$ has a unity element, (1) trivially follows from the definition of $\mathbf{G}(A)$. (1) implies (2). Since $\mathbf{B}(A)\cap(\mathbf{B}(A))^*$ is a zero-ring and every ideal of $\mathbf{B}(A)$ is idempotent, we obtain (3). Moreover, we have $\mathbf{G}(A)=\mathbf{G}(A)\cap(\mathbf{B}(A))^*=\mathbf{G}((\mathbf{B}(A))^*)$ by (2) and because of the hereditary property of $\mathbf{G}(A)$. Similarly, $\mathbf{B}(A)=\mathbf{B}(A)\cap(\mathbf{G}(A))^*$, because it can easily be proved that $\mathbf{B}(A)$ is hereditary.

For, if I is an arbitrary ideal of A, then $\mathbf{B}(I)\leqq I\cap\mathbf{B}(A)$, because every principal ideal of $\mathbf{B}(I)$ is also an ideal of A. Moreover, $I\cap\mathbf{B}(A)$ is a biregular ideal of I, and thus it follows that $I\cap\mathbf{B}(A)\leqq\mathbf{B}(I)$. □

THEOREM 44.5. Let $\mathbf{G}(A)$ be the Brown–McCoy radical and $\mathbf{B}(A)$ the maximal biregular ideal of a ring A. Let $A/\mathbf{G}(A)$ be a biregular ring. Then $\mathbf{B}(A)=(0)$ if and only if $(\mathbf{G}(A))^*\leqq\mathbf{G}(A)$.

PROOF. First, assume $\mathbf{B}(A)=(0)$.
Then

$$((\mathbf{G}(A))^*)^2/(((\mathbf{G}(A))^*)^2\cap\mathbf{G}(A)) \cong (((\mathbf{G}(A))^*)^2+\mathbf{G}(A))/\mathbf{G}(A).$$

By proposition 14.1, we have $((\mathbf{G}(A))^*)^2\cap\mathbf{G}(A)=(0)$ and thus

$$((\mathbf{G}(A))^*)^2 \cong (((\mathbf{G}(A))^*)^2+\mathbf{G}(A))/\mathbf{G}(A).$$

Being an ideal of a biregular ring, the ring on the right-hand side of this isomorphism is biregular itself and thus $((\mathbf{G}(A))^*)^2$ is biregular. Therefore $((\mathbf{G}(A))^*)^2\leqq$ $\leqq\mathbf{B}(A)$ and because of $\mathbf{B}(A)=(0)$ we have $((\mathbf{G}(A))^*)^2=(0)$. Hence $(\mathbf{G}(A))^*\leqq$ $\leqq\mathbf{G}(A)$. Conversely, if $(\mathbf{G}(A))^*\leqq\mathbf{G}(A)$, then we have $\mathbf{B}(A)\leqq(\mathbf{G}(A))^*\leqq\mathbf{G}(A)$ and since $\mathbf{B}(A)\cap\mathbf{G}(A)=(0)$ this implies $\mathbf{B}(A)=(0)$. □

PROPOSITION 44.6. Every ideal of the ring A contains a non-trivial biregular ideal if and only if $(\mathbf{B}(A))^*=(0)$.

PROOF. If $(\mathbf{B}(A))^*=(0)$, then $\mathbf{B}(A)I\neq(0)$ or $I\mathbf{B}(A)\neq(0)$ for every non-zero ideal I of A. Suppose $I\cdot\mathbf{B}(A)\neq(0)$. Then $I\cdot\mathbf{B}(A)=I\cap\mathbf{B}(A)\neq(0)$ is a biregular ideal of I.

If, however, $(\mathbf{B}(A))^*\neq(0)$, then $(\mathbf{B}(A))^*$ does not contain a biregular ideal because $(\mathbf{B}(A))^*\cap\mathbf{B}(A)=(0)$. □

PROPOSITION 44.7. If A is a biregular ring, then so is the full matrix ring A_n.

PROOF. Take $[a_{ij}]\in A_n$ with $a_{ij}\in A$. Since the principal ideals of A form a lattice, the finite sum $\sum(a_{ij})$ of the principal ideals (a_{ij}) has a unity element e. Now we take $E=[e_{ij}]$ with

$$e_{ij}=\begin{cases} e & \text{if } i=j,\\ 0 & \text{if } i\neq j.\end{cases}$$

The matrix $E=[e_{ij}]$ is contained in the centre of A_n and $E[a_{ij}]=[a_{ij}]E=[a_{ij}]$. We shall show that $[e_{ij}]$ is the unity element of the principal ideal $([a_{ij}])$. Let $[x]^{ij}$ be a matrix which has the element $x\in A$ in the (i,j)-th place, and 0's elsewhere. For arbitrary $x, y\in A$ and $B=[b_{ij}]\in A_n$, we have

$$[x]^{ip}B[y]^{qj} = [xb_{pq}y]^{ij}. \qquad (*)$$

Moreover, $E=\sum_{i=1}^{n}(e)^{ii}$ and since $(e)=\sum_{i,j}(a_{ij})$ we also have $e=\sum b_{ij}$, where

$$b_{ij} = \sum x_{ij}^{(k)} a_{ij} y_{ij}^{(k)} \in (a_{ij}).$$

But from equation ($*$) it follows that the matrices $[xa_{ij}y]^{tt}$ are contained in the principal ideal $([a_{ij}])$, and thus we have $E=[e_{ij}]\in([a_{ij}])$. Hence A_n is biregular. □

THEOREM 44.8. $\mathbf{B}(A_n)=(\mathbf{B}(A))_n$ for the maximal biregular ideal of a full matrix ring A_n.

PROOF. By proposition 44.7, we have $(\mathbf{B}(A))_n\leqq\mathbf{B}(A_n)$. Conversely, let $B==[b_{ij}]$ be an arbitrary matrix of $\mathbf{B}(A_n)$. We shall show that (b_{ij}) is a biregular ideal of A. Let E denote the unity element of the principal ideal (B). Since $B==EBE$, we have

$$b_{ij} = \sum_{p,q} e_{ip} b_{pq} e_{qj} \quad \text{with} \quad e_{ip},\ e_{qj}\in A.$$

From the above equation ($*$) we have $[b_{ij}]^{11}\in(B)$. Let c be an arbitrary element of the principal ideal (b_{ij}). Then $C=[c]^{11}\in(B)$. Since (B) is biregular, the principal ideal (C) contains a unity element $H=[h_{ij}]$. Then $H=\sum X_i C Y_i$ with $X_i, Y_i\in A_n$ and $h_{ij}\in(c)$. Since $C_1H=HC_1=C_1$ holds for every matrix C_1 in the ideal (C), we have $c_1h_{11}=h_{11}c_1$ for every $c_1\in(c)$. Hence h_{11} is the unity element of (c). Since c is an arbitrary element of the principal ideal $(a_{ij})\leqq A$ the ideal (a_{ij}) is a biregular ideal, and thus we have

$$\mathbf{B}(A_n) \leqq (\mathbf{B}(A))_n.$$

Therefore $\mathbf{B}(A_n)=(\mathbf{B}(A))_n$. □

THEOREM 44.9. $A=\mathbf{B}(A)\oplus(\mathbf{B}(A))^*$ holds for a ring A if and only if the maximal biregular ideal $\mathbf{B}((a))$ of every principal ideal (a) of A is a principal ideal itself.

PROOF. Suppose $A=\mathbf{B}(A)\oplus(\mathbf{B}(A))^*$. Then every element $a\in A$ possesses a decomposition $a=b+c$ with $b\in\mathbf{B}(A)$ and $c\in(\mathbf{B}(A))^*$. Hence we have

$$(a) = (b+c) \leqq (b)+(c).$$

If we multiply both sides with the unity element of (b), then we obtain $b=ea\in(a)$ and thus $c=a-b\in(a)$. Therefore, $(b)+(c)\leqq(a)$, and hence $(a)=(b)\oplus(c)$. Since

$\mathbf{B}(A)$ is hereditary,

$$\mathbf{B}((a)) = (a)\cap\mathbf{B}(A) = ((b)\oplus(c))\cap\mathbf{B}(A) = (b)\oplus((c)\cap\mathbf{B}(A)) = (b).$$

Hence $\mathbf{B}((A))$ is a principal ideal.

Conversely, if every ideal of the form $\mathbf{B}((a))$ is a principal ideal then $\mathbf{B}((a))$ contains a unity element, and thus $\mathbf{B}((a))$ is a direct summand of (a). Therefore $(a)=\mathbf{B}((a))\oplus(\mathbf{B}((a)))^*$. Since A is the sum of its principal ideals (a_α), we have

$$A = \sum_\alpha (a_\alpha) = \sum_\alpha \mathbf{B}((a_\alpha)) + \sum (\mathbf{B}((a_\alpha)))^*.$$

But on the right-hand side the two summands obviously have intersection (0). □

§ 45. ZEROID-PSEUDORADICALS

In this section we shall treat pseudoradicals which characterize in a certain sense the measure of singularity of a ring but which are not, in general, radicals in the sense of Amitsur and Kurosh. The sources for this section are Fuchs [3] and Szász [5]. It is easy to see that the smallest class of radical rings in the sense of Amitsur and Kurosh, which contains all pseudoradical rings in the sense of Fuchs, coincides with the *class of all rings.*

As is well known, important rings without divisors of zero, e.g. the discrete valuation rings, have a superfluously large Jacobson radical. Fuchs [3] wanted to define a pseudoradical for a ring, which equals the zero ideal in rings without divisors of zero.

We begin with the more general definitions of Szász [5].

Let V be a complete lattice with the zero element 0 and the greatest element i. Let a subset K be called an *upper class* of V if $x\in K$ and $x\leqq y$ imply $y\in K$. Then obviously $K=V$ if and only if $0\in K$. Let the empty set $\emptyset$ also be an upper class of V.

An upper class K is said to be an *upper class of finite character,* in short an *upper e-class,* if $\bigcup_{\gamma\in\Gamma} x_\gamma\in K$ implies the existence of a finite subset Δ of Γ with $\bigcup_{\delta\in\Delta} x_\delta\in K$. Both the set-theoretic intersection of a finite number of upper e-classes and the set-theoretic union of an arbitrary number of upper e-classes are also obviously upper e-classes of V.

Now, let K be a fixed upper e-class, and x a fixed element of V with $x\notin K$. An element $y\in V$ is called a *(K, x)-element*, if $x\cup y\notin K$. The element x itself is obviously a (K, x)-element.

An element $z\in V$ is said to be a *strong* *(K, x)-element*, if $y\cup z$ is also a (K, x)-element of V for every (K, x)-element y. Thus, every strong (K, x)-element is also a (K, x)-element of V.

THEOREM 45.1. The union $z_K(x)$ of all strong (K, x)-elements z is also a strong (K, x)-element. This coincides with the intersection of all maximal (K, x)-elements.

PROOF. If $z_1, z_2, \ldots, z_n$ are a finite number of strong (K, x)-elements, and if y is an arbitrary (K, x)-element of V, then since

$$(z_1 \cup z_2 \cup \ldots \cup z_n) \cup y = (\ldots((y \cup z_n) \cup z_{n-1}) \cup z_1),$$

$z_1 \cup z_2 \cup \ldots \cup z_n$ is a strong (K, x)-element of V. Now, let $z_K(x) = \bigcup_{\gamma \in \Gamma} z_\gamma$, where z_γ runs over all strong (K, x)-elements. If y is a (K, x)-element with $\bigcup_{\gamma \in \Gamma} z_\gamma \cup y \cup x \in K$, then it follows that $\bigcup_{\gamma \in \Gamma} w_\gamma \in K$ with $w_\gamma = z_\gamma \cup y \cup x$. Because of the finite character of K there exists a finite number of $\gamma_1, \gamma_2, \ldots, \gamma_n$ with

$$w_{\gamma_1} \cup w_{\gamma_2} \cup \ldots \cup w_{\gamma_n} \in K.$$

Hence $z_{\gamma_1} \cup \ldots \cup z_{\gamma_n} \cup y \cup x \in K$ and thus a contradiction follows. Therefore, the union $z_K(x)$ is also a strong (K, x)-element.

Because of the finite character of K, the union of every ascending chain of (K, x)-elements is a (K, x)-element as well. Hence Zorn's lemma can be applied. Therefore, there exists a maximal (K, x)-element m with $m \geqq y$ for every (K, x)-element $y \in V$. Let d_0 be the intersection of all maximal (K, x)-elements of V for a fixed element $x (\in V, \notin K)$. If y is an arbitrary (K, x)-element, and m is a maximal (K, x)-element with $y \leqq m$, then

$$d_0 \cup y \cup x \leqq m \cup m \cup x = m \cup x \notin K$$

and thus $d_0 \cup y \cup x \notin K$. Hence d_0 is a strong (K, x)-element of V. Therefore $d \leqq z_K(x)$. Conversely, if m is a maximal (K, x)-element, then $z_k(x) \cup m$ is also a (K, x)-element. By the maximality of m we have $m = z_K(x) \cup m$, and hence $z_K(x) \leqq m$. Therefore we also have $z_K(x) \leqq d_0$. Thus $d_0 = z_K(x)$. □

DEFINITION: Let $\mathfrak{M}$ be an arbitrary set of upper e-classes K_α of V and take $x \in V$. Then the intersection $z_{\mathfrak{M}}(x)$ of all $z_{K_\alpha}(x)$ with $K_\alpha \in \mathfrak{M}$ and $x \notin K_\alpha$ is said to be the *$\mathfrak{M}$-pseudoradical* of the element $x \in V$. We take $z_K(x) = i$ if $x \in K$ $(K \in \mathfrak{M})$.

Now let V be a complete lattice ordered groupoid, i.e. let V be both a complete lattice and a multiplicative groupoid, in which the following conditions hold:

(1) $$a \cdot b \leqq a \cap b;$$

(2) $$(a \cup b)c \leqq ac \cup bc;$$

(3) $$a(b \cup c) \leqq ab \cup ac.$$

We remark that this system of axioms (1), (2) and (3) differs from the usual.

A subset K of the lattice ordered groupoid V is said to be an *upper e-class groupoid* if the following conditions are fulfilled:

(1′) K is an upper e-class of the lattice V;
(2′) K is a subgroupoid of the groupoid V.

The empty set $\emptyset$ is also considered an upper e-class groupoid of V.

We see that K is obviously a dual ideal (a filter) of V because $k_1 \cdot k_2 \in K$, $k_1 k_2 \leqq k_1 \cap k_2$ ($k_1, k_2 \in K$) and because of the definition of the upper class. Both the set-theoretic union of every ascending chain of upper e-class groupoids and the set-theoretic intersection of a finite number of upper e-class groupoids are upper e-class groupoids of V.

Now let $\mathfrak{N}$ be an arbitrary set of upper e-class groupoids of the lattice ordered groupoid V. Then the $\mathfrak{N}$-pseudoradical $z_{\mathfrak{N}}(x)$ can be defined for every $x \in V$ by means of the above definition, and we have $x \leqq z_{\mathfrak{N}}(x) \leqq i$. We again take $z_K(x) = i$ for $x \in K$.

A *prime element* of V is an element p such that $ab \leqq p$ ($a, b \in V$) implies $a \leqq p$ or $b \leqq p$. Consequently, i is a prime element of V.

THEOREM 45.2. Let V be a complete lattice ordered groupoid satisfying axioms (1), (2) and (3). Let K be an upper e-class groupoid satisfying conditions (1′) and (2′) and x an element of V with $x \notin K$. Then every maximal (K, x)-element of V is a prime element. The pseudoradical $z_{\mathfrak{N}}(x) = \bigcap_{K \in \mathfrak{N}} z_K(x)$ is the intersection of prime elements for every set $\mathfrak{N}$ consisting of upper e-class groupoids.

PROOF. The second statement follows from the first. Now, let m be a maximal (K, x)-element. If $c_1 \not\leqq m$ and $c_2 \not\leqq m$ then we have the inequalities $c_1 \cup m > m$ and $c_2 \cup m > m$. Because of the maximality of m we have $c_1 \cup m \cup x \in K$ and $c_2 \cup m \cup x \in K$. By (2′), K is a subgroupoid of V. Therefore

$$m_0 = (c_1 \cup m \cup x)(c_2 \cup m \cup x) \in K.$$

By (2) and (3) $m_0 \leqq c_1 c_2 \cup m \cup x$ holds, and by the definition of the upper class we also have $c_1 c_2 \cup m \cup x \in K$. This means, however, that $c_1 c_2 \cup m$ is not a (K, x)-element. Hence $c_1 c_2 \cup m \neq m$ and $c_1 c_2 \not\leqq m$. Therefore m is a prime element of V. □

THEOREM 45.3. An ideal I of a ring A is a zeroid-pseudoradical $z_{\mathfrak{N}}(0)$ (in the lattice ordered semigroup of all ideals of the ring A) if and only if I is an intersection of prime ideals.

PROOF. It is clear that the ideals of a ring A form a lattice ordered semigroup satisfying the axioms (1), (2) and (3). Now, if the ideal I is a zeroid-pseudoradical, then, by theorem 45.2, I is an intersection of prime ideals of A.

Conversely, if the ideal I is an intersection of the prime ideals P_α of the ring A, then we shall explicitly give a set $\mathfrak{N}$ of upper e-class semigroups K_α of the lattice ordered semigroup V of all ideals of A such that $I = z_{\mathfrak{N}}(0)$. Let $C(P_\alpha)$ be the complementary set of the prime ideal P_α in A, i.e. $C(P_\alpha)=A\setminus P_\alpha$, and let $K(P_\alpha)$ be the set of those ideals $I_{\alpha\beta}$ of the ring A, for which $I_{\alpha\beta}\cap C(P_\alpha)$ is not empty. Then $K(P_\alpha)$ is obviously an upper e-class semigroup in the lattice ordered semigroup V. Furthermore, let $\mathfrak{N}$ be the set of all $K(P_\alpha)$ and let $I=\bigcap_\alpha P_\alpha$. Then we have $I= =z_{\mathfrak{N}}(0)$. □

By a *weak block* of a ring A we mean a subset B of A, such that $0\notin B$, and $I_1\cap B\neq\emptyset$ and $I_2\cap B\neq\emptyset$ imply $I_1\cdot I_2\cap B\neq\emptyset$ for arbitrary ideals L_1 and I_2 of A.

The complementary set $C(P)$ of a prime ideal P of A is obviously a weak block of A. If B is a weak block of A, then the complementary set $C(B)$ is not an ideal of A, in general. But if B is a weak block, and $C(B)$ is an ideal of A, then $C(B)$ is a prime ideal of A.

If B is a weak block of A, then the ideals I of A with $I\cap B\neq\emptyset$ form an upper e-class semigroup K in the lattice ordered semigroup V of all ideals of the ring A.

By a *block* B of the ring A we mean a weak block B such that $x\in B$ implies $x^n\in B$ for every exponent n. Every multiplicative semigroup H of A with $0\notin H$ is obviously a block of A and every block is trivially a weak block. If A is a full matrix ring S_n with $n\geqq 2$ over a division ring S, then the non-zero elements of A form a weak block which is not a block, and so it is not a semigroup. Moreover, there exist blocks which are not multiplicative semigroups H with $0\notin H$. However, if A is a commutative ring, and the complement $C(B)$ of a weak block B is an ideal of A, then B is necessarily a multiplicative semigroup with $0\notin B$.

It follows from theorem 45.3 that every zeroid-pseudoradical $z_{\mathfrak{N}}(0)$ contains the Baer lower nil radical. As for the relation of $z_{\mathfrak{N}}(0)$ to the Baer upper nil radical, we can assert the following:

THEOREM 45.4. If $\mathfrak{N}$ contains only those upper e-class subsemigroups $K(B)$ which are determined by blocks such that the elements $K(B)$ are ideals I of the ring A with $I\cap B\neq\emptyset$, then $z_{\mathfrak{N}}(0)$ contains the Baer upper nil radical U of A.

PROOF. Let I be an arbitrary $(K, 0)$-ideal of A with $K\in\mathfrak{N}$ and $I\in V$. We shall show that the upper nil radical U is a strong $(K, 0)$-element of V, i.e. $U+I\notin K(B)$. Assume to the contrary that $(U+I)\cap B\neq\emptyset$. Then there exist elements $u\in U$, $i\in I$ and $b\in B$ such that $u+i=b$, and since $u\in U$ $u^k=0$ for some K. The distributivity law yields

$$b^k = (u+i)^k\in B\cap(U+I),$$

because now B is a block of A. Since I is a $(K, 0)$-element, $I\cap B\neq\emptyset$ must hold which is a contradiction by the preceding. Therefore, $I+U\notin K$ and thus $U\leqq Z_{\mathfrak{N}}(0)$. □

Now we mention some example of *multiplicative semigroups M with* $0\notin M$ in a ring A which are also blocks of A.

I. Let A be a ring, G a right A-module, and N an A-submodule. Let H_1 and H_2 be arbitrary subsets of G with $0 \notin H_2$ and $H_1 \subset H_2$. Then each of the following, eventually empty subsets is a multiplicative semigroup M_j with $0 \notin M_j$ ($j=1, 2, 3, 4$):

(I$_1$) $$M_1 = \{x;\ x \in A,\ Nx = N\};$$

(I$_2$) $$M_2 = \{x;\ x \in A,\ nx = n \text{ for every } n \in N\};$$

(I$_3$) $$M_3 = \{x;\ x \in A,\ H_2 x = H_1\};$$

(I$_4$) $$M_4 = \{x;\ x \in A,\ H_2 x \subset H_1\}.$$

II. The *left units* (or the right units), i.e. those elements which are left divisors of every element of A, form a multiplicative semigroup M with $0 \notin M$.

III. The left *unity elements* (or the right unity elements) of a ring form a multiplicative semigroup M with $0 \notin M$.

IV. Every non-zero *idempotent element* e itself forms a multiplicative semigroup M with $0 \notin M$.

V. Both the set B_l of all *non-left divisors of zero* and the set B_r of all *non-right divisors of zero* of a ring A are multiplicative semigroups, which do not contain the zero element.

Now we define the zeroid-pseudoradical by

$$\mathbf{F}(A) = z_{\mathfrak{N}}(0),$$

where we choose $\mathfrak{N} = \{K_r, K_l\}$ with

$$K_r = \{I;\ I \in V,\ I \cap B_r \neq \emptyset\}$$

and

$$K_l = \{I;\ I \in V,\ I \cap B_l \neq \emptyset\}.$$

In other words, $\mathbf{F}(A)$ is a zeroid-pseudoradical $z_{\mathfrak{N}}(0)$ in the lattice ordered semigroup V of all ideals I of A for which $\mathfrak{N}$ contains only the upper e-class semigroups K_r and K_l, which are defined by B_r and B_l, respectively, where B_l denotes the set of all non-left divisors of zero and B_r the set of all non-right divisors of zero (cf. example V).

Thus, in the ring A $\mathbf{F}(A)$ is the largest ideal for which every element of $\mathbf{F}(A)+C$ is a left divisor of zero in A, whenever this holds for every element of C, and for which at the same time every element of $\mathbf{F}(A)+D$ is a right divisor of zero in A, whenever this holds for every element of D.

THEOREM 45.5. $\mathbf{F}(A)$ is an intersection of prime ideals.

PROOF follows from the definition of $\mathbf{F}(A)$ and from theorem 45.2, where a description of these prime ideals was given. □

If A is a ring without divisors of zero in which the non-units form an ideal, then the Jacobson radical $\mathbf{J}(A)$ coincides with this ideal. In these rings $\mathbf{F}(A)=(0)$, however obviously, in general $\mathbf{J}(A) \neq (0)$, as e.g. the ring of all rational numbers with odd denominators shows.

Moreover, the ring $A = \{a\}$ generated by a, with $a + a = a^3 + a^2$ $(a^2 \neq 0)$ is a ring with four elements, namely 0, a, a^2, $a + a^2$, in which $\mathbf{F}(A) = A$ but $\mathbf{J}(A) \neq A$, because $A/\mathbf{J}(A)$ is a field of two elements.

Since for the discrete direct sum A of an infinite number of arbitrary rings A_α, $\mathbf{F}(A) = A$ always holds, every ring A_α is a homomorphic image of a Fuchs zeroid-pseudoradical ring A. Hence the smallest such class of Amitsur–Kurosh radical rings which contains every Fuchs zeroid-pseudoradical ring, contains every ring. It is still an open question, whether $\mathbf{F}(A/\mathbf{F}(A)) = (0)$ in every ring. But we can prove:

Proposition 45.6. The factor ring $A/\mathbf{F}(A)$ does not contain non-zero nil ideals.

Proof. Let $C/\mathbf{F}(A)$ be a nil ideal of $A/\mathbf{F}(A)$ and B an arbitrary ideal of A, in which every element is a left divisor of zero. Then for every $c \in C$ there exists an exponent k with $c^k \in \mathbf{F}(A)$, and thus

$$(c+b)^k \in \mathbf{F}(A) + B.$$

Since every element of $\mathbf{F}(A) + B$ is a left divisor of zero in A, so is $c + b$, and hence we have $C \leqq \mathbf{F}(A)$ by the definition of $\mathbf{F}(A)$, because it can similarly be proved that when every element of D is a right divisor of zero in A, then so is every element of $C + D$. □

Theorem 45.7. If A is a ring with a right unity element and with minimum condition on principal right ideals, then the Fuchs zeroid-pseudoradical $\mathbf{F}(A)$ coincides with the Brown–McCoy radical $\mathbf{G}(A)$.

Proof. We shall show that every element of every proper two-sided ideal is a left divisor of zero in A. Otherwise there exists a proper ideal C, which contains a non-left divisor of zero c. In the set of those principal right ideals $(c)_r$ of A that are contained in C and for which c is not a left divisor of zero, there exists a minimal principal right ideal $(c_0)_r$. For this we obviously have $(c_0)_r = (c_0^2)_r$. Since A has a right unity element e, we have $c_0 = c_0^2 \cdot d$ where $d \in A$, and thus $c_0(c_0 \cdot d - e) = 0$. Since $C \neq A$, $e \notin C$. Consequently $c_0 d - e \neq 0$. But then $c_0(c_0 d - e) = 0$ contradicts the definition of c_0. Therefore every element of every proper ideal C is a left divisor of zero. If Z_l (or Z_r) is a pseudoradical in the lattice ordered semigroup V, which is determined by the block B_l (or B_r) of all non-left divisors of zero (or non-right divisors of zero) in A, then $Z_l(A) = \mathbf{G}(A)$ for the Brown–McCoy radical by theorem 45.1. By definition $\mathbf{F}(A) = Z_l \cap Z_r$. Since B_r and B_l are blocks in A, we have $U \leqq Z_r$ and $U \leqq Z_l$ for the upper nil radical U by theorem 45.4. Moreover, we have proved $\mathbf{U}(A) = \mathbf{G}(A)$ in Szász [2] for rings with minimum condition on principal right ideals and having a right unity element. Therefore, $\mathbf{G}(A)=\mathbf{U}(A)\leqq Z_l \cap Z_r=\mathbf{F}(A)\leqq Z_l=\mathbf{G}(A)$. Thus $\mathbf{F}(A)=\mathbf{G}(A)$. □

THEOREM 45.8. If A is a ring with unity element which is a regular ring in the sense of von Neuman, then the Fuchs zeroid-pseudoradical $\mathbf{F}(A)$ coincides with the Brown–McCoy radical $\mathbf{G}(A)$.

PROOF. Let M be an arbitrary proper maximal ideal of A and m an arbitrary element of M. Then there exists an element $x \in A$ with $m=m \cdot x \cdot m$. If e is the unity element of A, then $e \notin M$, and thus $xm-e\neq 0$. Hence, because of $m(xm-e)=0$, m is a left divisor of zero. Similarly, m is a right divisor of zero in A. Consequently, $\mathbf{G}(A)$ coincides with $\mathbf{F}(A)$. □

REMARK. In what follows, we give an example of a ring A with unity element which is regular in the sense of von Neumann and whose Fuchs zeroid-pseudoradical $\mathbf{F}(A)$ differs from zero. The existence of such a ring *contradicts* an incorrect statement in section 6 of Fuchs's paper [3]. In this ring A there also exist $\mathfrak{N}$-pseudoradicals $z_{\mathfrak{N}}(0)$, which differ from $\mathbf{F}(A)$. Hence the $\mathfrak{N}$-pseudoradicals $z_{\mathfrak{N}}(0)$ and $z_{\mathfrak{N}}(I)$ for an ideal I of A are proper generalizations of the Fuchs zeroid-pseudoradical of a ring and the van Leeuwen pseudoradical [1] of an ideal of a ring, respectively.

EXAMPLE. Let M be a left vector space of an uncountable dimension $d=\aleph_\nu$ over an arbitrary division ring S. Let A, as a right operator domain for M, be the ring of all S-endomorphisms of the S-vector space M. Then the ring A has a unity element 1, and A is by the methods of Johnson and Kiokemeister [1] regular in the sense of von Neumann. As in Chapter IV of the book [2] of Jacobson, (0), ..., $I_{\mathfrak{n}}$, ..., A are all ideals of A. Here the ideal $I_{\mathfrak{n}}$ for an infinite cardinal $\mathfrak{n}(\leqq\aleph_\nu)$ denotes the set of all S-endomorphisms $a\in A$, for which $\dim_S(Ma)<\mathfrak{n}$. The ideals of A form a chain, and thus $I_{\aleph_\nu}\neq(0)$ is the only maximal (proper) ideal of A. Since every ideal of A is idempotent because of the regularity in the sense of von Neumann, all upper e-class semigroups K_α of the lattice ordered semigroup V of all ideals of A can be explicitly determined, because now every upper e-class of the lattice V is also an upper e-class semigroup of the lattice ordered semigroup V:

$$\begin{aligned}
K_{-1} &= \{(0), I_{\aleph_0}, \dots, I_{\aleph_\nu}, A\}\\
K_0 &= \{I_{\aleph_0}, I_{\aleph_1}, \dots, I_{\aleph_\nu}, A\}\\
K_1 &= \{I_{\aleph_1}, \dots, I_{\aleph_\nu}, A\},\\
&\vdots\\
K_\nu &= \{I_{\aleph_\nu}, A\}\\
K_{\nu+1} &= \{A\}\\
K_{\nu+2} &= \emptyset.
\end{aligned}$$

For instance, we have $z_{K_0}(0)=(0)$, $z_{K_1}(0)=I_{\aleph_\nu}$, ... and for the Brown–McCoy radical, as well as for the zeroid-pseudoradical we have

$$\mathbf{F}(A) = \mathbf{G}(A) = z_{K_{\nu+1}}(0) = I_{\aleph_\nu} \neq (0).$$

Now, let $\mathfrak{N}$ denote a non-empty subset of the set $\{K_{-1}, K_0, K_1, K_{\nu+2}\}$ of upper e-class semigroups of V. Then $z_{\mathfrak{N}}(0)$ certainly differs from $\mathbf{F}(A)=I_{\aleph_\nu}$. □

Finally, we treat the Fuchs zeroid-pseudoradical of a full matrix ring over a uniform ring. Here a ring A is called *uniform*, if for a finite number of elements $a_1, a_2, ..., a_n$ of an ideal K consisting of left or right divisors of zero, there exists a non-zero element c of A with $a_1c=...=a_nc=0$ or $ca_1=...\ ca_n=0$. In particular, every ring with a two-sided annihilator or every ring without divisors of zero is a uniform ring.

PROPOSITION 45.9. Every element of an ideal I of the full matrix ring A_n over a uniform ring A is a left divisor of zero in A_n if and only if there exists an ideal K of A with $I \leqq K_n$ such that every element of K is a left divisor of zero in A.

PROOF. First assume that $I \leqq K_n$ where every element of K is a left divisor of zero in A. If $[a_{ij}]$ is a matrix of I, then there exists a non-zero element c of A with $a_{ij}c=0$ for every a_{ij} because of the uniformity of A. Then the diagonal matrix $[c]$ annihilates the matrix $[a_{ij}]$ on the right.

Conversely, assume now that every element of an ideal I of A_n is a left divisor of zero in A_n. Let K be the ideal of A generated by all elements of all matrices in I. Then we have $I \leqq K_n$. It remains to show that every element of K is a left divisor of zero in A. Take $a^{(\nu)} \in K$ with $a^{(\nu)} = a_1^{(\nu)} + ... + a_m^{(\nu)}$, where we may suppose that $a_i^{(\nu)}$ is the (j_{i_ν}, k_{i_ν})-th element of the matrix $b_i^{(\nu)}$. If $(x)_{jk}$ denotes the matrix with x in the (j, k)-th place and 0's elsewhere, then the matrices

$$[c_i^\nu] = \sum_{s=1}^{n} [x_\nu]_{sj_{i_\nu}} [b_i^{(\nu)}][y_\nu]_{k_{i_\nu}s}$$

are diagonal matrices of the form $\langle x_\nu b_i^{(\nu)} y_\nu, ..., x_\nu b_i^{(\nu)} y_\nu \rangle$ for arbitrary elements $x_\nu, y_\nu \in A$. Every matrix $[c_i^\nu]$ is contained in I and $[c]$ has a right annihilator matrix $(\gamma_{ij}) \in A_n$ since

$$[c^\nu] = [c_1^{(\nu)}] + ... + [c_{m\nu}^{(\nu)}] = \langle x_\nu b^{(\nu)} y_\nu, ..., x_\nu b^{(\nu)} y_\nu \rangle$$

and

$$[c] = [c^{(1)}] + ... + [c^{(t)}] \in I.$$

Then every element γ_{ij} of (γ_{ij}) annihilates the element $x_\nu b^\nu y_\nu$ on the right, and thus every element of AKA is a left divisor of zero. But then every element of K^3, and hence of K too, is a left divisor of zero in A. □

THEOREM 45.10. $\mathbf{F}(A_n)=(\mathbf{F}(A))_n$ where $\mathbf{F}(A_n)$ is the Fuchs zeroid-pseudoradical of a full matrix ring A_n over a uniform ring A.

PROOF. Let A be a uniform ring, $I=\mathbf{F}(A_n)$ and K an ideal of A, which consists only of left divisors of zero in A. Then, by proposition 45.9, every element of the ideal K_n is a left divisor of zero. Moreover, every element of $I+K_n$ is a left divisor of zero in A_n by theorem 45.1 and the definition of $\mathbf{F}(A_n)$. Thus, by proposition 45.9, there exists an ideal B of A such that every element of B is a left divisor of zero in A and $I+K_n \leqq B_n$. Since K_n has been chosen arbitrarily, we have $I \leqq (Z_l)_n$ and, similarly, $I \leqq (Z_r)_n$. Therefore $\mathbf{F}(A_n) \leqq (\mathbf{F}(A))_n$.

Conversely, if C is an ideal of A_n, which consists only of left divisors of zero in A_n, then, by proposition 45.9, there exists an ideal D of A with $C \leqq D_n$ such that every element of D is a left divisor of zero in A. Then we have

$$(\mathbf{F}(A))_n + C \leqq (\mathbf{F}(A))_n + D_n = (\mathbf{F}(A)+D)_n.$$

Since every element of $\mathbf{F}(A)+D$ is a left divisor of zero in A, every element of $(\mathbf{F}(A))_n + C$ is also a left divisor of zero in A_n. Thus $(\mathbf{F}(A))_n \leqq Z_l(A_n)$. Analogously we can show that $(\mathbf{F}(A))_n \leqq Z_r(A_n)$. Thus, we have proved $(\mathbf{F}(A))_n = \mathbf{F}(A_n)$. □

PROBLEM 63. Find a necessary and sufficient condition for $A=\mathbf{M}(A)\oplus(\mathbf{M}(A))^*$ to be valid for the maximal von Neumann regular ideal $\mathbf{M}(A)$ of A and its two-sided annihilator $(\mathbf{M}(A))^*$. (Give an analogue of theorem 44.9 for biregular rings.)

PROBLEM 64. Investigate the radicals of the factor ring $A/\mathbf{B}(A)$, where $\mathbf{B}(A)$ is the maximal biregular ideal of a ring A (e.g. does $\mathbf{B}(A/\mathbf{B}(A))=(0)$ hold?).

PROBLEM 65. Investigate the zeroid-pseudoradicals defined by the blocks mentioned in I, II, III and IV of § 45, in a full matrix ring, in an MHR-ring (with unity element), and in a von Neumann regular ring (with unity element), respectively.

PROBLEM 66 (Fuchs). Does $\mathbf{F}(A/\mathbf{F}(A))=(0)$ hold in every ring A for the Fuchs zeroid-pseudoradical $\mathbf{F}(A)$?

PROBLEM 67. Is the Brown–McCoy radical $\mathbf{G}(A)$ contained in the Fuchs zeroid-pseudoradical $\mathbf{F}(A)$ for every MHR-ring A?

PROBLEM 68.† Determine all simple rings without divisors of zero. (a ring of this kind is not necessarily a division ring.)·

PROBLEM 69. Investigate the Behrens radical of a full matrix ring (the *Behrens radical* is the upper radical determined by the class of subdirectly irreducible rings which contain a non-zero idempotent element in the heart; cf. Behrens [1]).

Problem 70.† Investigate the various concrete radicals (nil, Jacobson, Brown–McCoy, etc.) and the various concrete zeroid-pseudoradicals in the following classes of rings:

(a) linearly compact rings (or these in a narrow sense) (cf. Zelinsky [1], Leptin [1, 2], Wiegandt [1, 2, 3]),

(b) rings, in which every (finitely generated) right ideal is a principal right ideal, or in which at the same time every (finitely generated) left ideal is also a principal left ideal (cf. Szász [12]),

(c) rings, in which every proper right ideal differs from its two-sided idealizer, or in which every proper subring differs from its right idealizer (cf. Freidman [1, 2] etc.),

(d) rings with a distributive lattice of right ideals,

(e) rings which have an uncountably infinite number of left ideals, but only a countably infinite number of right ideals (cf. Szász [3]),

(f) Banach algebras,

(g) group algebras over a field,

(h) semigroup algebras over a field.

Problem 71. Investigate the direct limit and the inverse limit of radical rings (for concrete, or for general radicals or zeroid-pseudoradicals). (For this problem we refer the reader to Zelinsky [1] who dealt with the Jacobson radical.)

Problem 72. Investigate the Kronecker product of radical algebras over a field for concrete or for general radicals or zeroid pseudoradicals.

Problem 73. Which wreath sums of radical rings are again radical rings? (For the notion of the wreath sum, cf. Kaloujnine and Krasner [1].)

Problem 74.† For which rings with minimum condition on principal right ideals is the maximal torsion ideal not a ring direct summand of the ring (cf. Szász [40])?

Problem 75.† For which right Noetherian rings A are the following statements valid simultaneously?

(1) A/P is a simple ring with non-zero right socle for every prime ideal P.

(2) The minimum condition for principal right ideals holds in the left annihilator of every homomorphic image.

Problem 76.† Which prime rings can be embedded into a prime ring with unity element?

Problem 77. In which rings is every subring an endomorphic image?

Problem 78. In which rings is every endomorphic image a ring direct summand?

PROBLEM 79. In which rings A is nA a ring direct summand of A for every integer n?

PROBLEM 80. Which rings have a unique minimal subring?

PROBLEM 81. In which ring are the distinct subrings always non-isomorphic?

PROBLEM 82.† In which rings A is every element a left multiplicator (i.e. to every $a \in A$ there exists a rational integer n with $(a+n)A=(0)$)?

PROBLEM 83. In which rings is the commutator ideal finite or can be finitely generated?

PROBLEM 84. In which rings A has the additive group Z^+ of the centre Z, a finite group-theoretic index with respect to A^+?

PROBLEM 85. In which group algebras over a field has the ideal, generated by the commutator group of the group, a finite dimension?

PROBLEM 86. In which Jacobson radical-rings is every automorphism φ a quasi-inner automorphism (i.e. $\varphi: x \to (1-a)x(1-a)^{-1}$ for a suitable $a \in A$)?

PROBLEM 87. Does there exist a simple, finitely generated but infinite ring?

PROBLEM 88.† Let $\hat{a}=\{(1-x)a(1-x)^{-1}; x \in A\}$ in a Jacobson radical-ring. When is every class $\hat{a}$ finite, and when is the number of the classes $\hat{a}$ finite?

PROBLEM 89. In which rings with minimum condition on principal right ideals is this condition inherited by every subring?

PROBLEM 90.† Which rings have a distributive lattice of right ideals (of subrings)?

PROBLEM 91.† In which rings A does $aA+Aa=A$ hold for every $a \in A$?

PROBLEM 92. In which rings A does $(1-a)A+A(1-a)=A$ hold for every $a \in A$, where (in the case $1 \notin A$) 1 is an operator?

PROBLEM 93.† Is A a Jacobson radical ring, if $A=\bigcup_{m=1}^{n} R_m$, $\bigcap_{m=1}^{n} R_m=(0)$ and $R_k^*=\bigcup_{\substack{m \neq k \\ m=1}}^{n} R_m \neq A$ holds for every k and for certain right ideals R_m of A ($\bigcup$ and $\bigcap$ are to be understood set-theoretically)?

PROBLEM 94. Which rings A have a finite subset E such that $R \cap E \neq (0)$ for every non-zero right ideal R of A?

PROBLEM 95† (Kertész). Which right Artinian rings are also left Artinian?

PROBLEM 96† (Kertész). Has the ring A necessarily a two-sided unity element, if the maximal trivial A-submodule M_0 of M (i.e. $M_0A=(0)$) is a direct summand of M in every right A-module M?

PROBLEM 97 (Andrunakievich). Let $\mathbf{N}(A)$ denote the sum of all nil right ideals of A. Does $\mathbf{N}(A/\mathbf{N}(A))=(0)$ hold?

PROBLEM 98. Does the maximum condition hold for two-sided ideals in a Jacobson radical ring if every commutative subring is Noetherian?

PROBLEM 99.† In which rings A does the centralizer of the centralizer of every subring B of A coincide with B?

PROBLEM 100. Does there exist an antisimple ring which is not a nil ring?

PROBLEM 101. Does there exist a nil ring which is not anti-simple?

PROBLEM 102 (Roos). We can show that the class $(\mathbf{E}_5)_r$ of all rings with the condition $a\in aA+AaA$ forms an Amitsur–Kurosh radical class. A similar fact holds for $(\mathbf{E}_5)_l$ with $a\in Aa+AaA$ for every $a\in A$. Let $\mathbf{E}_6$ be the class of all rings A with $a\in aA+Aa+AaA$ for every $a\in A$ (which is also a radical class). Does $\overline{(\mathbf{E}_5)_r\cup(\mathbf{E}_5)_l}=\mathbf{E}_6$ hold, where $\overline{\mathbf{X}}$ denotes the lower radical class determined by the class $\mathbf{X}$ ($\overline{(\mathbf{E}_5)_l\cup(\mathbf{E}_5)_r}\subseteq\mathbf{E}_6$ is almost trivial)?

PROBLEM 103. Let S be the adjoint semigroup ($x\circ y=x+y-xy$) of a ring A. Call u and v right congruent mod $H(a)$ with $a\in A$ in S, if there exist exponents l and m such that $a^{(l)}\circ u=a^{(m)}\circ v$ holds. (Here $a^{(2)}=a\circ a$ and $a^{(k+1)}=a^{(k)}\circ a$). Let $G(a)$ be the two-sided congruence generated by $H(a)$ in $S\times S$. For which rings A does $S=e\cup B$ hold with $e\notin B$, $e\circ e=e$, $e\circ b=b\circ e=b$ for every $b\in B$ and $(b_1,b_2)\in G(b_3)$ for every $b_1,b_2,b_3\in B$? (According to the above definition of $G(a)$, $G(a)\subseteq S\times S$ holds for the set-theoretic product $S\times S$.)

PROBLEM 104 (Kertész). Which nilpotent rings are the Jacobson radicals of right Artinian rings?

PROBLEM 105. Which transfinitely nilpotent nil rings are the Jacobson radicals of rings with minimum condition on principal right ideals?

PROBLEM 106. In which rings are all proper, finitely generated right ideals isomorphic among themselves?

PROBLEM 107. In which rings are all right quasi-regular proper, finitely generated right ideals isomorphic among themselves?

PROBLEM 108. What is the Fuchs zeroid-pseudoradical of a von Neumann regular ring without (one-sided or two-sided) unity element?

PROBLEM 109. What is the Fuchs zeroid-pseudoradical of a ring with minimum condition on principal right ideals and without (one-sided or two-sided) unity element?

PROBLEM 110. What is the Fuchs zeroid-pseudoradical of a full matrix ring over a non-uniform ring?

PROBLEM 111 (Rédei). Does there exist an infinite but non-nilpotent nil ring, all proper subrings of which are nilpotent?

PROBLEM 112 (S. Lajos). Find characterizations of different classes of rings in term of the circle operation ($x \circ y = x + y - xy$).

PROBLEM. 113 (S. Lajos). Characterize the semigroups which can occur as adjoint semigroups of rings.

APPENDIX

The following problems have been entirely or partially solved since the completion of the manuscript:

Problem No.	Solved by
1.	Yu-Lee Lee (Manhattan)
3.	N. Divinsky (Vancouver)–J. Krempa–A. Suliński (Warsaw)
24.	P. M. Cohn (London)
27.	I. Kaplansky (Chicago)
38 and 39.	B. J. Gardner (Hobart)
40.	W. Stephenson (London)
48.	F. A. Szász (Budapest)
54.	F. Hansen (Bochum)
70. (d)	W. Stephenson (London)
74.	C. Ayoub (University Park, Pennsylvania) and Dinh van Huyn (Halle)
76.	E. Kiss (Budapest)
82.	F. A. Szász (Budapest)
95.	A. Widiger (Halle)
96.	E. Kiss (Budapest)
99.	E. Kiss and L. Rónyai (Budapest)
	Partially solved by
36.	R. Baer (Zürich)
45.	P. M. Cohn (London)
68.	P. M. Cohn (London)
75.	F. A. Szász (Budapest) and I. A. Amin (Cairo)
88.	P. M. Cohn (London)
90.	Á. Szendrei (Szeged) solved the part in brackets
91.	P. M. Cohn (London)
93.	Tran Trong Hue and F. A. Szász (Budapest)

BIBLIOGRAPHY

ABIAN, A.

[1] On the nilpotency of nil algebras, Amer. Math. Monthly *74* (1967), 33–34.

[2] Direct product decomposition of commutative semisimple rings, Proc. Amer. Math. Soc. *24* (1970), 502–507.

ABIAN, A.–MCWORTER, W. A.

[1] On the index of nilpotency of some nil algebras, Boll. Un. Mat. Ital. (3) *18* (1963), 252–255.

ABRHAN, I.

[1] О множествах нильпотентных элементов и радикалах прямого произведения полугрупп, Mat. Časopis *18* (1968), 25–33.

[2] Note on the set of nilpotent elements and on radicals of semigroups, Mat. Časopis *21* (1971), 124–130.

AHSAN, J.

[1] Rings all of whose cyclic modules are quasi-injective, Proc. London Math. Soc. (3) *27* (1973), 425–439.

ALBU, T.

[1] Modules décomposables de Dickson, C. R. Acad. Sci. Paris, Sér. A–B, *273* (1971), A369–A372.

[2] Un critère de décomposabilité des modules de torsion, Bull. Math. Soc. Sci. Math. R. S. Roumanie *15* (1971), 3–8.

ALIN, J. S.–ARMENDARIZ, E. P.

[1] TTF-classes over perfect rings, J. Austral. Math. Soc. *11* (1970), 499–503.

ALLEN, H. P.

[1] Invariant radical splittings: a Hopf approach, J. Pure Appl. Algebra *3* (1973), 1–19.

ALMEIDA COSTA, A.

[1] Sur les idéaux nucléaires d'un demi-anneau, Univ. Lisboa Rev. Fac. Ci. A (2) *11* (1965/66), 277–293.

Almeida Costa, A.–Noronha Galvão, M. L.

[1] Sur les demi-anneaux réticulés, Univ. Lisboa Rev. Fac. Ci. A (2) *14* (1972/73), 195–208.

Amitsur, S. A.

[1] Nil *PI*-rings, Proc. Amer. Math. Soc. *2* (1951), 538–540.
[2] An embedding of *PI*-rings, Proc. Amer. Math. Soc. *3* (1952), 3–9.
[3] The problem of Kurosh–Levitzki–Jacobson, Riveon Lemmatematika *5* (1952), 41–48.
[4] A general theory of radicals I, Radicals in complete lattices, Amer. J. Math. *74* (1952), 774–786.
[5] A general theory of radicals II, Radicals in rings and bicategories, Amer. J. Math. *76* (1954), 100–125.
[6] A general theory of radicals III, Applications, Amer. J. Math. *76* (1954), 126–136.
[7] The identities of *PI*-rings, Proc. Amer. Math. Soc. *4* (1953), 27–34.
[8] On rings with identities, J. London Math. Soc. *30* (1955), 464–470.
[9] The radical ring generated by a single element, Riveon Lemmatematika *9* (1955), 41–44.
[10] Radicals of polynomial rings, Canad. J. Math. *8* (1956), 355–361.
[11] Algebras over infinite fields, Proc. Amer. Math. Soc. *7* (1956), 35–48.
[12] A generalization of Hilbert's Nullstellensatz, Proc. Amer. Math. Soc. *8* (1957), 649–656.
[13] The radical of field extensions, Bull. Res. Council Israel, Sect. F7F (1957/1958), 1–10.
[14] Rings with a pivotal monomial, Proc. Amer. Math. Soc. *9* (1958), 635–642.
[15] On the semisimplicity of group algebras, Michigan Math. J. *6* (1959), 251–253.
[16] Nil semi-groups of rings with a polynomial identity, Nagoya Math. J. *27* (1966), 103–111.
[17] Prime rings having polynomial identities with arbitrary coefficients, Proc. London Math. Soc. (3) *17* (1967), 470–486.
[18] Embeddings in matrix rings, Pacific J. Math. *36* (1971), 21–29.
[19] Rings of quotients and Morita contexts, J. Algebra *17* (1971), 273–298.
[20] Nil radicals. Historical notes and some new results, Colloquia Math. Soc. J. Bolyai, 6. Rings, Modules and Radicals, Bolyai J. Mat. Társulat, Budapest 1973, 47–65.

Anderson, T.

[1] The Levitzki radical in varieties of algebras, Math. Ann. *194* (1971), 27–34.

Anderson, T.–Divinsky, N.–Suliński, A.

[1] Hereditary radicals in associative and alternative rings, Canad. J. Math. *17* (1965), 594–603.
[2] Lower radical properties for associative and alternative rings, J. London Math. Soc. *41* (1966), 417–424.

Anderson, T.–Heinicke, A. G.

[1] A note on the Jacobson and Brown–McCoy radicals, Canad. Math. Bull. *11* (1968), 737–738.

(Andriyanov, V. I.) Андриянов, В. И.

[1] Смешанные гамильтоновы нилькольца, Мат. Зап. Урал. Гос. Унив. *5* (1966), № 3, 15–30.
[2] Смешанные Г-кольца, Уч. Зап. Свердл. Гос. Пед. Ин-та *54* (1967), 12–21.

(ANDRIYANOV, V. I.–FREIDMAN, P. A.) Андриянов, В. И.–Фрейдман, П. А.

[1] О гамильтоновых кольцах, Уч. Зап. Свердл. Гос. Пед. Ин-та *31* (1965), 3–23.

(ANDRUNAKIEVICH, V. A.) Андрунакиевич, В. А.

[1] Полурадикальные и радикальные кольца, Докл. Акад. Наук СССР *55* (1947), 3–5.
[2] Полурадикальные кольца, Изв. Акад. Наук СССР *12* (1948), 129–178.
[3] К определению радикала кольца, Изв. Акад. Наук СССР *16* (1952), 217–224.
[4] Радикал обобщенных Q-колец, Изв. Акад. Наук СССР *18* (1954), 419–426.
[5] Кольца с условием минимальности для идеалов, Докл. Акад. Наук СССР *98* (1954), 323–326.
[6] Бирегулярные кольца, Мат. Сборник *39* (81) (1956), 447–464.
[7] Кольца с аннуляторным условием, Изв. Акад. Наук СССР *20* (1956), 547–568.
[8] Антипростые и сильно идемпотентные кольца, Изв. Акад. Наук СССР *21* (1957), 125–144.
[9] Модулярные идеалы, радикалы и полупростота колец, Успехи Мат. Наук *12* (1957), 133–139.
[10] К теории радикалов ассоциативных колец, Докл. Акад. Наук СССР *113* (1957), 487–490.
[11] Радикалы ассоциативных колец I, Мат. Сборник *44* (86) (1958), 179–212.
[12] Радикалы ассоциативных колец II, Мат. Сборник *55* (97) (1961), 329–346.
[13] Об одной характеристике наднильпотентного радикала, Успехи Мат. Наук *161* (1961), 127–130.
[14] О полурадикальных идеалах, Сибирск. Мат. Жур. *1* (1960), 547–554.
[15] Радикалы и разложение кольца, Докл. Акад. Наук СССР *145* (1962), 9–12.
[16] Первичные модули и радикал Бэра, Сибирск. Мат. Жур. *2* (1961), 801–806.
[17] Сильно регулярные кольца, Бул. Акад. Штиинце РСС Молд. (1964), № 6, 75–77.
[18] Радикалы слабо ассоциативных колец, Докл. Акад. Наук СССР *134* (1960), 1271–1274.

(ANDRUNAKIEVICH, V. A.–RYABUKHIN, JU.M.) Андрунакиевич, В. А.–Рябухин, Ю. М.

[1] Специальные модули и специальные радикалы, Докл. Акад. Наук СССР *147* (1962)' 1274–1277.
[2] Специальные модули и специальные радикалы, Памяти Н. Г. Чеботарёва, Издат. Казан. Унив. Казань 1964, 7–16.
[3] Модули и радикалы, Докл. Акад. Наук СССР *156* (1964), 991–994.
[4] Связанные кольца, Докл. Акад. Наук СССР *162* (1965), 1219–1222.
[5] Связанные кольца, Исследования по общей алгебре, Кишинёв 1965, 3–24.
[6] Наднильпотентные и подидемпотентные радикалы алгебр и радикалы модулей, Мат. Исслед. *3* (1968), № 2, 5–15.
[7] Прямые суммы алгебр с делением, Докл. Акад. Наук СССР *189* (1969), 927–929.
[8] Кольца без нильпотентных элементов и вполне простые идеалы, Докл. Акад. Наук СССР *180* (1968), 9–11.
[9] Аддитивная теория идеалов в системах с частными, Изв. Акад. Наук СССР, Сер. Мат. *31* (1967), 1057–1090.
[10] О существовании радикала Брауна–Маккоя в алгебрах Ли, Докл. Акад. Наук СССР *179* (1968), 503–506.

[11] Вполне идемпотентные кольца и обобщенный нильрадикал, Докл. Акад. Наук СССР *201* (1971), 1015–1018.

[12] Кручения в алгебрах, Докл. Акад. Наук СССР *208* (1973), 265–268.

[13] Радикалы алгебр и структурная теория, Наука, Москва 1979.

Arendt, B. D.

[1] Semisimple bands, Trans. Amer. Math. Soc. *143* (1969), 133–143.

Arens, R. F.–Kaplansky, I.

[1] Topological representation of algebras, Trans. Amer. Math. Soc. *63* (1948), 457–481.

Arhangel'skiĭ, A. V.–Wiegandt, R.

[1] A general classification of connectednesses and disconnectednesses in topology, General Top. Appl. *5* (1975), 9–33.

Armendariz, E. P.

[1] On radical extensions of rings, J. Austral. Math. Soc. *7* (1967), 552–554.

[2] Hereditary subradicals of the lower Baer radical, Publ. Math. Debrecen *15* (1968), 91–93.

[3] Closure properties in radical theory, Pacific J. Math. *26* (1968), 1–7.

[4] Quasi-injective modules and stable torsion classes, Pacific J. Math. *31* (1969), 277–280.

[5] On finite-dimensional torsion-free modules and rings, Proc. Amer. Math. Soc. *24* (1970), 566–571.

Armendariz, E. P.–Fisher, J. W.

[1] Regular *PI*-rings, Proc. Amer. Math. Soc. *39* (1973), 247–251.

Armendariz, E. P.–Leavitt, W. G.

[1] Non-hereditary semisimple classes, Proc. Amer. Math. Soc. *18* (1967), 1114–1117.

[2] The hereditary property in the lower radical construction, Canad. J. Math. *20* (1968), 474–476.

(Arnautov, V. I.) Арнаутов, В. И.

[1] Топологический радикал Бэра и разложение кольца, Сибирск. Мат. Жур. *5* (1964), 1209–1227.

[2] Дополнительные радикалы в топологических кольцах, I–II, Мат. Исслед. *3* (1968), № 2, 16–30; *4* (1969), № 1, 3–15.

[3] К теории топологических колец, Сибирск. Мат. Жур. *6* (1965), 249–261.

[4] Радикалы в кольцах с базисом групповых окрестностей нуля, Мат. Исслед. *3* (1968), № 4, 3–17.

[5] Алгебраические радикалы в топологических кольцах, Мат. Исслед. *4* (1969), № 2, 116–122.

[6] Вложение радикалов, Мат. Исслед. *4* (1969), № 3, 3–20.

[7] Радикалы в ограниченных кольцах, Бул. Акад. Штиинце РСС Молд. Сер. Физ.-Техн. и Мат. Н. *14* (1969), № 2, 3–8.

(Arnautov, V. I.–Vodinchar, M. I) Арнаутов, В. И.–Водинчар, М. И.

[1] Радикалы топологических колец, Мат. Исслед. *3* (1968), № 2, 31–61.

(ARSHINOV, M. N.–SADOVSKI, L. E.) Аршинов, М. Н.–Садовский, Л. Е.

[1] Некоторые теоретико-структурные свойства групп и полугрупп, Успехи Мат. Наук *27* (1972), № 6, 139–180.

ARTIN, E.–NESBITT, C.–THRALL, R. M.

[1] Rings with minimum condition, Univ. Michigan, Michigan 1944.

ASANO, S.

[1] On the radical of quasi-Frobenius algebras, Kodai Math. Sem. Rep. *13* (1961), 135–151.

[2] Remarks concerning two quasi-Frobenius rings with isomorphic radicals, Kodai Math. Sem. Rep. *13* (1961), 227–234.

(ASATIANI, R. V.) Асатиани, Р. В.

[1] Радикалы стабильного типа в группах автоморфизмов модулей, Сакарт. ССР Мецн. Акад. Моамбе *47* (1967), 13–18.

ASAWA, K.

[1] Über Ringe mit Vielfachenkettensatz, Proc. Imp. Acad. Tokyo *15* (1939), 288–291.

AUBERT, K. E.

[1] Sur le radical de McCoy, C. R. Acad. Sci. Paris *237* (1953), 10–12.

AULT, J. C.–WATTERS, J. F.

[1] Circle groups of nilpotent rings, Amer. Math. Monthly *80* (1973), 48–52.

(BABICH, A. M.) Бабич, А. М.

[1] О радикале Левицкого, Докл. Акад. Наук СССР *126* (1959), 242–243.

BAER, R.

[1] Radicals ideals, Amer. J. Math. *65* (1943), 537–568.

[2] Metaideals, in: Report of a Conference on Linear Algebras, Nat. Acad. Sci. Washington D. C. 1957, 33–52.

BAER, R.–HEINEKEN, H.

[1] Radical groups of finite Abelian subgroup rank, Illinois J. Math. *16* (1972), 533–580.

BALLIEU, R.

[1] Anneaux finis, systèmes hypercomplexes de rang trois sur un corps commutatif, Ann. Soc. Sci. Brux. Sér. 1, *61* (1947), 222–227.

BARBUT, E.

[1] On nil semirings with ascending chain conditions, Fund. Math. *68* (1970), 261–264.

BARNES, D. W.

[1] On the radical of a ring with minimum condition, J. Austral. Math. Soc. *5* (1965), 234–236.

BARROS, C. M. DE

[1] Une caractérisation des anneaux fortement réguliers, Rev. Colombiana Mat. *2* (1968), 1–5.

BARTOLOZZI, F.

[1] Anelli artiniani con radicale a quadrato nullo, Rend. Circ. Mat. Palermo (2) *19* (1970), 113–122.

BASS, H.

[1] Finitistic dimension and a homological generalization of semi-primary rings, Trans. Amer. Math. Soc. *95* (1960), 466–488.

BAUMSLAG, G.

[1] A universal approach to groups and rings, Lecture Notes, Courant Institute of Math. Sciences, New York Univ. New York 1963.

BEACHY, J. A.

[1] Cotorsion radicals and projective modules, Bull. Austral. Math. Soc. *5* (1971), 241–253.
[2] On quasi-artinian rings, J. London Math. Soc. *3* (1971), 449–452.
[3] On maximal torsion radicals, Canad. J. Math. *25* (1973), 712–726.

BECHEANU, M.

[1] Remarque sur les radicaux spéciaux, Rev. Roumaine Math. Pur. Appl. *10* (1965), 357–360.

BECHTELL, H.

[1] Nongenerators of rings, Proc. Amer. Math. Soc. *19* (1968), 241–245.

BEHRENS, E. A.

[1] Nichtassoziative Ringe, Math. Ann. *127* (1954), 441–452.
[2] Distributiv darstellbare Ringe, Math. Z. *73* (1960), 409–432.

BEIDLEMAN, J. C.

[1] Quasiregularity in near rings, Math. Z. *89* (1965), 224–229.
[2] A radical for near ring modules, Michigan Math. J. *12* (1965), 377–383.
[3] On the theory of radicals in distributively generated near-rings I–II, Math. Ann. *173* (1967), 89–101, 200–218.
[4] A note on regular near-rings, J. Indian Math. Soc. *33* (1969), 207–209.

BELLUCE, L. P.–JAIN, S. K.

[1] A remark on primitive rings and *I*-pivotal monomials, J. Math. Sci. *1* (1966), 49–51.

BERGMAN, G. M.

[1] A ring primitive on the right but not on the left, Proc. Amer. Math. Soc. *15* (1964), 473–475.

BERNHARDT, R. L.

[1] Splitting hereditary torsion theories over semiperfect rings, Proc. Amer. Math. Soc. *22* (1969), 681–687.

[2] On splitting in hereditary torsion theories, Pacific J. Math. *39* (1971), 31–38.

BETSCH, G.

[1] Ein Radikal für Fastringe, Math. Z. *78* (1962), 86–90.

[2] Struktursätze für Fastringe, Diss. Univ. Tübingen 1963.

BICAN, L.

[1] The lattice of radical filters of a commutative Noetherian ring, Comment. Math. Univ. Carolinae *12* (1971), 53–59.

BIGGS, R. G.

[1] A new bound for nil *U*-rings, Canad. J. Math. *22* (1970), 403–407.

BIRKHOFF, G.

[1] Subdirect unions in universal algebra, Bull. Amer. Math. Soc. *50* (1944), 764–768.

BJÖRK, J. E.

[1] Rings satisfying a chain condition on principal ideals, J. Reine Angew. Math. *236* (1969), 112–119.

[2] On quasimodular maximal right ideals of primitive rings generated by their minimal right ideals, Acta Math. Acad. Sci. Hung. *21* (1970), 193–197.

[3] Conditions which imply that subrings of artinian rings are artinian, J. Reine Angew. Math. *247* (1971), 123–138.

[4] Radical properties of perfect modules, J. Reine Angew. Math. *253* (1972), 78–86.

[5] Noetherian and artinian chain conditions of associative rings, Arch. Math. (Basel) *24* (1973), 366–378.

BLACKETT, D. W.

[1] Simple and semisimple near rings, Proc. Amer. Math. Soc. *4* (1953), 772–785.

BLAIR, R. L.

[1] Ideal lattices and the structure of rings, Trans. Amer. Math. Soc. *75* (1953), 135–153.

[2] A note on *f*-regularity in rings, Proc. Amer. Math. Soc. *6* (1955), 511–515.

BLAND, P. E.

[1] Perfect torsion theories, Proc. Amer. Math. Soc. *41* (1973), 349–355.

BLOCK, R. E.

[1] A unification of the theories of Jordan and alternative algebras, Amer. J. Math. *94* (1972), 389–412.

BOSÁK, J.

[1] О радикалах полугрупп, Mat. Fyz. Časopis *12* (1962), 230–234.

[2] On radicals of semigroups, Mat. Fyz. Časopis *18* (1968), 204–211.

Bostok, F. A.–Patterson, E. M.

[1] A generalization of Divinsky's radical, Proc. Glasgow Math. Soc. *6* (1963), 75–87.

Bourbaki, N.

[1] Éléments de Mathématique, Fascicule 23, Algèbre, Chapitre 8, Modules et anneaux semisimples, Actualité Sci. Industr. 1261, Hermann, Paris 1958.

Bourne, S.

[1] The Jacobson radical of a semiring, Proc. Nat. Acad. Sci. USA *37* (1951), 163–170.

Bourne, S.–Zassenhaus, H.

[1] On the semiradical of a semiring, Proc. Nat. Acad. Sci. USA *44* (1958), 907–914.

(Bovdi, A. A.) Бовди, А. А.

[1] О радикале групповой алгебры, Докл. и Сообщ. Ужгородск. Унив. Серия Физ.-Мат. Наук *4* (1961), 84–85.

Brameret, M. P.

[1] Treillis d'idéaux et structure d'anneaux, C. R. Acad. Sci. Paris *255* (1962), 1434–1435.

Brauer, R.

[1] On the nilpotency of the radical of a ring, Bull. Amer. Math. Soc. *48* (1942), 752-758.

Bridger, M.

[1] An ideal criterion for torsion freeness, Proc. Amer. Math. Soc. *33* (1972), 285–291.

Brown, B.

[1] An extension of the Jacobson radical, Proc. Amer. Math. Soc. *2* (1951), 114–117.

Brown, B.–McCoy, N. H.

[1] Radicals and subdirect sums, Amer. J. Math. *69* (1947), 46–58.
[2] The radical of a ring, Duke Math. J. *15* (1948), 495–599.
[3] The maximal regular ideal of a ring, Proc. Amer. Math. Soc. *1* (1950), 165–171.
[4] Some theorems on groups with applications to ring theory, Trans. Amer. Math. Soc. *69* (1950), 302–311.

Brown, W. C.

[1] A splitting theorem for algebras over commutative von Neumann regular rings, Proc. Amer. Math. Soc. *36* (1972), 369–374.

Büke, A.

[1] Nilpotente Algebren von Index 3 über einem Körper der Charakteristik 2, Rev. Fac. Sci. Univ. Istambul *622* (1957), 84–89.

(BUSHUYEV, V. F.) Бушуев, В. Ф.

[1] О радикалах в кольцах Рисса, Уч. Зап. Ивановск. Гос. Пед. Ин-та *61* (1969), 84–113.

[2] Специальные радикалы в кольцах Рисса, Уч. Зап. Ивановск. Гос. Пед. Ин-та *61* (1969), 114–131.

(BUTSARKINA, L. P.) Буцаркина, Л. П.

[1] О кольцах первой ступени с нильпотентными образующими, Мат. Зап. Урал. Гос. Унив. *3* (1962), 25–29.

CAHEN, P.-J.

[1] Torsion theory and associated primes, Proc. Amer. Math. Soc. *38* (1973), 471–476.

CALAIS, JOSETTE

[1] Demi-groupes quasi-inversifs, C. R. Acad. Sci. Paris *252* (1961), 2357–2359.

CAMPBELL, J. M.

[1] Torsion theories and coherent rings, Bull. Austral. Math. Soc. *8* (1973), 233–239.

CARNS, G. L.–CHAO, C.

[1] On the radical of the group algebra of a p-group over a modular field, Proc. Amer. Math. Soc. *33* (1972), 323–328.

CARREAU, F.

[1] Théorie générale des radicaux, Ph. D. Thesis, Univ. Montreal 1967.

[2] Sous-catégories réflexives et la théorie générale des radicaux, Fund. Math. *71* (1971), 223–242.

CARTAN, H.–EILENBERG, S.

[1] Homological Algebra, Princeton University Press, Princeton 1956.

CATEFORIS, V. C.

[1] On regular self-injective rings, Pacific J. Math. *30* (1969), 39–45.

CATTANEO, G. M.

[1] Sulle sottoalgebre semisemplici di un anello semplice e artiniano, Rend. Mat. (6) *2* (1969), 15–21.

CHALABI, A.

[1] Modules over group algebras and their application in the study of semi-simplicity, Math. Ann. *201* (1973), 57–63.

CHANDRA, H.

[1] On the radical of a Lie algebra, Proc. Amer. Math. Soc. *1* (1950), 14–17.

CHANDRAN, V. R.

[1] A note on a problem of F. Szász, Acta Math. Acad. Sci. Hung. *24* (1973), 85–86.

CHATTERS, A. W.

[1] A decomposition theorem for Noetherian hereditary rings, Bull. London Math. Soc. *4* (1972), 125–126.

CHATTERS, A. W.–HEINICKE, A. G.

[1] Localization at a torsion theory in hereditary Noetherian rings, Proc. London Math. Soc. (3) *27* (1973), 193–204.

CHEATHAM, T.

[1] Direct sums of torsion-free covers, Canad. J. Math. *25* (1973), 1002—1005.

(CHEREMISIN, A. I.) Черемисин, А. И.

[1] О кольцах Рисса, Сибирск. Мат. Жур. *11* (1970), 199–212.

[2] Структурная теория Ф-колец, Сибирск. Мат. Жур. *11* (1970), 879–895; Исправления, ib. *14* (1973), 450–451.

CHEW, B. S.–NEGGERS, J.

[1] Local rings with left vanishing radical, J. London Math. Soc. (2) *4* (1971), 374–378.

CHEW, K. L.–ANG, C. H.

[1] Rings with ω-nilpotent radicals, Nanta Math. *5* (1971), № 1, 1–11.

CHIBA, K.–TOMINAGA, H.

[1] Note on strongly regular rings and P_1-rings, Proc. Japan Acad. *51* (1975), 259–261.

CHUNG, L. W.–LUH, J.

[1] Characterization of semisimple Artinian rings, Math. Japon. *21* (1976), 226–228.

CLARK, W. E.

[1] Generalized radical rings, Canad. J. Math. *20* (1968), 88–94.

CLARK, W. E.–LEWIN, I.

[1] On minimal ideals in the circle composition semigroup of a ring, Publ. Math. Debrecen *14* (1967), 99–104.

CLARKE, F. M.

[1] Note on quasiregularity and the Perlis–Jacobson radical, Portugaliae Math. *11* (1952), 89–94.

CLARKE, R. J.

[1] On the radical of the centre of a group algebra, J. London Math. Soc. (2) *1* (1969), 565–572.

[2] On the radical of the group algebra of a *p*-nilpotent group, J. Austral. Math. Soc. *13* (1972), 119–123.

CLIFFORD, A. H.

[1] Semigroups admitting relative inverses, Ann. of Math. *42* (1941), 1037–1049.
[2] Semigroups without nilpotent ideals, Amer. J. Math. *71* (1949), 46–58.
[3] Bands of semigroups, Proc. Amer. Math. Soc. *5* (1954), 499–504.
[4] Radicals in semigroups, Semigroup Forum *1* (1970), 103–127.

CLIFFORD, A. H.–MILLER, D. D.

[1] Semigroups having zeroid elements, Amer. J. Math. *70* (1948), 117–125.

COHN, P. M.

[1] On a class of simple rings, Mathematica *5* (1958), 103–117.
[2] Simple rings without zero-divisors and Lie division rings, Mathematica *6* (1959), 14–18.
[3] Morita equivalence and duality, Queen Mary Coll. Math. Notes, Queen Mary College, London 1970.
[4] Free radical rings, Colloquia Math. Soc. J. Bolyai, 6. Rings, Modules and Radicals, Bolyai J. Mat. Társulat, Budapest 1973, 135–145.
[5] The embedding of radical rings in simple radical rings, Bull. London Math. Soc. *3* (1971), 185–188; Corrections: ib. *4* (1972), 54 and ib. *5* (1973), 322.

CONRAD, P. F.

[1] The relationship between the radical of a lattice-ordered group and complete distributivity, Pacific J. Math. *14* (1964), 494–499.

COSTA, A. A.

[1] Sistemas Hipercomplexos e Representacoes, Centro de Estudos Matematicos, Faculdade de Ciencias do Porto 1948.
[2] On semiprimary rings, Centro Estudos Mat. Fac. Ci. Porto *14* (1945), 1–38.
[3] Sobre nilideais e ideais quasiregulares, Anais Fac. Ci. Porto *34* (1950), 1–28.

COURTER, R.

[1] Finite direct sums of complete matrix rings over perfect completely primary rings, Canad. J. Math. *21* (1969), 430–446.

DELANGHE, R.

[1] On the center and the radical of a Clifford algebra, Simon Stevin *42* (1968/69), 123–131.

DESKINS, W. E.

[1] A radical for near rings, Proc. Amer. Math. Soc. *5* (1954), 825–827.
[2] On the radical of a group algebra, Pacific J. Math. *8* (1958), 693–697.

DICKSON, S. E.

[1] Direct decompositions of radicals, Proc. Confer. Categorical Algebra, La Jolla, California 1965, 366–374.
[2] A note on hypernilpotent radical properties for associative rings, Canad. J. Math. *19* (1967), 447–448.

[3] A torsion theory for abelian categories, Trans. Amer. Math. Soc. *121* (1966), 223–225.
[4] Noetherian splitting rings are artinian, J. London Math. Soc. *42* (1967), 732–736.
[5] Decomposition of modules II. Rings without chain conditions, Math. Z. *104* (1968), 349–357.

DIEM, J. E.

[1] A radical for lattice ordered rings, Pacific J. Math. *25* (1968), 71–82.

DIVINSKY, N.

[1] Rings and Radicals, Allen, London 1965.
[2] Pseudo-regularity, Canad. J. Math. *7* (1955), 401–410.
[3] *D*-regularity, Proc. Amer. Math. Soc. *9* (1958), 62–71.
[4] On simple semiradical and radical algebras, J. London Math. Soc. *34* (1959), 225–250.
[5] General radicals that coincide with the classical radical on rings with D. C. C. Canad. J. Math. *13* (1961), 639–644.
[6] Duality between radical and semisimple classes of associative rings, Scripta Math. *29* (1973), 409–416.

DIVINSKY, N.–KREMPA, J.–SULIŃSKI, A.

[1] Strong radical properties of alternative and associative rings, J. Algebra *17* (1971), 369–388.

DIVINSKY, N.–SULIŃSKI, A.

[1] Kurosh radicals of rings with operators, Canad. J. Math. *17* (1965), 278–280.

DIXON, P. G.

[1] Semiprime Banach algebras, J. London Math. Soc. (2) *6* (1973), 676–678.

DJABALI, M.

[1] Demi-groupes nilpotents et radical d'un anneau noethérien ou artinien, C. R. Acad. Sci. Paris, Sér. A–B, *264* (1967), A493–A495.

DLAB, V.

[1] The concept of a torsion module, Amer. Math. Monthly *75* (1968), 973–976.
[2] Distinguished submodules, J. Austral. Math. Soc. *8* (1968), 661–670.
[3] A characterization of perfect rings, Pacific J. Math. *33* (1970), 79–88.
[4] On a class of perfect rings, Canad. J. Math. *22* (1970), 79–88.
[5] Structure of perfect rings, Bull. Austral. Math. Soc. *2* (1970), 117–124.

(DOMANOV, O. I.) Доманов, О. И.

[1] О вложениях неассоциативных полурадикальных колец в радикальные, Мат. Исслед. *5* (1970), № 4, 84–90.

(DOROFEEV, G. V.) Дорофеев, Г. В.

[1] О нильпотентности правоальтернативных колец, Алгебра и Логика *9* (1970), 302–305.
[2] О локально нильпотентном радикале неассоциативных колец, Алгебра и Логика *10* (1971), 355–364.

DRAZIN, M. P.

[1] Rings with nil commutator ideals, Rend. Arch. Math. Palermo (2) *6* (1957), 51–64.

[2] A generalization of polynomial identities in rings, Proc. Amer. Math. Soc. *8* (1957), 352–361.

[3] The nilpotence of nil subrings, Amer. J. Math. *79* (1957), 67–72.

DUBISCH, R.–PERLIS, S.

[1] The radical of an alternative algebra, Amer. J. Math. *70* (1948), 540–546.

[2] On total nilpotent algebra, Amer. J. Math. *73* (1951), 439–452.

(DYMENT, Z. M.) Дымент, З. М.

[1] О максимальных коммутативных подалгебрах полной матричной алгебры, Весці Акад. Навук БССР, Сер. Фіз.-Мат. Х. (1968), № 1, 17–24.

[2] О максимальных коммутативных нильпотентных подалгебрах класса нильпотентности из полной матричной алгебры, Весці Акад. Навук БССР, Сер. Фіз.-Мат. Х. (1971), № 6, 46–52.

EAGON, J. A.

[1] Finitely generated domains over Jacobson semi-simple rings are Jacobson semi-simple, Amer. Math. Monthly *74* (1967), 1091–1092.

ECKSTEIN, F.

[1] Complete semisimple rings with ideal neighbourhoods of zero, Arch. Math. Basel *18* (1967), 587–590.

[2] Semi-direct splitting of the radical, Abh. Math. Sem. Univ. Hamburg *32* (1968), 61–72.

[3] Semigroup methods in ring theory, J. Algebra *12* (1969), 177–190.

EGGERT, N. H.

[1] Quasi regular groups of finite commutative nilpotent algebras, Pacific J. Math. *36* (1971), 631–634.

EHRLICH, G.

[1] Unit-regular rings, Portugal Math. *27* (1968), 209–212.

ELDRIDGE, K. E.

[1] On ring structures determined by groups, Proc. Amer. Math. Soc. *23* (1969), 472–577.

(ELIZAROV, V. P.) Елизаров, В. П.

[1] Радикальное кольцо частных, Сибирск. Мат. Жур. *3* (1962), 360–367.

[2] О теоремах Голди, Сибирск. Мат. Жур. *10* (1969), 58–63.

[3] Сильные предкручения и сильные фильтры, модули и кольца частных, Сибирск. Мат. Жур. *14* (1973), 549–559.

ELLIGER, S.

[1] Interdirekte Summen von Moduln, J. Algebra *18* (1971), 271–303.

ENESCU, I.

[1] Algèbres associatives sur un corps noncommutatif, Bul. Inst. Politehn. Iaşi, Sect. I, *15* (19) (1969), № 1–2, 9–15.

EQBAL, A.–WIEGANDT, R.

[1] On lower radicals of semigroups, *57* (1973), 163–167.

ERICKSON, T. S.–MONTGOMERY, S.

[1] The prime radical in special Jordan rings, Trans. Amer. Math. Soc. *156* (1971), 155–164.

ERTEL, G.

[1] Structure des $\mathfrak{F}$-algèbres, C. R. Acad. Sci. Paris, Sér. A–B, *272* (1971), A520–A522, A570–A572.

[2] Structure des $\mathfrak{F}$-algèbres dans le cas où le corps de base est de caractéristique 2, C. R. Acad. Sci. Paris, Sér. A–B, *274* (1972), A381–A383.

FAITH, C. C.

[1] Rings with minimum condition on principal ideals, I, Archiv. Math. *10* (1959), 327–330; Erratum, Archiv. Math. *10* (1959), 480.

[2] Rings with minimum condition on principal ideals, II, Archiv. Math. *12* (1961), 179–181.

[3] Radical extension of rings, Proc. Amer. Math. Soc. *12* (1961), 274–283.

[4] Lectures on injective modules and quotient rings, Lecture Notes in Math. Vol. 49, Springer-Verlag, Berlin–Heidelberg–New York 1967.

[5] Algebra: Rings, Modules and Categories I, Springer-Verlag, Berlin–Heidelberg–New York 1973.

FAITH, C. C.–UTUMI, Y.

[1] On a new proof of Litoff's theorem, Acta Math. Acad. Sci. Hung. *14* (1963), 369–372·

(FAIZULLAEV, ZH.) Файзуллаев, Ж.

[1] Об одном описании полупростого класса модулей, Докл. Акад. Фанхои РСС Точикистон *15* (1972), № 7, 15–17.

FARÊS, N.–HAGE, I.

[1] Sur une classe d'anneaux semi-simples et un type d'idempotents, C. R. Acad. Sci. Paris, Sér. A–B, *274* (1972), A817–A820.

FERRERO, G.

[1] Gli stems *p*-singolari con radicale proprin, Inst. Lombardo Accad. Sci. Lett. Rend. A *104* (1970), 91–105.

FIELDHOUSE, D. J.

[1] Regular modules and rings, Illinois J. Math. *16* (1972), 217–221.

[2] Regular rings and modules, J. Austral. Math. Soc. *13* (1972), 477–491.

[3] Regular modules over semi-local rings, Colloquia Math. Soc. J. Bolyai, 6. Rings, Modules and Radicals, Bolyai J. Mat. Társulat, Budapest 1973, 193–196.

FISCHER, E.–STRUIK, R. R.

[1] Nil algebras and periodic groups, Amer. Math. Monthly *75* (1968), 611–623.

FISHER, J. W.

[1] On the nilpotency of nil subrings, Canad. J. Math. *22* (1970), 1211–1216.
[2] Nil subrings with bounded indices of nilpotency, J. Algebra *19* (1971), 509–516.
[3] Nil subrings of endomorphism rings of modules, Proc. Amer. Math. Soc. *34* (1972), 75–78.
[4] Structure of semiprime *PI*-rings, Proc. Amer. Math. Soc. *39* (1973), 465–467.

FISHER, J. W.–SNIDER, R. L.

[1] Prime von Neumann regular rings and primitive group algebras, Proc. Amer. Math. Soc. *44* (1974), 244–250.

FLANIGAN, F. J.

[1] Radical behavior and the Wedderburn family, Bull. Amer. Math. Soc. *79* (1973), 66–70.

FORMANEK, E.

[1] A problem of Passman on semisimplicity, Bull. London Math. Soc. *4* (1972), 375–376.

FORSYTHE, A.–MCCOY, N. H.

[1] On the commutativity of certain rings, Bull. Amer. Math. Soc. *52* (1946), 523-526.

FORT, J.

[1] Radical tertaire d'un sousmodule et sousmodules tertaires dans un module sur un anneau non nécessairement commutatif, C. R. Acad. Sci. Paris *254* (1962), 1900–1902.

FOSTER, D. M.

[1] Radicals and bimodules, Proc. Amer. Math. Soc. *38* (1973), 47–52.

FOUNTAIN, J. B.

[1] Nilpotent principal ideal rings, Proc. London Math. Soc. (3) *20* (1970), 348–364.

(FREIDMAN, P. A.) Фрейдман, П. А.

[1] О кольцах с идеализаторным условием I, Изв. Высш. Учеб. Завед. Мат. *2* (1960), 213–222.
[2] О кольцах с идеализаторным условием II, Учен. Зап. Урал. Гос. Унив. *23* (1959), 35–47.
[3] О кольцах с идеализаторным условием III, Учен. Зап. Урал. Гос. Унив. *24* (1960), 49–61.
[4] К теории радикала ассоциативного кольца, Изв. Высш. Учеб. Завед. Мат. *3* (1958), 225–232.
[5] О кольцах разрешимого типа, Автореферат дисс. Свердловск 1960.
[6] О некоторых условиях конечности в ассоциативных кольца, Мат. Зап. Урал. Гос. Унив. *3* (1961), 77–84.

[7] О кольцах все подкольца которых являются метаидеалами конечного индекса, Мат. Сборник *65* (1964), 313–323.
[8] Кольца с правым идеализаторным условием, Мат. Зап. Урал. Гос. Унив. *4* (1963), 51–58.
[9] Кольца с правым идеализаторным условием, Изв. Высш. Учеб. Завед. Мат. *2* (15) (1960), 213–222.
[10] О кольцах с идеализаторным условием для бесконечных подколец, Мат. Зап. Уральск. Ун-та *7* (1970), 213–221.
[11] Кольца, все коммутативные подкольца которых артиновы, Изв. Высш. Учеб. Завед. Мат. (1973), № 10, 83–89.

(Freidman, P. A.–Andriyanov, V. I.) Фрейдман, П. А.–Андриянов, Б. И.
[1] Гамильтонские кольца, Симпоз. Алгебр. Тарту, Гос. Унив. 1960, 120–121.

Fried, E.
[1] Beiträge zur Theorie der Frobenius-Algebren, Math. Ann. *155* (1964), 265–269.
[2] A teljes mátrixgyűrű egy tulajdonsága (A property of the full matrix ring), Mat. Lapok *20* (1969), 63–70.

Fried, E.–Wiegandt, R.
[1] Connectednesses and disconnectednesses of graphs, Algebra Univ. *5* (1975), 411–428.

Fröhlich, A.
[1] Radical modules over a Dedekind domain, Nagoya Math. J. *27* (1966), 643–662.

Fuchs, L.
[1] Abelian Groups, Akadémiai Kiadó, Budapest 1958.
[2] A remark on the Jacobson radical, Acta Sci. Math. Szeged *14* (1952), 157–168.
[3] On a new type of radical, Acta Sci. Math. Szeged *16* (1955), 43–53.
[4] Torsion preradicals and ascending Loewy series of modules, J. Reine Angew. Math. *239/240* (1969), 169–179.

Fuchs, L.–Halperin, I.
[1] On the embedding of a regular ring in a regular ring with identity, Fund. Math. *54* (1964), 285–290.

Fuchs, L.–Szele, T.
[1] Contribution to the theory of semisimple rings, Acta Math. Acad. Sci. Hung. *3* (1952), 233–239.

Fuelberth, J. D.
[1] On commutative splitting rings, Proc. London Math. Soc. (3) *20* (1970), 393–408.

Gabriel, P.–Oberst, U.
[1] Spektralkategorien und reguläre Ringe im von Neumannschen Sinn, Math. Z. *92* (1966), 389–395.

GARDNER, B. J.

[1] Torsion classes and pure subgroups, Pacific J. Math. *33* (1970), 109–116.
[2] A note on types, Bull. Austral. Math. Soc. *2* (1970), 275–276.
[3] Radicals of abelian groups and associative rings, Univ. Tasmania, Techn. Rep. № 36, March 1971.
[4] Rings whose modules form few torsion classes, Bull. Austral. Math. Soc. *4* (1971), 355–359.
[5] A note on radicals and polynomial rings, Math. Scand. *31* (1972), 83–88.
[6] Some closure properties for torsion classes of abelian groups, Pacific J. Math. *42* (1972), 45–61.
[7] Two notes on radicals of Abelian groups, Comment. Math. Univ. Carolinae *13* (1972), 419–430.
[8] Radicals of abelian groups and associative rings, Acta Math. Acad. Sci. Hung. *24* (1973), 259–268.
[9] Generalized pure-hereditary radical classes of abelian groups, Comment. Math. Univ. Carolinae *14* (1973), 187–195.
[10] Some remarks on radicals of rings with chain conditions, Acta Math. Acad. Sci. Hung. *25* (1974), 263–268.
[11] On the scarcity of nilpotent radical classes, Ann. Univ. Sci. Budapest R. Eötvös, Sect. Math. *16* (1973), 23–24.
[12] Some radical constructions for associative rings, J. Austral. Math. Soc. *18* (1974), 442–446.
[13] Some aspects of T-nilpotence, Pacific J. Math. *53* (1974), 117–130.

(GELFAND, I. M.–SHILOV, G. E.) Гельфанд, И. М.–Шилов, Г. Е.

[1] О различных методах введения топологии в множестве максимальных идеалов нормированного кольца, Мат. Сборник *9* (51) (1941), 25–38.

(GEMINTERN, V. I.) Геминтерн, В. И.

[1] О некоторых типах матричных колец, Сибирск. Мат. Жур. *5* (1964), 310–318.
[2] Полупервичные кольца с атомной структурой левых идеалов, Сибирск. Мат. Жур. *8* (1967), 947–951.

GERSTENHABER, M.

[1] On nil algebras and linear varieties of nilpotent matrices I, Amer. J. Math. *80* (1958), 614–622.
[2] On nil algebras and linear varieties of nilpotent matrices II, Duke Math. J. *27* (1960), 21–31.
[3] On nil algebras and linear varieties of nilpotent matrices III, Ann. of Math. *70* (1959), 167–205.
[4] On nil algebras and linear varieties of nilpotent matrices IV, Ann. of Math. *75* (1942), 382–418.

GILMER, R. W. JR.

[1] The pseudo-radical of a commutative ring, Pacific J. Math. *19* (1966), 275–284.

GOLAN, J. S.–TEPLY, M. L.

[1] Torsion-free covers, Israel J. Math. *15* (1973), 237–256.

GOLDIE, A. W.

[1] Decompositions of semisimple rings, J. London Math. Soc. *3* (1956), 40–48.

[2] The structure of prime rings under ascending chain condition, Proc. London Math. Soc. *8* (1958), 589–608.

[3] The structure of prime rings with maximum condition, Proc. Nat. Acad. Sci. USA *44* (1958), 584–586.

[4] Semiprime rings with maximum condition, Proc. London Math. Soc. *10* (1960), 201–220.

GOLDMAN, O.

[1] A characterization of semisimple rings with the descending chain condition, Bull. Amer. Math. Soc. *52* (1946), 1021–1027; Addition, Bull. Amer. Math. Soc. *53* (1947), 956.

[2] Semisimple extensions of rings, Bull. Amer. Math. Soc. *52* (1946), 1028–1032.

(GOLOD, E. S.) Голод, Е. С.

[1] О нильалгебрах и финитно аппроксимируемых p-группах, Изв. Акад. Наук СССР *28* (1964), 273–276.

(GOLOD, E. S.–SHAFAREVICH, I. R.) Голод, Е. С.–Шафаревич, И. Р.

[1] О башне полей классов, Изв. Акад. Наук СССР *28* (1964), 261–272.

(GOLOVINA, L. I.) Головина, Л. И.

[1] Коммутативные радикальные кольца, Изв. Акад. Наук СССР *14* (1950), 449–472.

GÓMEZ PARDO, J. L.

[1] Teorías torsión en categorías exactas, Algebra 9. Univ. de Santiago de Compostela, Depart. de Algebra y Fundamentos, España 1972.

GOODEARL, K. R.

[1] Singular torsion and the splitting properties, Memoirs Amer. Math. Soc. *124* (1972).

[2] Prime ideals in regular self-injective rings, Canad. J. Math. *25* (1973), 829–839.

[3] Von Neumann Regular Rings, Pitman, London 1978.

GOPALA RAO, N. R.

[1] On regular rings, Indian J. Math. *10* (1968), 95–100.

(GORBACHUK, E. L.) Горбачук, Е. Л.

[1] Расщепляемость кручения и предкручения в категории правых Λ-модулей, Мат. Заметки *2* (1967), 681–688.

[2] Радикалы в модулях над разными кольцами, Мат. Исслед. *7* (1972), № 1, 44–59.

[3] Коммутативные кольца, над которыми все кручения расщепляемы, Мат. Исслед. *7* (1972), № 2, 81–90.

GORDON, R.–LENAGAN, T. H.–ROBSON, J. C.

[1] Krull dimension-nilpotency and Gabriel dimension, Bull. Amer. Math. Soc. *79* (1973), 716–719.

GORDON, R.–ROBSON, J. C.

[1] Krull dimension, Memoirs Amer. Math. Soc. *133* (1973).

GORMAN, H. E.

[1] Radical regularity in differential rings, Canad. J. Math. *23* (1971), 197–201.
[2] Differential rings and modules, Scripta Math. *29* (1973), 25–35.

GOULDING, T. L.–ORTIZ, A. H.

[1] Structure of semiprime $(p; q)$ radicals, Pacific J. Math. *37* (1971), 97–99.

GOURSAUD, J. M.

[1] Sur les anneaux de groupes semi-parfaits, Canad. J. Math. *25* (1973), 922–928.

(GOVOROV, V. E.) Говоров, В. Е.

[1] Производные радикалов модулей, Мат. Заметки *3* (1968), 633–642.
[2] О радикалах и функторах, Сибирск. Мат. Жур. *9* (1968), 1419–1421.

(GOYAN, I. M.) Гоян, И. М.

[1] Радикал Бэра почти колец, Бул. Акад. Штиинце РСС Молд. (1966), № 4, 32–38.

GRAY, M.

[1] Radical subcategories, Pacific J. Math. *23* (1967), 79–89.
[2] A radical approach to algebra, Addison–Wesley Publ. Comp. Reading/Mass.–London 1970.

GREEN, J. A.–STONEHEWER, S. E.

[1] The radicals of some group algebras, J. Algebra *13* (1969), 137–142.

(GRIGOR, R. S.) Григор, Р. С.

[1] К теории радикалов полугрупп I, Мат. Исслед. *6* (1971), № 4, 37–55.

GRILLET, M. P.

[1] Semisimple *A*-semigroups and semirings, Fund. Math. *76* (1972), 109–116.

(GRINGLAZ, L. YA.) Гринглаз, Л. Я.

[1] Локально нильпотентные почти кольца, Мат. Зап. Урал. Гос. Унив. *5* (1965), 35–42.
[2] О некоторых радикалах стабильного типа, Труды Рижск. Алг. Сем. Рига 1969, 29–50.
[3] О локально стабильном радикале представления, Мат. Зап. Уральск. Ун-та *8* (1973), № 3, 35–42.

GROBLER, J. J.

[1] Die Theorie der Primideale in Pseudoringen, Tydskr. Natuurwet. *11* (1971), 99–107.

GROSSHANS, F.

[1] Semi-simple algebraic groups defined over a real closed field, Amer. J. Math. *94* (1972), 473–485.

GRÜNENFELDER, L. A.

[1] Hopf-Algebren und Coradikal, Math. Z. *116* (1970), 166–182.

GUPTA, R. N.

[1] On the existence of the Köthe radical and its coincidence with the lower nil radical for a class of rings, Nieuw Archiv voor Wiskunde (3) *16* (1968), 90–93.
[2] On primitivity of matrix rings, Amer. Math. Monthly *75* (1968), 636–637.
[3] Characterization of rings whose classical quotient rings are perfect rings, Publ. Math. Debrecen *17* (1970), 215–222.

HAIMO, F.

[1] Endomorphism radicals which characterize some divisible groups, Ann. Univ. Sci. Budapest R. Eötvös, Sect. Math. *10* (1967), 25–29.
[2] The Jacobson radical of some endomorphism rings, Studies on Abelian Groups (Symp. Montpellier 1967), Springer-Verlag, Berlin–Heidelberg–New York 1968, 143–146.
[3] Radicals with integrity and row-finite matrices, Proc. Amer. Math. Soc. *24* (1970), 144–147.
[4] Radical and antiradical groups, Rocky Mountain J. Math. *3* (1973), 91–106.

HALBERSTADT, E.

[1] Le radical d'un annéide régulier, C. R. Acad. Sci. Paris, Sér. A–B, *270* (1970), A361–A363.
[2] Structure des annéides réguliers artiniens et semisimples, C. R. Acad. Sci. Paris, Sér. A–B, *270* (1970), A435–A437.

HALL, M.

[1] The position of the radical in an algebra, Trans. Amer. Math. Soc. *48* (1940), 391–404.

HALL, T. E.

[1] The radical of the algebra of any finite semigroup over any field, J. Austral. Math. Soc. *11* (1970), 350–352.

HAMPTON, C. R.–PASSMAN, D. S.

[1] On the semisimplicity of group rings of solvable groups, Trans. Amer. Math. Soc. *173* (1972), 289–301.

HANSEN, F.

[1] Das Jacobsonsche Radikal, modulare Rechtsideale und Frattinische Unterstrukturen in Ringen, Diss. Dokt. Naturwiss. Abt. Math. Ruhr-Univ. Bochum 1972.

[2] Lösung von Problemen und Verallgemeinerung von Sätzen aus der Theorie des Jacobsonschen Radikals, Acta Math. Acad. Sci. Hung. *23* (1972), 299–307.

[3] Lösung eines Problems bezüglich der modularen Rechtsideale, Acta Math. Acad. Sci. Hung. *24* (1973), 151–153.

HANSEN, G. W.

[1] The characterization of semi-simple Artinian rings in terms of generalized inverses, Thesis, Brigham Univ. 1968.

HARADA, M.

[1] On semi-simple abelian categories, Osaka J. Math. *7* (1970), 89-95.

HARTLEY, B.

[1] Locally nilpotent ideals of a Lie algebra, Proc. Cambridge Phil. Soc. *63* (1967), 257–272.

HARTNEY, J. F. T.

[1] On the radical theory of a distributively generated near-ring, Math. Scand. *23* (1968), 214–220.

HAUGER, G.

[1] Präradikale in Kategorien. Diss. Univ. München 1969.

[2] Präradikale und Morphismenklassen, Manuscripta Math. *2* (1970), 397–402.

HEATHERLY, H. E.

[1] Distributive near-rings, Quart. J. Math. Oxford (2) *24* (1973), 63–70.

HEIDEMA, J.

[1] The lattice of radical ideals in a commutative ring is distributive, Tydskr. Natuurwet. *10* (1970), 245–249.

HEINICKE, A. G.

[1] A note on lower radical constructions for associative rings, Canad. Math. Bull. *11* (1968), 23–30.

[2] Some results in the theory of radicals of associative rings, Ph. D. Thesis, Univ. British Columbia, Vancouver 1968.

[3] Subdirect sums, hereditary radicals and structure spaces, Proc. Amer. Math. Soc. *25* (1970), 29–33.

[4] Hereditary radical ideals in a ring, J. London Math. Soc. (2) *2* (1970), 539–543.

HELWIG, K. H.

[1] Halbeinfache Jordan-Algebren, Math. Z. *109* (1969), 1–28.

HELZER, G.

[1] Semi-primary split rings, Pacific J. Math. *44* (1973), 541–552.

HENTZEL, I. R.

[1] A note on modules and radicals, Proc. Amer. Math. Soc. *19* (1968), 1385–1386.
[2] Nil semi-simple (−1,1) rings, J. Algebra *22* (1972), 442–450.

HENTZEL, I. R.–SLATER, M.

[1] On the Andrunakievich Lemma for alternative rings, J. Algebra *27* (1973), 243–256.

HERSTEIN, I. N.

[1] Theory of Rings, Univ. Chicago 1961.
[2] Topics in Ring Theory, Univ. Chicago 1965.
[3] A theorem of Levitzki, Proc. Amer. Math. Soc. *13* (1962), 213–214.
[4] A counter example in Noetherian Rings, Proc. Nat. Acad. Sci. USA *54* (1965), 1036–1037.
[5] Noncommutative Rings, Carus Math. Monogr. № 15, Mathematical Association of America, New York 1968.

HERSTEIN, I. N.–SMALL, L.

[1] Nil rings satisfying certain chain conditions, Canad. J. Math. *16* (1964), 771–776; An addendum, Canad. J. Math. *18* (1966), 300–302.

HIGMAN, G.

[1] On a conjecture of Nagata, Proc. Cambridge Philos. Soc. *52* (1956), 1–4.

HILL, D. A.

[1] Semi-perfect *q*-rings, Math. Ann. *200* (1973), 113–121.

HILLE, E.

[1] Functional Analysis and Semigroups, Amer. Math. Soc. Colloq. Publ. Vol. 31, New York 1948.

HOCHSTER, M.

[1] Expanded radical ideals and semiregular ideals, Pacific J. Math. *44* (1973), 553–568.

HOEHNKE. H.-J.

[1] Nilpotenzkriterien, Math. Ann. *132* (1957), 401–404.
[2] Lösung eines Problems von Ch. Hopkins, J. Reine Angew. Math. *198* (1957), 112–120.
[3] Zur Strukturtheorie der Halbgruppen, Math. Nachr. *26* (1963), 8–13.
[4] Eine Charakterisierung des 0-Radikals einer Halbgruppe, Publ. Math. Debrecen *11* (1964), 72–73.
[5] Über das untere und obere Radikal einer Halbgruppe, Math. Z. *89* (1965), 300–311.
[6] Structure of semigroups, Canad. J. Math. *18* (1966), 449–491.
[7] Über antiautomorphe und involutorische primitive Halbgruppen, Czechoslovak Math. J. *15* (1965), 50–62.
[8] Radikale in allgemeinen Algebren, Math. Nachr. *32* (1966), 347–383.
[9] Die Radikale und das Prinzip des maximalen homomorphen Bildes in Bikategorien, Archiv. Math. Brno *3* (1967), 191–207.
[10] Das Brown–McCoysche 0-Radikal für Algebren und seine Anwendung in der Theorie der Halbgruppen, Fund. Math. *66* (1970), 155–175.

HOEHNKE, H.-J.–SEIDEL, H.

[1] Über das 0-Radikal einer Halbgruppe, Monatsber. Deutsch. Akad. Wiss. Berlin *5* (1963), 667–670.

HOFFMAN, A. E.

[1] Minimal hereditary semisimple classes, J. Sci. Nat. Math. Government College, Lahore (West Pakistan) *9* (1969), 29–32.
[2] Direct sum closure properties of radicals, J. Natur. Sci. Math. *10* (1970), 53–58.
[3] Radicals of semigroups with zero, Notices Amer. Math. Soc. April 1973, 703–715.

HOFFMAN, A. E.–LEAVITT, W. G.

[1] A note on the termination of the lower radical construction, J. London Math. Soc. *43* (1968), 617–618.
[2] Properties inherited by the lower radical, Portugaliae Math. *27* (1968), 63–66.

HOFMANN, K. H.

[1] Über das Nilradikal lokal kompakter Gruppen, Math. Z. *91* (1966), 206–215.

HOLCOMBE, M.

[1] Representations of *2*-primitive near-rings and the theory of near-algebras, Proc. Royal Irish Acad. *73* (1973), Sect. A, № 13.

HOO, C. S.–SHUM, K. P.

[1] On algebraic radicals in mobs, Colloq. Math. *25* (1972), 25–35.
[2] On the nilpotent elements of semigroups, Colloq. Math. *25* (1972), 211–224.

HOORN, W. G. VAN

[1] Some generalisations of the Jacobson radical for seminear-rings and semirings, Math. Z. *118* (1970), 69–82.

HOPKINS, CH.

[1] Nilrings with minimal condition for admissible left ideals, Duke Math. J. *4* (1938), 664-667.
[2] Rings with minimal condition for left ideals, Ann. of Math. *40* (1939), 721–730.

HOTZEL, E.

[1] On simple rings with minimal one-sided ideals, Preprint, 1970.

HSÜ, Y. H.

[1] L^2-radical and L^2-direct sum, Chinese Math. *8* (1966), 538–549.

HUE, TRAN TRONG–SZÁSZ, F. A.

[1] On the homomorphic closure of some semisimple classes, Math. Seminar Notes, Kobe Univ. *7* (1979), 281–290.

[2] A sufficient covering condition for a left duo ring to be nilpotent, Math. Seminar Notes, Kobe Univ. *7* (1979), 179–182.
[3] On the radical classes and the transfree images of rings, Acta Math. Acad. Sci. Hungar. (to appear).
[4] The action of the lower radical construction and the semisimple closure operator on the class of unequivocal rings, Acta Math. Acad. Sci. Hungar. (to appear).
[5] Some notes on the upper and lower radicals, Periodica Math. Hungar. (to appear).
[6] On the radical classes determined by regularities, Acta Sci. Math. Szeged (to appear).

HUNGER, K.

[1] Nilpotence of nil subrings implies more general nilpotence, Archiv. Math. *18* (1967), 136–139.

HUNGERFORD, TH. W.

[1] On the structure of principal ideal rings, Pacific J. Math. *25* (1968), 543–547.

HUQ, S. A.

[1] Commutator, nilpotency and solvability in categories, Quart. J. Math. Oxford *19* (1968), 363–389.
[2] Semivarieties and subfunctors of the identity functor, Pacific J. Math. *29* (1969), 303–309.

HUTCHINSON, J. J.–ZELMANOWITZ, J.

[1] Subdirect sum decompositions of endomorphism rings, Pacific J. Math. *47* (1973), 129–134.

IKEDA, M.

[1] Über die einstufig nichtkommutativen Ringe, Nagoya Math. J. *27* (1966), 371–379.

INGRAHAM, E. C.

[1] On the occasional equality of the lower and Jacobson radicals in Noetherian rings, Arch. Math. Basel *20* (1969), 267–269.

(IOFFE, L. SH.) Иоффе, Л. Ш.

[1] Радикал модуля, Сибирск. Мат. Жур. *5* (1964), 820–826.
[2] *S*-радикал модуля, Труды Моск. Ин-та Инж. Ж.-Д. Трансп. *352* (1969), 233–242.

ION, I. D.

[1] Radicals of projective limits of associative rings, Stud. Cerc. Math. *16* (1964), 1141–1145.

ISEKI, K.

[1] Sur le *G*-radical d'un anneau topologique, C. R. Acad. Sci. Paris *234* (1952), 1938–1939.
[2] On the Brown–McCoy radical in topological rings, Annais Acad. Brasil Ci. *25* (1953), 29–86.

ISEKI, K.–MIYANAGA, Y.

[1] On a radical in a semiring, Proc. Japan. Acad. *32* (1956), 562–563.

IVAN, J.

[1] The radical and semisimplicity of direct product of algebras, Mat. Fyz. Časopis Slovensk. Akad. Vied. *7* (1957), 158–167.

IVANOV, G.

[1] Rings with zero singular radical, J. Algebra *16* (1970), 340–346.

JACOBSON, N.

[1] The Theory of Rings, Mathematical surveys, No. 2. American Mathematical Society, New York 1943.
[2] Structure of Rings. American Mathematical Society Colloquium Publications, Vol. 37, Providence 1956.
[3] The radical and semisimplicity of arbitrary rings, Amer. J. Math. *67* (1945), 300–320.
[4] Structure theory of simple rings without finiteness assumptions, Trans. Amer. Math. Soc. *57* (1945), 228–245.
[5] A topology for the set of primitive ideals in an arbitrary ring, Proc. Nat. Acad. Sci. USA *31* (1945), 333–338.
[6] Structure theory for algebraic algebras of bounded degree, Ann. of Math. *46* (1945), 695–707.
[7] On the theory of primitive rings, Ann. of Math. *48* (1947), 8–21.
[8] Une généralisation du théorème d'Engel, C. R. Acad. Sci. Paris *234* (1952), 579–581.
[9] Structure theory for a class of Jordan algebras, Proc. Nat. Acad. Sci. USA *55* (1966), 243–251.

JAEGERMANN, M.

[1] Morita contexts and radicals, Bull. Acad. Polonaise Sci. *20* (1972), 619–623.

JAEGERMANN, M.–KREMPA, J.

[1] Rings in which ideals are annihilators, Fund. Math. *76* (1972), 95–107.

JAMBOR, P.

[1] On generation of torsion theories, Comment. Math. Univ. Carolinae *13* (1972), 79–98; Correction: ib. *14* (1973), 61.

JANESKI, J.–WEISSGLASS, J.

[1] Regularity of semilattice sums of rings, Proc. Amer. Math. Soc. *39* (1973), 479–482.

JANS, J. P.

[1] Compact rings with open radical, Duke Math. J. *24* (1957), 573–577.
[2] Rings and Homology, Holt, Rinehart and Winston, New York–Chicago 1964.
[3] Some aspects of torsion, Pacific J. Math. *15* (1965), 1249–1259.
[4] Torsion associated with duality, Tohoku Math. J. *24* (1972), 449–452.

JATEGAONKAR, A. V.

[1] A counter-example in ring theory and homological algebra, J. Algebra *12* (1969), 418–440.

JENKINS, T. L.

[1] The upper radical determined by regular rings and rings with D. C. C. Boll. Un. Mat. Ital. *1* (1968), 514–516.
[2] The upper radical determined by regular rings, Amer. Math. Monthly *74* (1967), 1240.
[3] The theory of radicals and of radical rings, Ph. D. Thesis, Univ. Nebraska 1966.
[4] Another characterization of the Jacobson radical, Amer. Math. Monthly *74* (1967), 1237.
[5] A maximal ideal radical class, J. Nat. Sci. Math. *7* (1967), 191–195.
[6] On Andrunakievic's lemma, J. Nat. Sci. Math. *12* (1972), 129–130.

JENKINS, T. L.–KREILING, D.

[1] Semisimple classes and upper-type radical classes of narings, Proc. Amer. Math. Soc. *26* (1970), 378–382.

JENNER, W. E.

[1] The radical of a non-associative ring, Proc. Amer. Math. Soc. *1* (1950), 348–351.

JENNINGS, S. A.

[1] A note on chain condition in nilpotent rings and groups, Bull. Amer. Math. Soc. *50* (1944), 574–763.
[2] On rings whose associated Lie rings are nilpotent, Bull. Amer. Math. Soc. *53* (1947), 593–597.
[3] Radical rings with nilpotent associated groups, Trans. Royal Soc. Canada, Sectio Math. (3) *49* (1955), 31–38.
[4] The group ring of a class of infinite nilpotent groups, Canad. J. Math. *7* (1955), 169–187.

(JENSEN, CHR.–SKORNYAKOV, L. A.) Йенсен, Хр.–Скорняков, Л. А.

[1] О прямых пределах полупростых колец, Мат. Заметки *7* (1970), 461–468.

JIZUKA, K.

[1] On the Jacobson radical of a semiring, Tohoku Math. J. *11* (1959), 409–421.

JIZUKA, K.–NAKABURA, K.

[1] A note on the semiradical of a semiring, Kumamoto J. Sci. Ser. A, *4* (1959), 1–3.

JOHNSON, B. E.

[1] The Wedderburn decomposition of Banach algebras with finite dimensional radical, Amer. J. Math. *90* (1968), 866–876.

JOHNSON, D. G.

[1] A structure theory for a class of lattice-ordered rings, Acta Math. *104* (1960), 163–215.

JOHNSON, M. J.

[1] Radicals of endomorphism near-rings, Rocky Mountain J. Math. *3* (1973), 1–7.

JOHNSON, R. E.

[1] Equivalence rings, Duke Math. J. *15* (1948), 787–793.

[2] Prime rings, Duke Math. J. *18* (1951), 799–809.

[3] Representations of prime rings, Trans. Amer. Math. Soc. *74* (1953), 351–356.

[4] Prime matrix rings, Proc. Amer. Math. Soc. *16* (1965), 1099–1105.

JOHNSON, R. E.–KIOKEMEISTER, F.

[1] The endomorphisms of the total operator domain of an infinite module, Trans. Amer. Math. Soc. *62* (1947), 904–930.

JONES, A.–LUMER, G.

[1] A note on radical rings, Bol. Fac. Ingen. Agriments Montevideo *5* (1956), 11–15.

JOOSTE, T. W. DE

[1] Prime ideals and nil radicals in ringoids, Math. Z. *105* (1968), 21–35.

JUMP, K.

[1] On the radical of a pseudo-ring, Math. Z. *116* (1970), 69–70; Correction: ib. *119* (1971), 94.

KALOUJNINE, L.–KRASNER, M.

[1] Produit complet des groupes de permutations et problème d'extension des groupes, Acta Sci. Math. Szeged *13* (1950), 208–230.

KANBARA, H.

[1] Note on Krull–Remak–Schmidt–Azumaya's theorem, Osaka J. Math. *9* (1972), 409–413.

KANDÔ, T.

[1] Strong regularity in arbitrary rings, Nagoya Math. J. *4* (1952), 51–53.

KAPLANSKY, I.

[1] Infinite Abelian Groups, Univ. Michigan 1954.

[2] On problem of Kurosh and Jacobson, Bull. Amer. Math. Soc. *52* (1946), 496–500.

[3] Rings with a polynomial identity, Bull. Amer. Math. Soc. *54* (1948), 575–580.

[4] Topological rings, Bull. Amer. Math. Soc. *54* (1948), 809–826.

[5] Topological representation of algebras, Trans. Amer. Math. Soc. *68* (1950), 62–75.

[6] Fields and Rings, Univ. Chicago Press, Chicago 1969.

(KASHU, A. I.) Кашу, А. И.

[1] Квази-терциарные идеалы, Мат. Исслед. *2* (1967), № 2, 96–113.

[2] Квазипримарные идеалы в нетеровых кольцах, Мат. Исслед. *3* (1968), № 2, 113–122.

[3] Радикалы в модулях, Мат. Исслед. *3* (1968), № 4, 78–107.

[4] Об аксиоматике кручений в Λ-модулях на языке левых идеалов кольца Λ, Мат. Исслед. *3* (1968), № 4, 166–174.

[5] Радикалы в категории Л-модулей, определенные с помощью элементов, Бул. Акад. Штиинце РСС Молд. 1969, № 1, 24–30.

[6] Радикальные замыкания в категории модулей, Мат. Исслед. *5* (1970), № 4.

[7] О делимости в модулях, Мат. Исслед. *6* (1971), № 2, 74–84.

[8] Некоторые характеризации кручений и устойчивых радикалов модулей, Мат. Исслед. *8* (1973), № 2, 176–182.

KAYE, S. M.

[1] Ring theoretic properties of matrix rings, Canad. Math. Bull. *10* (1967), 365–374.

(KAZIMIRSKI, P. S.–DROGOMIZHSKA, M. M.) Казімірський, П. С.–Дрогомижська, М. М.

[1] Certain properties of rings whose zero divisors belong to the Perlis–Jacobson radical, Доповіді Акад. Наук Україн. ССР Сер. A, (1971), 778–780.

KEGEL, O.

[1] Zur Nilpotenz gewisser assoziativer Ringe, Math. Ann. *149* (1962), 258–260.

[2] On rings that are sums of two subrings, J. Algebra *1* (1964), 103–109.

KEIMEL, K.

[1] Radicals in lattice-ordered rings, Colloquia Math. Soc. J. Bolyai, 6. Rings, Modules and Radicals, Bolyai J. Mat. Társulat, Budapest 1973, 237–253.

KELLY, G. M.

[1] On the radical of a category, J. Austral. Math. Soc. *4* (1964), 299–307.

(KEMKHADZE, SH. S.) Кемхадзе, Ш. С.

[1] К определению радикала в теории групп, Батумис Сех. Пед. Инст. Шром. *12* (1965), 246–253.

[2] Квазинильпотентные группы и связанные с ними радикалы, Тбилиси Унив. Шром. *117* (1966), 103–127.

[3] Локальные теоремы Мальцева и радикальные классы групп, Уч. Зап. Иванов. Гос. Пед. Ин-та *106* (1972), 106–108.

KERTÉSZ, A.

[1] Vorlesungen über Artinsche Ringe, Akadémiai Kiadó, Budapest–Teubner, Leipzig 1968.

[2] On radical-free rings of endomorphisms, Acta Univ. Debrecen *5* (1959), 159–161.

[3] Beiträge zur Theorie der Operatormoduln, Acta Math. Acad. Sci. Hung. *8* (1957), 235–257.

[4] Vizsgálatok az operátormodulusok elméletében (Notes on operator modules) III, Magy. Tud. Akad. Mat. Fiz. Oszt. Közl. *9* (1959), 105–120.

[5] A characterization of the Jacobson radical, Proc. Amer. Math. Soc. *14* (1963), 595–597.

[6] Gyűrűk Jacobson-féle radikáljáról (On the Jacobson radical of rings), Magy. Tud. Akad. Mat. Fiz. Oszt. Közl. *16* (1966), 445–461.

[7] Modules and semisimple rings II, Publ. Math. Debrecen *4* (1955/56), 229–236.

[8] Eine Charakterisierung der halbeinfachen Ringe, Acta Math. Acad. Sci. Hung. *9* (1958), 343–344.

[9] Über artinsche Ringe, Wiss. Z. Humboldt-Univ. Berlin, Math. Nat. R. *12* (1963), 823–826.
[10] Zur Frage der Spaltbarkeit von Ringen, Bull. Acad. Polonaise Sci. *12* (1964), 91–93.
[11] Noethersche Ringe die artinsch sind, Acta Sci. Math. Szeged *31* (1970), 219–228.
[12] Ein Radikalbegriff für Moduln, Colloquia Math. Soc. J. Bolyai, 6. Rings, Modules and Radicals, Bolyai J. Mat. Társulat, Budapest 1973, 255–257.
[13] Eine neue Charakterisierung der halbeinfachen Ringe, Acta Sci. Math. Szeged *34* (1973), 169–170.

KERTÉSZ, A.–STEINFELD, O.
[1] A féligegyszerű gyűrűk jellemzéseiről (On characterizations of semisimple rings), Magy. Tud. Akad. Mat. Fiz. Oszt. Közl. *9* (1959), 301–314.
[2] On the symmetry of semisimple rings, Amer. Math. Monthly *67* (1960), 450-452.

KERTÉSZ, A.–WIDIGER, A.
[1] Artinsche Ringe mit artinschem Radikal, J. Reine Angew. Math. *242* (1970), 8–15.

KEZLAN, TH. P.
[1] Some rings with nil commutator ideals, Michigan Math. J. *12* (1965), 105–111.

(KHAKHUTAISHVILI, A. S.) Хахутаишвили, А. С.
[1] Наследственные радикалы в классе нильпотентных групп, Латв. Мат. Ежегодник *7* (1970), 277–282.
[2] О радикально эквивалентных группах и радикалах в сплетениях групп, Латв. Мат. Ежегодник *9* (1971), 265–276.
[3] К теории радикалов в классе конечных групп, Латв. Мат. Ежегодник *11* (1972), 191–204.

KIEŁPIŃSKY, R.
[1] Direct sums of torsions, Bull. Acad. Polonaise Sci. Série math. astr. phys. *19* (1971), 281–286.

KLASA, J.
[1] Semisimplicity and von Neumann's regularity, Semigroup Forum *2* (1971), 354–361.

KLEIN, A. A.
[1] A note about two properties of matrix rings, Israel J. Math. *8* (1970), 90–92.
[2] Matrix rings of finite degree nilpotency, Pacific J. Math. *36* (1971), 387–391.

KNOTT, R. P.
[1] Group algebras of some generalised soluble groups, Proc. Edinburgh Math. Soc. *18* (1972), 1–5.
[2] The Brown–McCoy radical of certain group algebras, J. London Math. Soc. (2) *6* (1973), 617–625.

KOETHE, G.

[1] Die Struktur der Ringe, deren Restklassenring nach dem Radikal vollständig reduzibel ist, Math. Z. *32* (1930), 161–186.

KOH, K.

[1] On a structure theorem for a semisimple ring-like object, J. Algebra *10* (1968), 360–367.
[2] On simple rings with maximal annihilator right ideals, Canad. Math. Bull. *8* (1965), 667–668.
[3] On a semiprimary ring, Proc. Amer. Math. Soc. *19* (1968), 205–208.
[4] On some characteristic properties of self-injective rings, Proc. Amer. Math. Soc. *19* (1968), 209–213.
[5] On one-sided ideals of a prime type, Proc. Amer. Math. Soc. *28* (1971), 321–329.

KOH, K.–LUH, J.

[1] Maximal regular right ideal space of a primitive ring, Trans. Amer. Math. Soc. *170* (1972), 269–277.

KOH, K.–MEWBORN, A. C.

[1] Prime rings with maximal annihilator and maximal complement right ideals, Proc. Amer. Math. Soc. *16* (1965), 1073–1076.
[2] A class of prime rings, Canad. Math. Bull. *9* (1966), 63–72.
[3] The weak radical of a ring, Proc. Amer. Math. Soc. *18* (1967), 554–559.

(KOIFMAN, L. A.) Койфман, Л. А.

[1] Радикал модуля, Сивирск. Мат. Жур. *7* (1966), 1204–1207.
[2] Почти-полупростые кольца, Мат. Сборник *83* (1970), 120–150.
[3] Кольца, над которыми каждый модуль имеет максимальный подмодуль, Мат. Заметки *7* (1970), 359–367.

(KOSTRIKIN, A. I.) Кострикин, А. И.

[1] О локальной нильпотентности колец Ли, удовлетворяющих условию Энгеля, Докл. Акад. Наук СССР *118* (1958), 1074–1077.

(KOSTRIKIN, A. I.–SHAFAREVICH, I. R.) Кострикин, А. И.–Шафаревич, И. Р.

[1] Группы гомологий нильпотентных алгебр, Докл. Акад. Наук СССР *115* (1957), 1066–1069.

KOVÁCS, L. G.

[1] A note on regular rings, Publ. Math. Debrecen *4* (1956), 465–468.

KREILING, D.–JENKINS, T. L.

[1] On the construction of the Kurosh lower radical class for associative rings, J. Austral. Math. Soc. *13* (1972), 362–364.

KREMPA, J.

[1] Logical connections between some open problems concerning nil rings, Fund. Math. *76* (1972), 121–130.

[2] On radical properties of polynomial rings, Bull. Acad. Polonaise Sci., Série math. astr. phys. *20* (1972), 545–548.

[3] Radicals of semigroup rings, Fund. Math. *85* (1974), 57–71.

[4] Nil matrix rings (manuscript).

KROENER, C. W.

[1] On radical rings, Rend. Sem. Mat. Univ. Padova *41* (1968), 261–275.

KRULL, W.

[1] Jacobsonsches Radikal und Hilbertscher Nullstellensatz, Proc. Internat. Congr. Math. Vol. 2, Cambridge 1950, 56–64.

[2] Jacobsonsche Ringe, Hilbertscher Nullstellensatz, Dimensionstheorie, Math. Z. *54* (1951), 354–387.

KRUSE, R. L.

[1] A note on the adjoint group of a finitely generated radical ring, J. London Math. Soc. (2) *1* (1969), 743–744.

[2] On the circle group of a nilpotent ring, Amer. Math. Monthly *77* (1970), 168–170.

KRUSE, R. L.–PRICE, D. T.

[1] Nilpotent Rings, Gordon and Breach Science Publ. New York–London–Paris 1969.

[2] On the subring structure of finite nilpotent rings, Pacific J. Math. *31* (1969), 103–117.

[3] On the classification of nilpotent rings, Math. Z. *113* (1970), 215–223.

KUCZKOWSKI, J. E.

[1] On radicals in a certain class of semigroups, Mat. Časopis *20* (1970), 278–280.

KUPISCH, H.

[1] Beiträge zur Theorie nichthalbeinfacher Ringe mit Minimalbedingung, J. Reine Angew. Math. *201* (1959), 100–112.

[2] Über eine Klasse von Ringen mit Minimalbedingung I, Arch. Math. *17* (1966), 20–35.

KURATA, Y.

[1] On an n-fold torsion theory in the category $_RM$, J. Algebra *22* (1972), 559–572.

KURATA, Y.–KURATA, S.

[1] A generalization of prime ideals in rings II, Proc. Japan Acad. *45* (1969), 75–78.

(KUROCHKIN, V. M.) Курочкин, В. М.

[1] Расщепление алгебр в полупрямую сумму радикала и полупростой подалгебры, Московск. Гос. Унив. Учен. Зап. *148* (1951), № 4, 192–203.

(Kurosh, A. G.) Курош, А. Г.

[1] Теория групп, Москва 1967.

[2] Проблемы теории колец, связанных с проблемой Бернсайда о периодических группах, Изв. Акад. Наук СССР *5* (1941), 233–240.

[3] Радикалы колец и алгебр, Мат. Сборник *33* (75) (1953), 13–26.

[4] Radicaux en théorie des groupes, Bull. Soc. Math. Belgique *14* (1962), 307–310.

[5] Радикалы в теории групп, Сибирск. Мат. Жур. *3* (1962), 912–931.

(Kuzmin, E. N.) Кузьмин, Е. Н.

[1] Локально нильпотентный радикал алгебр Мальцева, удовлетворяющих n-му условию Энгеля, Докл. Акад. Наук СССР *177* (1967), 508–510.

Lajos, S.

[1] Generalized ideals in semigroups, Acta Sci. Math. Szeged *22* (1961), 217–222.

[2] On quasiideals of regular rings, Proc. Japan Acad. *38* (1962), 210–211.

[3] A note on completely regular semigroups, Acta Sci. Math. Szeged *28* (1967), 261–265.

[4] On regular duo rings, Proc. Japan Acad. *45* (1969), 157–158.

[5] On semilattices of groups, Proc. Japan Acad. *45* (1969), 383–384.

[6] On the bi-ideals in semigroups, Proc. Japan Acad. *45* (1969), 710–712.

[7] On regular rings and semigroups, K. Marx Univ. of Economics, Dept. Math. Budapest № 4 (1975), 1–25.

[8] Characterizations of completely regular elements in semigroups, Acta Sci. Math. Szeged *40* (1978), 297–300.

Lajos, S.–Szász, F. A.

[1] Characterizations of strongly regular rings I–II, Proc. Japan Acad. *46* (1970), 38–39, 287–289.

[2] On the bi-ideals in associative rings, Proc. Japan Acad. *46* (1970), 505–507.

[3] Bi-ideals in associative rings, K. Marx Univ. of Economics, Dept. Math. Budapest 1970, № 4.

[4] Some characterizations of two-sided regular rings, Acta Sci. Math. Szeged *31* (1970), 223–228.

[5] Bi-ideals in associative rings, Acta Sci. Math. Szeged *32* (1971), 185–193.

[6] On (m, n)-ideals in associative rings, Publ. Math. Debrecen *25* (1978), 265–273.

Lallement, G.–Petrich, M.

[1] Structure of a class of regular semigroups and rings, Bull. Amer. Math. Soc. *73* (1967), 419–422.

Lambek, J.

[1] On the structure of semi-prime rings and their rings of quotients, Canad. J. Math. *13* (1961), 392–417.

[2] Lectures on Rings and Modules, Blaisdell Publ. Co. Waltham/Mass.–Toronto–London 1966.

[3] Torsion theories, additive semantics and rings of quotients, Lecture Notes in Math. Vol. 177, Springer-Verlag, Berlin–Heidelberg–New York 1971.

LAMBEK, J.–MICHLER, G.

[1] The torsion theory at a prime ideal of a right Noetherian ring, J. Algebra *25* (1973), 364–389.

LANSKI, C.

[1] Nil subrings of Goldie rings are nilpotent, Canad. J. Math. *21* (1969), 904–907.
[2] Solvable normal subgroups and nilpotent ideals in matrix rings, Math. Ann. *202* (1973), 251–264.

LARSEN, M. D.–MIRBAGHERI, A.

[1] A note on the intersection of the powers of the Jacobson radical, Rocky Mountain J. Math. *1* (1971), 617–621.

LA TORRE, D. R.

[1] A note on the Jacobson radical of a hemiring, Publ. Math. Debrecen *14* (1967), 9–13.
[2] The Brown–McCoy radicals of a hemiring, Publ. Math. Debrecen *14* (1967), 15–28.
[3] An internal characterization of the 0-radical of a semigroup, Math. Nachr. *45* (1970), 279–281.
[4] Modular congruences and the Brown–McCoy radical for semigroups, Proc. Amer. Math. Soc. *29* (1971), 427–433.

(LATYSHEV, V. N.) Латышев, В. Н.

[1] О делителях нуля и нильэлементах в алгебре Ли, Сибирск. Мат. Жур. *4* (1963), 830–836.
[2] Два замечания о *PI*-алгебрах, Сибирск. Мат. Жур. *4* (1963), 1120–1121.

LAXTON, R. R.

[1] A radical and its theory for distributively generated near-rings, J. London Math. Soc. *38* (1963), 40–49.
[2] Prime ideals and the ideal radical of a distributively generated near-ring, Math. Z. *83* (1964), 8–17.
[3] Note on the radical of a near-ring, J. London Math. Soc. (2) *6* (1972), 12–14.

LAXTON, R. R.–MACHIN, A.

[1] On the decomposition of near-rings, Abh. Math. Sem. Univ. Hamburg *38* (1972), 221–230.

LEAVITT, W. G.

[1] Sets of radical classes, Publ. Math. Debrecen *14* (1967), 321–324.
[2] Strongly hereditary radicals, Proc. Amer. Math. Soc. *21* (1969), 703–705.
[3] Hereditary semisimple classes, Glasgow Math. J. *11* (1970),7–8.
[4] Note on two problems of A. Kertész, Publ. Math. Debrecen *6* (1959), 83-85.
[5] Type radicals, Glasgow Math. J. *9* (1968), 22–29.
[6] Radical and semisimple classes with specified propertes, Proc. Amer. Math. Soc. *24* (1970), 680–687.
[7] Lower radical constructions, Colloquia Math. Soc. J. Bolyai, 6. Rings, Modules and Radicals, Bolyai J. Mat. Társulat, Budapest 1973, 319–323.

[8] The General Theory of Radicals, Note of University of Nebraska, Lincoln 1972.
[9] General radical theory in rings, Rocky Mountain J. of Math. *3* (1973), 431–439.
[10] The intersection property of an upper radical, Archiv Math. *24* (1973), 486–492.
[11] A note on upper radicals in rings, J. Austral. Math. Soc. A22 (1976), 264–273.

LEAVITT, W. G.–JENKINS, T. L.
[1] Non-hereditariness of the maximal ideal radical class, J. Nat. Sci. Math. Government College, Lahore (W. Pakistan) *7* (1967), 103–105.

LEAVITT, W. G.–LEE, Y. L.
[1] A radical coinciding with the lower radical in associative and alternative rings, Pacific J. Math. *30* (1969), 459–462.

LEAVITT, W. G.–TANGEMAN, R. L.
[1] On Sasiada's ring, J. London Math. Soc. (2) *1* (1969), 61–63; Corrigendum, ib. (2) *2* (1970), 32.

LEDUC, P. Y.
[1] Radical et primitivité des catégories, Cahiers Top. Géom. Diff. *10* (1968), 347–350.
[2] Catégories localement artiniennes sans radical, Canad. Math. Bull. *12* (1969), 675–676.

LEE, Y. L.
[1] On the construction of upper radical properties, Proc. Amer. Math. Soc. *19* (1968), 1165–1166.
[2] A characterization of Baer lower radical property, Kyungpook Math. J. *7* (1967), 45–46.
[3] On the construction of lower radical properties, Pacific J. Math. *28* (1969), 393–395.

LEE, Y. L.–PROPES, R. E.
[1] On intersections and unions of radical classes, J. Austral. Math. Soc. *13* (1972), 354–356.

LEEUWEN, L. C. A. VAN
[1] On the zeroid radical of a ring, Indag. Math. *21* (1959), 428–433.
[2] On a generalization of the Jenkins radical, Arch. Math. Basel *22* (1971), 155–160.
[3] Upper radicals determined by simple rings, Colloquia Math. Soc. J. Bolyai, 6. Rings, Modules and Radicals, Bolyai J. Mat. Társulat, Budapest 1973, 337–349.

LEEUWEN, L. C. A. VAN–JENKINS, T. L.
[1] On the hereditariness of the upper radical, Archiv Math. *25* (1974), 135–137
[2] A note on radical semisimple classes, Publ. Math. Debrecen *21* (1974), 179–184.
[3] A supernilpotent nonspecial radical class, Bull. Austral. Math. Soc. *9* (1973), 343–348.
[4] Upper radicals and simple rings, Periodica Math. Hung. *6* (1975), 69–74.

LENAGAN, T. H.
[1] The nil radical of a ring with Krull dimension, Bull. London Math. Soc. *5* (1973), 307–311.

LENNOX, J. C.

[1] The Fitting-Gaschütz-Hall relation in certain soluble groups by finite groups, J. Algebra *24* (1973), 219–225.

LEPTIN, H.

[1] Linear kompakte Moduln und Ringe I, Math. Z. *62* (1955), 241–267.
[2] Linear kompakte Moduln und Ringe II, Math. Z. *66* (1957), 289–327.

LERON, U.–VAPNE, A.

[1] Polynomial identities of related rings, Israel J. Math. *8* (1970), 127–137.

LESIEUR, L.

[1] Idéal complètement premier d'un anneau noethérien à gauche intègre, Colloquia Math. Soc. J. Bolyai, 6. Rings, Modules and Radicals, Bolyai J. Mat. Társulat, Budapest 1973. 337–349.

LESIEUR, L.–CROISOT, R.

[1] Sur le notion de radical, Acad. Royale Belgique Cl. Sci. *44* (1958), 75–93.
[2] Une propriété charactéristique des idéaux tertiaires, C. R. Acad. Sci. Paris *246* (1958), 517–520.
[3] Structure des anneaux premiers noetheriens à gauche, C. R. Acad. Sci. Paris *248* (1959), 2545–2547.
[4] Algèbre Noetherienne non-commutatif, Mém. des Sci. Math. Paris 1963.

LEVICH, E. M.–KHAKHUTAISHVILI, A. S.) Левич, Е. М.–Хахутаишвили, А. С.

[1] О наследственно радикальных классах двуступенно разрешимых групп, Латв. Мат. Ежегодник *9* (1971), 161–167.

LEVITZKI, J.

[1] On rings which satisfy the minimum condition for the right hand ideals, Comp. Math. *1* (1939), 214–222.
[2] On the radical of a general ring, Bull. Amer. Math. Soc. *49* (1943), 462–466.
[3] A characteristic condition for semi-primary rings, Duke Math. J. *11* (1944), 367–368.
[4] Solution of a problem of G. Köthe, Amer. J. Math. *67* (1945), 437–442.
[5] On three problems concerning nil rings, Bull. Amer. Math. Soc. *51* (1945), 913–919.
[6] On a problem of A. Kurosh, Bull. Amer. Math. Soc. *52* (1946), 1033–1035.
[7] On powers with transfinite exponents I, Riveon Lemmatematika *1* (1946), 8–13.
[8] On powers with transfinite exponents II, Riveon Lemmatematika *2* (1947), 1–7.
[9] A lemma on the radical with application, Riveon Lemmatematika *3* (1949), 20–24.
[10] On multiplicative systems, Comp. Math. *8* (1950), 76–80.
[11] A theorem on polynomial identities, Proc. Amer. Math. Soc. *1* (1950), 334–341.
[12] Prime ideals and the lower radical, Amer. J. Math. *73* (1951), 25–29.
[13] A note on prime ideals, Riveon Lemmatematika *5* (1951), 1–4.
[14] Contributions to the theory of nilrings, Riveon Lemmatematika *7* (1954), 50–70.
[15] On nil subrings, Israel J. Math. *1* (1963), 215–216.

LEWAND, R.

[1] Hereditary radicals in Jordan rings, Proc. Amer. Math. Soc. *33* (1972), 302–306.

LICOIU, D.

[1] Catégories semi-primaires, Rev. Roum. Math. Pures Appl. *17* (1972), 1649–1652.

LIEBERT, W.

[1] Endomorphism rings of reduced complete torsion-free modules over complete discrete valuation rings, Proc. Amer. Math. Soc. *36* (1972), 375–378.
[2] Endomorphism rings of reduced torsion-free modules over complete discrete valuation rings, Trans. Amer. Math. Soc. *169* (1972), 347–363.
[3] The Jacobson radical of some endomorphism rings, J. Reine Angew. Math. *262/263* (1973), 166–170.

(LIKHTMAN, A. I.) Лихтман, А. И.

[1] О кольцах, радикальных над коммутативным подкольцом, Мат. Сборник *83* (1970), 513–523.

(LIPKINA, Z. S.) Липкина, З. С.

[1] Строение компактных колец, Сибирск. Мат. Жур. *14* (1973), 1346–1348.

(LIVSHITS, A. KH.) Лившиц, А. Х.

[1] Теоретико-категорные основы двойственности радикальности и полупростоты, Сибирск. Мат. Жур. *5* (1964), 319–336.

LONCUOR, C.

[1] Radical d'une algèbre d'un produit direct de groupes finis, Bull. Soc. Math. Belg. *23* (1971), 423–435.

LUH, J.

[1] On the concepts of radical of semigroups having kernel, Portugaliae Math. *19* (1960), 189–198.
[2] A characterization of regular rings, Proc. Japan Acad. *39* (1963), 741–742.
[3] A remark on quasiideals of a regular ring, Proc. Japan Acad. *40* (1964), 660–661.
[4] On the structure of *J*-rings, Amer. Math. Monthly *74* (1967), 164–166.

MACLANE, S.

[1] Duality for groups, Bull. Amer. Math. Soc. *56* (1950), 485–516.

MACRAE, R. E.

[1] On the nilpotence of nil algebras, Proc. Cambridge Philos. Soc. *59* (1963), 679–680.

(MAKHARADZE, L. M.) Махарадзе, Л. М.

[1] Локально нильпотентные идеалы в топологических кольцах, Мат. Сборник *41* (1957), 395–414.

[2] Топологические нильпотентные кольца с условием минимальности, Успехи Мат. Наук *12* (1957), 181–186.
[3] Локально нильпотентный радикал в локально ограниченных кольцах, Труды Выч. Центр. Акад. Наук Груз. ССР *2* (1961), 21–22.
[4] О нильпотентности топологических колец, Сообщ. Акад. Наук Груз. ССР *40* (1965), 257–269.
[5] Обобщенный локально нильпотентный радикал в топологических кольцах, Сакарт. ССР Мецн. Акад. Моамбе *59* (1970), 25–28.

(MALTSEV, A. I.) Мальцев, А. И.
[1] О разложении алгебры в прямую сумму радикала и полупростой подалгебры, Докл. Акад. Наук СССР *36* (1942), 46–50.
[2] Об ассоциативных кольцах, радикальных над своими подкольцами, Мат. Заметки *11* (1972), 33–40.
[3] О тождествах полных матричных алгебр, Сибирск. Мат. Жур. *14* (1973), 1349–1350.

MARANDA, I. M.
[1] Injective structures, Trans. Amer. Math. Soc. *110* (1964), 98–135.

(MARKOVICHEV, A. S.) Марковичев, А. С.
[1] Нил-системы и радикал в альтернативных артиновых кольцах, Мат. Заметки *11* (1972), 299–306.

MARTINDALE, W. S.
[1] Prime rings satisfying a generalized polynomial identity, J. Algebra *12* (1969), 576–584.

MARTY, R.
[1] Radical, socle et relativisation, Studies on Abelian Groups (Symp. Montpellier 1967), Springer-Verlag, Berlin–Heidelberg–New York 1968, 287–300.

MARUBAYASHI, H.
[1] On potent rings and modules, Osaka J. Math. *9* (1972), 231–259.

MARUBAYASHI, H.–MURATA, K.
[1] A note on radicals of ideals in nonassociative rings, Proc. Japan Acad. *45* (1969), 131–134.

MASSAGLI, R. A.
[1] On a new radical in a topological ring, Pacific J. Math. *45* (1973), 577–584.

MATLIS, E.
[1] Modules with descending chain condition, Trans, Amer. Math. Soc. *97* (1966), 495–508.

McCOY, N. H.
[1] The Theory of Rings, Macmillan, New York–London 1964.
[2] Subdirect sums of rings, Bull. Amer. Math. Soc. *53* (1947), 856–877.
[3] Prime ideals in general rings, Amer. J. Math. *71* (1948), 823–833.

[4] The prime radical of a polynomial ring, Publ. Math. Debrecen *4* (1956), 161–162.
[5] Completely prime and completely semi-prime ideals, Colloquia Math. Soc. J. Bolyai, 6. Rings, Modules and Radicals, Bolyai J. Mat. Társulat, Budapest 1973, 147–152.

McCrimmon, K.

[1] The radical of a Jordan algebra, Proc. Nat. Acad. Sci. USA *62* (1969), 671–678.
[2] A characterization of the radical of a Jordan algebra, J. Algebra *18* (1971), 103–111.
[3] A characterization of the Jacobson–Smiley radical, J. Algebra *18* (1971), 565–573.

McKnight, J. D. Jr.–Musser, G. L.

[1] Special $(p; q)$ radicals, Canad. J. Math. *24* (1972), 38–44.

McLean, K. R.

[1] Commutative artinian principal ideal rings, Proc. London Math. Soc. (3) *36* (1973), 249–272.

McWorter, W. A.

[1] Some simple properties of simple nil rings, Canad. Math. Bull. *9* (1966), 197–200.

Mewborn, A. C.

[1] Regular rings and Baer rings, Math. Z. *121* (1971), 211–219.

Meyberg, K.

[1] Identitäten und das Radikal in Jordan-Tripelsystemen, Math. Ann. *197* (1972), 203–220.

Michler, G.

[1] Kleine Ideale, Radikale und die Eins in Ringen, Publ. Math. Debrecen *12* (1965), 231–252.
[2] Radikale und Sockel, Math. Ann. *167* (1966), 1–48.
[3] Radicals and structure spaces, J. Algebra *4* (1966), 199–219.
[4] Charakterisierung einer Klasse von Noetherschen Ringen, Math. Z. *100* (1967), 163–182.
[5] On maximal nilpotent subrings of right Noetherian rings, Glasgow Math. J. *8* (1967), 89–101.

(Mikaelyan, G. S.) Микаелян, Г. С.

[1] О Q-картеровских подгруппах локально конечных групп с Q-радикалом конечного индекса, Айк. ССР Гит. Акад. Тег. Мат. *7* (1972), 413–423.

(Mikheev, I. M.) Михеев, И. М.

[1] Локально правонильпотентный радикал в классе правоальтернативных колец, Алгебра и Логика *11* (1972), 174–185.

Miller, R. W.

[1] Endomorphism rings of finitely generated projective modules, Pacific J. Math. *47* (1973), 199–220.

MILLER, R. W.–TURNIDGE, D. R.

[1] Co-Artinian rings and Morita duality, Israel J. Math. *15* (1973), 12–26.

MIRBAGHERI, A.

[1] Some results on finiteness of radical algebras, J. Austral. Math. Soc. *11* (1970), 291–296.

[2] The nilpotency of left Noetherian radical rings, Portugal. Math. *29* (1970), 151–155.

MITCHELL, B.

[1] Rings with several objects, Advances in Math. *8* (1972), 1–161.

MLITZ, R.

[1] Ein Radikal für universale Algebren und seine Anwendung auf Polynomringe mit Komposition, Monatsh. Math. *75* (1971), 144–152.

[2] Verbandstheoretische Erzeugung von Radikalen in Multioperatorgruppen, Colloquia Math. Soc. J. Bolyai, 6. Rings, Modules and Radicals, Bolyai J. Mat. Társulat, Budapest 1973, 351–356.

MOEN, S.

[1] Free derivation modules and a criterion for regularity, Proc. Amer. Math. Soc. *39* (1973), 221–227.

MOORE, H. G.–YAQUB, A.

[1] A generalization of Boolean rings, Portugal. Math. *27* (1968), 213–216.

MORITA, K.

[1] Duality theorems for modules and its application to the theory of rings with minimum condition, Sci. Rep. Tokyo Kyoiku Daigaku *6* (1958), 83–142.

[2] Localizations in categories of modules I, Math. Z. *114* (1970), 121–144.

[3] Localizations in categories of modules II, J. Reine Angew. Math. *242* (1970), 163–169.

[4] Localizations in categories of modules III, Math. Z. *119* (1971), 313–320.

MORRISON, D. R.

[1] Biregular rings and the ideal lattice isomorphisms, Proc. Amer. Math. Soc. *6* (1952), 46–49.

MOTOSE, K.

[1] A note on perfect rings and semi-perfect rings, J. Fac. Sci. Shinshu Univ. *4* (1969), 67–69.

[2] On the radical of a group ring, Math. J. Okayama Univ. *15* (1971), 35–36.

[3] Note on results of D. A. R. Wallace, J. Fac. Sci. Shinshu Univ. 7 (1972), 119–122.

MÜLLER, W.

[1] Darstellungstheoretische Eigenschaften der symmetrischen Gruppe, 1. Über die Größe des Radikals in Gruppenringen über der symmetrischen Gruppe, Diss. Univ. München 1969.

[2] Über die Größe des Radikals in Gruppenringen über der symmetrischen Gruppe, Manuscripta Math. *4* (1971), 39–60.

Murata, K.–Kurata, Y.–Marubayashi, H.

[1] A generalization of prime ideals in rings, Osaka J. Math. *6* (1969), 291–301.

Musser, G. L.

[1] Linear semiprime $(p; q)$ radicals, Pacific J. Math. *37* (1971), 749–757.

Myung, H. C.

[1] A generalization of the prime radical in nonassociative rings, Pacific J. Math. *42* (1972), 187–193.

[2] A characterization of the Jacobson radical in ternary algebras, Proc. Amer. Math. Soc. *38* (1973), 228–234.

Nagata, M.

[1] On the theory of radicals in a ring, J. Math. Soc. Japan *3* (1951), 330–344.

[2] On the nilpotency of nil algebras, J. Math. Soc. Japan *4* (1952), 296–301.

Nakajima, A.

[1] A radical and a subcategory in an exact category, Math. J. Okayama Univ. *14* (1969/70), 59–67.

Nakano, T.

[1] A nearly semi-simple ring, Comment. Math. Univ. St. Pauli *7* (1959), 27–33.

Năstăsescu, C.–Popescu, N.

[1] Anneaux semi-artiniens, Bull. Soc. Math. France *96* (1968), 357–368.

(Neroslavski, O. M.) Нерославский, О. М.

[1] Интерпретация Ext_R^2 (KM), группы гомологий и когомологий радикального кольца и присоединенной группы, Весці Акад. Навук БССР, Сер. Фiз.-Мат. Н. 1971, № 4, 20–26.

[2] Структуры, связанные с радикальными кольцами, Весці Акад. Навук БССР, Сер. Фiз.-Мат. Н. 1973, № 2, 5–10.

Neumann, J. von

[1] On regular rings, Proc. Nat. Acad. Sci. USA *22* (1936), 707–713.

Nicholson, W. K.

[1] Rings whose elements are quasi-regular or regular, Aequationes Math. *9* (1973), 64–70.

Niewieczerzal, D.–Terlikowska, B.

[1] A note on alternative semiprime rings, Bull. Acad. Polonaise Sci. *20* (1972), 265–268.

Niţă, C.

[1] Sur les anneaux A, tels que tout A-module simple est isomorphe à un idéal, C. R. Acad. Sci. Paris, Sér. A–B, *268* (1969), A88–A91.

[2] Remarques sur les anneaux N-réguliers, C. R. Acad. Sci. Paris, Sér. A–B, *271* (1970), A345–A348.
[3] Anneaux N-réguliers de dimension finie, Revue Roum. Math. Pures Appl. *18* (1973), 923–926.

Noronha Galvão, M. L.–Almeida Costa, A.
[1] Sur le demi-anneau des nombres naturels, An. Fac. Sci. Univ. Porto *48* (1965), 35–39.

O'Donnell, S. R.
[1] A general injectivity for modules, Math. Z. *124* (1972), 9–19.

Oehmke, R. H.
[1] On maximal congruences and finite semisimple semigroups, Trans. Amer. Math. Soc. *125* (1966), 223–237.
[2] Quasi-regularity in semigroups, Algèbre et Théorie des Nombres, Sém. P. Dubreil–M.-L. Dubreil-Jacotin–L. Lesieur–C. Pisot *23* (1969/70), № 1.

Okuzumi, M.
[1] Automorphisms of a free nilpotent algebra, Kodai Math. Sem. Rep. *20* (1968), 374–384.

Ornstein, A. J.
[1] Rings with restricted minimum condition, Proc. Amer. Math. Soc. *19* (1968), 1145–1150.

Ortiz, A. H.
[1] A construction in general radical theory, Canad. J. Math. *22* (1970), 1097–1100.
[2] An intersection theorem for a class of Brown–McCoy radicals, Tamkang J. Math. *2* (1971), 117–121.
[3] On the structure of semiprime rings, Proc. Amer. Math. Soc. *38* (1973), 22–26.

Osborn, J. M.
[1] Representations and radicals of Jordan algebras, Scripta Math. *29* (1973), 297–329.

Osofsky, B. L.
[1] Rings all of whose finitely generated modules are injective, Pacific J. Math. *14* (1964), 645–650.
[2] Noncommutative rings whose cyclic modules have cyclic injective hulls, Pacific J. Math. *25* (1968), 331–340.
[3] Non-injective cyclic modules, Proc. Amer. Math. Soc. *19* (1968), 1383–1384.

Palmer, T. W.
[1] The reducing ideal is a radical, Pacific J. Math. *43* (1972), 207–220.

Panella, G.
[1] Un teorema di Golod-Safarevic e alcune sue conseguence, Confer. Sem. Mat. Univ. Bari *104* (1966), 17.
[2] Osservazione riguardante gli anelli che verificano una condizione di massimo, Atti Accad. Naz. Lincei Rend. Cl. Sci. Fis. Mat. Natur. (8) *42* (1967), 755–756.

PAPP, Z.

[1] *S*-modules and torsion theories over Artinian rings, Arch. Math. Basel *23* (1972), 598–602.

(PARFENOV, V. A.) Парфенов, В. А.

[1] О слабо разрешимом радикале алгебр Ли, Сибирск. Мат. Жур. *12* (1971), 171–176.

PASSMAN, D. S.

[1] Nil ideals in group rings, Michigan Math. J. *9* (1962), 375–384.
[2] On the semisimplicity of modular group algebras, Proc. Amer. Math. Soc. *20* (1969), 515–519.
[3] On the semisimplicity of modular group algebras II, Canad. J. Math. *21* (1969), 1137–1145.
[4] On the semisimplicity of twisted group algebras, Proc. Amer. Math. Soc. *25* (1970), 161–166.
[5] Radicals of twisted group rings I, II, Proc. London Math. Soc. (3) *20* (1970), 409–437; (3) *22* (1971), 633–651.
[6] On the semisimplicity of group of linear groups, I, II, Pacific J. Math. *47* (1973), 221–228.

PATTERSON, E. M.

[1] On the radicals of certain rings of infinite matrices, Proc. Roy. Soc. Edinburgh *65* (1960), 263–271.
[2] On the radicals of row-finite matrices, Proc. Roy. Soc. Edinburgh *66* (1961), 42–46.
[3] The Jacobson radical of a pseudo-ring, Math. Z. *89* (1965), 348–364.

(PAVLOV, I. A.) Павлов, И. А.

[1] О коммутативных нильпотентных алгебрах матриц I, II, Докл. Акад. Наук БССР *11* (1967), 870–872; *12* (1968), 393–396.
[2] Коммутативные нильпотентные алгебры матриц класса *n*-3 I, II, Весцi Акад. Навук БССР, Сер. Фiз.-Мат. Н., 1968, № 4, 39–46; № 5, 10–16.

(PEKELIS, A. S.) Пекелис, А. С.

[1] Структурные изоморфизмы радикальных групп, Докл. Акад. Наук СССР *204* (1972). 288–291.

PEREMANS, W.

[1] The radical of a ring, Math. Centrum Amsterdam, Rapport ZW, Vol. 013, 1950.

PERLIS, S.

[1] A characterization of the radical of an algebra, Bull. Amer. Math. Soc. *48* (1942), 128–132.

PHAM, D.

[1] Sur les anneaux indexables, C. R. Acad. Sci. Paris *245* (1957), 1683–1685.

PIACENTINI CATTANEO, G. M.

[1] Alcune osservazioni sulle potenze del radicale di un anello, Atti Accad. Naz. Lincei Rend. Cl. Sci. Fis. Mat. Natur. (8) *53* (1972), 15–17.

PIERCE, R. S.

[1] Radicals in function rings, Duke Math. J. *23* (1956), 253–261.

(PLOTKIN, B. I.) Плоткин, Б. И.

[1] О нильрадикале группы, Докл. Акад. Наук СССР *98* (1954), 341–343.
[2] О радикалах в группах, Успехи Мат. Наук *10* (1955), № 3, 179–180.
[3] Радикальные группы, Мат. Сборник *37* (1955), 507–526.
[4] Радикальные и полупростые группы и алгебры Ли, Труды 3-го Всес. Мат. Съезда, 1, Акад. Наук СССР, Москва 1956, 33.
[5] Радикальные и полупростые группы, Труды Моск. Мат. Об-ва *6* (1957), 299–336.
[6] Радикал и нильэлементы в группах, Изв. Высш. Учеб. Завед. Мат. (1958), № 1, 130–135.
[7] Об одном радикале группы автоморфизмов группы с условием максимальностя, Докл. Акад. Наук СССР *130* (1960), 977–981.
[8] Радикальные группы, у которых радикал обладает возрастающим центральным рядом, Уч. Зап. Уральск. Ун-та *23* (1960), № 3, 40–43.
[9] Радикалы в групповых парах, Докл. Акад. Наук СССР *140* (1961), 1019–1022.
[10] Радикалы, связанные с представлениями групп, Докл. Акад. Наук СССР *144* (1962), 52–55.
[11] О некоторых радикалах групп автоморфизмов, Успехи Мат. Наук *17* (1962), № 4, 165–171.
[12] О полупростых группах, Сибирск. Мат. Жур. *9* (1968), 623–631.
[13] О полугруппе радикальных классов групп, Сибирск. Мат. Жур. *10* (1969), 1091–1108.
[14] Стабильность, нильпотентность и радикалы в группах автоморфизмов модулей, Труды Рижск. Алг. Сем., Рига 1969, 209–252.
[15] О функториалах, радикалах и корадикалах в группах, Мат. Зап. Уральск. Ун-та *7* (1970), № 3, 150–182.
[16] Радикалы и многообразия в представлениях групп, Латв. Мат. Ежегодник *10* (1972), 75–132.

(PLOTKIN, B. I.–KEMKHADZE, SH. S.) Плоткин, Б. И.–Кемхадзе, Ш. С.

[1] Об одной схеме построения радикалов в группах, Сибирск. Мат. Жур. *6* (1965), 1197–1201.

(POLIN, S. V.) Полин, С. В.

[1] Радикалы в Ω-почтикольцах, I, II, Изв. Высш. Учеб. Завед. Мат. (1972), № 1, 64–75; № 2, 63–71.

PONDĚLÍČEK, B.

[1] A note on radicals of semigroups, Mat. Časopis *23* (1973), 14–16.

POPESCU, N.–VRACIU, C.

[1] Some remarks about semi-artinian rings, Revue Roum. Math. Pures Appl. *18* (1973), 1413–1422.

POSNER, E. C.

[1] Left valuation rings and simple radical rings, Trans. Amer. Math. Soc. *107* (1963), 458–465.
[2] Prime rings satisfying a polynomial identity, Proc. Amer. Math. Soc. *11* (1960), 180–183.
[3] Primitive matrix rings, Archiv. Math. *12* (1961), 97–101.

POSNER, E. C.–SCHNEIDER, H.

[1] Hyperplanes and prime rings, Archiv. Math. *11* (1960), 322–326.

POWERS, D.

[1] Semisimple algebras of finite dimension over the real field, Scientia (Valparaíso) *134* (1967), 17–22.

PROPES, R. E.

[1] A characterization of the Behrens radical, Kyungpook Math. J. *10* (1970), 49–52.
[2] Homomorphically closed partitions, Kyungpook Math. J. *11* (1971), 17–20.
[3] Radicals of PID's and Dedekind domains, Canad. J. Math. *24* (1972), 566–572.

PROPES, R. E.–LEE, Y. L.

[1] Radical properties and partitions of rings, Kyungpook Math. J. *7* (1967), 37–39.

PUTCHA, M. S.–YAQUB, A.

[1] Rings satisfying monomial constraints, Proc. Amer. Math. Soc. *39* (1973), 10–18.

RAMAKOTAIAH, D.

[1] Radicals for near-rings, Math. Z. *96* (1957), 45–56.
[2] Radicals for near-rings, Math. Z. *96* (1967), 45–56.
[3] A radical for near-rings, Arch. Math. Basel *23* (1972), 482–483.

RAMAMURTHI, V. S.

[1] On splitting cotorsion radicals, Proc. Amer. Math. Soc. *39* (1973), 457–461.
[2] Weakly regular rings, Canad. Math. Bull. *16* (1973), 317–321.

RANGA RAO, M. L.

[1] Azumaya, semisimple and ideal algebras, Bull. Amer. Math. Soc. *78* (1972), 588–592.

RANGASWAMY, K. M.–VANAJA, N.

[1] A note on modules over regular rings, Bull. Austral. Math. Soc. *4* (1971), 57–62.
[2] Regular and Baer rings, Proc. Amer. Math. Soc. *42* (1974), 354–358.

RASHID, M. A.–WIEGANDT, R.

[1] The hereditariness of the upper radical, Acta Math. Acad. Sci. Hung. *24* (1973), 343–347.

RÉDEI, L.

[1] Algebra, I, Akademische Verlagsgesellschaft, Leipzig 1959.

[2] Die einstufig nichtkommutativen endlichen Ringe, Acta Math. Acad. Sci. Hung. *8* (1957), 401–422.

RENAULT, G.

[1] Radicaux des anneaux de groupes, Colloquia Math. Soc. J. Bolyai, 6. Rings, Modules and Radicals, Bolyai J. Mat. Társulat, Budapest 1973, 397–421.

RIBENBOIM, P.

[1] Rings and modules, Interscience Tracts in Pure and Applied Mathematics, № 24, Interscience Publishers (John Wiley and Sons, Inc.), New York–London–Sydney 1969.

RICH, M.

[1] Some radical properties of *s*-rings, Proc. Amer. Math. Soc. *30* (1971), 40–42.

RIEDL, CHR.

[1] Radikale für Fastmoduln, Fastringe und Kompositionringe, Diss. Univ. Wien 1966.

RIEFFEL, M. A.

[1] A general Wedderburn theorem, Proc. Nat. Acad. Sci. USA *54* (1965), 1513.

ROBERTSON, E. F.

[1] Generalizations of nilpotency in rings, Proc. Edinburgh Math. Soc. *18* (1973), 265–272.

ROBINSON, D. S.

[1] Groups in which normality is a transitive relation, Proc. Cambridge Philos. Soc. *60* (1964) 21–38.

ROBSON, J. C.

[1] Do simple rings have unity elements? J. Algebra *5* (1967), 140–143.

[2] Pri-rings and ipli-rings, Quart. J. Math. Oxford *18* (1967), 125–145.

[3] Artinian quotient rings, Proc. London Math. Soc. (3) *17* (1967), 600–616.

(ROIZ, E. N.) Ройз, Е. Н.

[1] О компрессивном радикале полугрупп, Мат. Сборник *92* (1973), 530–540.

[2] О радикалах Джекобсона и Бэра колец и их мультипликативных полугрупп, Изв. Высш. Учеб. Завед. Мат. (1975), № 10, 71–79.

ROLEWICZ, S.

[1] Some remarks on radical in commutative B_0-algebras, Bull. Acad. Polonaise Sci. Série math. astr. phys. *15* (1967), 153–155.

ROOS, C.

[1] The radical property of nonassociative rings such that every homomorphic image has no nonzero left annihilating ideals, Math. Nachr. *64* (1974), 385–391.

ROQUETTE, P.

[1] Abspaltung des Radikales in vollständigen lokalen Ringen, Abh. Math. Sem. Univ. Hamburg *23* (1959), 75–113.

ROSA, B. DE LA

[1] Ideals and radicals, Diss. Techn. Hochschule Delft 1970.
[2] A radical class which is fully determined by a lattice isomorphism, Acta Sci. Math. Szeged *33* (1972), 337–341.

ROSEN, M. I.

[1] The Jacobson radical of a group algebra, Michigan Math. J. *13* (1966), 477–480.

ROSSA, R. F.

[1] More properties inherited by the lower radical, Proc. Amer. Math. Soc. *33* (1972), 247–249.

RUBIN, R. A.

[1] Absolutely torsion-free rings, Pacific J. Math. *46* (1973), 503–514.

RUTTER, E. A., JR.

[1] Torsion theories over semiperfect rings, Proc. Amer. Math. Soc. *34* (1972), 389–395.

(RYABUKHIN, YU. M.) Рябухин, Ю. М.

[1] О вложении радикалов, Бул. Акад. Штиинце РСС Молд. (1963), № 11, 34–40.
[2] Радикалы в категориях, Бул. Акад. Штиинце РСС Молд. (1964), № 6, 58–74.
[3] О наднильпотентных и специальных радикалах, Исслед. по Алгебре и Матем. Анализу, Акад. Наук Молдав. ССР, Кишинёв 1965, 65–72.
[4] Полустрого наследственные радикалы в примитивных классах колец, Исслед. по общ. алгебре, Кишинёв 1965, 112–122.
[5] О нижних радикалах колец, Матем. Заметки Акад. Наук СССР *2* (1967), № 3, 239–244.
[6] К теории нижнего ниль-радикала колец, Алгебра и Логика, Акад. Наук СССР *6* (1967), № 4, 83–92.
[7] Радикалы в Ω-группах, I, Общая теория, Мат. Исслед. *3* (1968), № 2, 123–160.
[8] О радикале в категориях, Мат. Исслед. *2* (1967), № 3, 107–165.
[9] К теории радикалов неассоциативных колец, Мат. Исслед. *3* (1968), № 1, 86–99.
[10] О счетном простом квазирегулярном кольце, Мат. Заметки *4* (1968), 399–403.
[11] Радикалы в Ω-группах II. Идеально наследственные радикалы, Мат. Исслед. *3* (1968), № 4, 108–135.
[12] Об одном классе локально нильпотентных колец, Алгебра и Логика *7* (1968), № 5, 100–108.
[13] Радикалы в Ω-группах III. Специальные и квазиспециальные радикалы, Мат. Исслед. *4* (1969), № 1, 110–131.

[14] Алгебры без нильпотентных элементов, Докл. Акад. Наук СССР *181* (1969), 43–46.
[15] Алгебры без нильпотентных элементов, I, II, Алгебра и Логика *8* (1969), 181–214; 215–240.
[16] К проблеме существования простого нилькольца, Сибирск. Мат. Жур. *10* (1969), 950–956.

SADIQ, ZIA, M.–WIEGANDT, R.
[1] On the terminating of the lower radical construction, Ann. Univ. Sci. Budapest R. Eötvös, Sect. Math. (manuscript).

SAI, Y.
[1] On regular categories, Osaka J. Math. *7* (1970), 301–306.

SANDOMIERSKI, F. L.
[1] Semisimple maximal quotient rings, Trans. Amer. Math. Soc. *128* (1967), 112–120.

SANDS, A. D.
[1] Primitive rings of infinite matrices, Proc. Edinburgh Math. Soc. *14* (1964), 47–53.
[2] Prime ideals in matrix rings, Proc. Glasgow Math. Soc. *2* (1966), 194–195.
[3] The radical of a certain infinite matrix ring, Ann. Univ. Sci. Budapest R. Eötvös, Sect. Math. *12* (1969), 143–145.
[4] On radicals of infinite matrix rings, Proc. Edinburgh Math. Soc. (2) *16* (1968/69), 195–203.
[5] Radicals and Morita contexts, J. Algebra *24* (1973), 335–345.

(SANOV, I. N.) Санов, И. Н.
[1] Алгорифм Эвклида и односторонние разложения на простые множители для матричных колец, Сибирск. Мат. Жур. *8* (1967), 846–852.

SANTOSUOSSO, G.
[1] Sul trasporto ad un ultraprodotto di anelli di proprietà dei suoi fattori, Rend. Mat. (6) *1* (1968), 82–99.

SĄSIADA, E.
[1] Solution of the problem of existence of a simple radical ring, Bull. Acad. Polonaise Sci. Série math. astr. phys. *9* (1961), 257.

SĄSIADA, E.–COHN, P. M.
[1] An example of a simple radical ring, J. Algebra *5* (1967), 373–377.

SĄSIADA, E.–SULIŃSKI, A.
[1] A note on Jacobson radical, Bull. Acad. Polonaise Sci. Série math. astr. phys. *10* (1962), 421–423.

SATYANARAYANA, M.
[1] Semisimple rings, Amer. Math. Monthly *74* (1967), 1086.
[2] On dual semigroups, Publ. Math. Debrecen *20* (1973), *45*–51.

Satyanarayana, M.–Al-Amiri, H.

[1] Completely prime radical and primary ideal representations in non-commutative rings, Math. Z. *121* (1971), 181–187.

Schelter, W.–Roberts, P.

[1] Flat modules and torsion theories, Math. Z. *129* (1972), 331–334.

Schneider, H.–Weissglass, J.

[1] Group rings, semigroup rings and their radicals, J. Algebra *5* (1967), 1–15.

Schwarz, Š.

[1] Zur Theorie der Halbgruppen, Sbornik Prac Prirodovedeckej Fakulty Slovenskej Univerzity v Bratislave *6* (1943), 1–64.

[2] On semigroups having a kernel, Czechoslovak Math. J. *1* (76) (1951), 256–301.

Scott, S. D.

[1] Formation radicals for near-rings, Proc. London Math. Soc. (3) *25* (1972), 441–464.

Scozzafava, R.

[1] The radical of a locally compact semialgebra, Studia Math. Acad. Sci. Hung. *8* (1973), 39–41.

Segal, I. E.

[1] The group algebra of a locally compact group. Trans. Amer. Math. Soc. *61* (1947), 69–105.

Seidel, H.

[1] Eine Charakterisierung des 0-Radikales einer Halbgruppe, Math. Nachr. *34* (1967), 163–166.

[2] Über das Radikal einer Halbgruppe, Math. Nachr. *29* (1965), 255–263.

(Sesekin, N. F.) Сесекин, Н. Ф.

[1] О критериях нильпотентности радикалов Фиттинга бесконечных групп и некоторые следствия, Мат. Зап. Уральск. Ун-та *6* (1967), № 1, 89–93.

Sexauer, N. E.–Warnock, J. E.

[1] The radical of the row finite matrices over an arbitrary ring. Trans. Amer. Math. Soc. *139* (1969), 287–295.

(Shayn, B. M.) Шайн, Б. М.

[1] Теория радикалов полугрупп, Резюме сообш. и докл. VIII Всес. Коллок. Общ. Алг. Рига 1967.

[2] A note on radicals in regular semigroups, Semigroup Forum *3* (1971/72), 84–85.

Shanny, R. F.

[1] Regular endomorphism rings of free modules, J. London Math. Soc. (2) *4* (1971), 353–354.

(SHATALOVA, M. A.) Шаталова, М. А.

[1] Об одном классе структурно упорядоченных колец, Уч. Зап. Моск. Обл. Пед. Ин-та *185* (1967), 125–134.

[2] К теории радикалов в структурно упорядоченных кольцах, Мат. Заметки *4* (1968), 639–648.

(SHCHERBATSKI, I. K.) Щербацкий, И. К.

[1] О совпадении примарностей, Бул. Акад. Штиинце РСС Молд. (1973), № 1, 25–30.

(SHCHUKIN, K. K.) Щукин, К. К.

[1] К теории радикалов групп, Сибирск. Мат. Жур. *3* (1962), 932–942.

[2] Радикалы в теории групп, Сибирск. Мат. Жур. *3* (1962), 912–931.

[3] К теории радикалов в группах, Докл. Акад. Наук СССР *142* (1962), 1047–1049.

[4] О вербальных радикалах групп, Уч. Зап. Кишиневск. Гос. Ун-та *82* (1965), 97–99.

[5] Минимальный радикал и слабо неразложимые группы, Мат. Заметки *13* (1973), 447–456.

(SHESTAKOV, I. P.) Щестаков, И. П.

[1] Конечномерные алгебры с ниль-базисом, Алгебра и Логика *10* (1971), 87–99.

(SHEVRIN, L. N.) Шеврин, Л. Н.

[1] К общей теории полугрупп, Мат. Сборник *53* (1961), 367–386.

[2] Нильполугруппы с некоторыми условиями конечности, Мат. Сборник *55* (1961), 473–480.

(SHIRSHOV, A. I.) Ширшов, А. И.

[1] О некоторых неассоциативных нилькольцах и алгебраических алгебрах, Мат. Сборник *4* (83) (1957), 381–394.

(SHIRYAEV, V. M.) Ширяев, В. М.

[1] Об инверсных полугруппах с данным *G*-радикалом, Докл. Акад. Наук БССР 14 (1970), 782–785.

SHOCK, R. C.

[1] Essentially nilpotent rings, Israel J. Math. *9* (1971), 180–185.

[2] Nil subrings in finiteness conditions, Amer. Math. Monthly *78* (1971), 741–748.

[3] A note on the prime radical, J. Math. Soc. Japan *24* (1972), 347–376.

(SHULGEIFER, E. G.) Шульгейфер, Е. Г.

[1] Разложение на простые множители в структурах с умножением, Украинский Мат. Жур. *2* (1950), 100–114.

[2] К общей теории радикалов в категориях, Мат. Сборник *51* (93) (1960), 487–500.

[3] Функторная характеризация строгих радикалов в категориях, Сибирск. Мат. Жур. *7* (1966), 1412–1421.

[4] Локализации и сильно наследственные строгие радикалы в категориях, Труды Моск. Мат. Об-ва *19* (1968), 271–301.

SIERPIŃSKA, A.

[1] Radicals of rings of polynomials in non-commutative indeterminates, Bull. Acad. Polonaise Sci. *21* (1973), 805–808.

(SIMONYAN, L. A.) Симонян, Л. А.

[1] О двух радикалах алгебр Ли, Докл. Акад. Наук СССР *157* (1964), 281–283.
[2] Некоторые радикалы алгебр Ли, Сибирск. Мат. Жур. *6* (1965), 1101–1107.

SINGH, S.

[1] On tertiary radicals of an ideal in a ring, Riv. Mat. Univ. Parma (2) *8* (1967), 291–296.

SINHA, I.

[1] On the radical of subrings of rings, Math. Student *34* (1966), 185–190.

SINHA, I.–SRIVASTAVA, J.

[1] Relative projectivity and a property of Jacobson radical, Publ. Math. Debrecen *18* (1971), 37–41.

(SKORNYAKOV, L. A.) Скорняков, Л. А.

[1] *T*-гомоморфизмы колец, Мат. Сборник *42* (1957), 425–440.
[2] Дедекиндовы структуры с дополнениями и регулярные кольца, Физматгиз, Москва 1961.
[3] Локально бикомпактные бирегулярные кольца, Мат. Сборник *62* (1963), 3–13.
[4] Гомологическая классификация колец, Труды 4-го Всес. Мат. Съезда, 1961, т. 2, 22–32, Наука, Ленинград 1964.
[5] Цепные слева кольца, Памяти Н. Г. Чеботарева, Издат. Казан Унив. Казань 1964, 74–88.
[6] О коновских кольцах, Алгебра и Логика *3* (1965), 5–30.
[7] Einfache lokal bikompakte Ringe, Math. Z. *87* (1965), 241–251.
[8] О цепных слева кольцах, Изв. Высш. Учеб. Завед. Мат. (1966), № 4, 114–117.
[9] Гомологическая классификация колец, Мат. Вестник *4* (1967), 415–434.
[10] Когда все модули – полуцепные? Мат. Заметки *5* (1969), 173–182.
[11] Цепные слева полугруппы, Сибирск. Мат. Жур. *11* (1970), 168–182.
[12] О расщеплении инъективных модулей, Мат. Исслед. *5* (1970), № 2, 183–186.
[13] О максимально-идеальных покрытиях колец, Мат. Зап. Уральск. Ун-та *7* (1970), 208–212.
[14] О радикалах Ω-колец, Избранные вопросы алгебры и логики, Новосибирск 1973, 283–299.

SLATER, M.

[1] Prime alternative rings, J. Algebra *21* (1972), 394–409.

(SLINKO, A. M.) Слинько, А. М.

[1] Об эквивалентности некоторых нильпотентностей в правоальтернативных кольцах, Алгебра и Логика *9* (1970), 342–348.
[2] О радикалах йордановых колец, Алгебра и Логика *11* (1972), 206–215.

[3] О радикале Джекобсона и абсолютных делителях нуля специальных йордановых алгебр, Алгебра и Логика *11* (1972), 711–723.

[4] Замечание о радикалах и дифференцированиях колец, Сибирск. Мат. Жур. *13* (1972), 1395–1397.

SLOVER, R.

[1] The Jacobson radical of row-finite matrices, J. Algebra *12* (1969), 345–359.

[2] A note on the radical of row-finite matrices, Glasgow Math. J. *13* (1972), 80–81.

SMILEY, M. F.

[1] The radical of an alternative ring, Ann. Math. *49* (1948), 702–709.

[2] An application of a radical of Brown and McCoy to non-associative rings, Amer. J. Math. *72* (1950), 93–100.

SMITH, J. R.

[1] Local domains with topologically *T*-nilpotent radical, Pacific J. Math. *30* (1969), 233–245.

SNIDER, R. L.

[1] Complemented hereditary radicals, Bull. Austral. Math. Soc. *4* (1971), 307–320.

[2] Lattices of radicals, Pacific J. Math. *40* (1972), 207–220.

SORANI, G.

[1] Radical subrings of matrix rings, Amer. Math. Monthly *73* (1966), 989–991.

SPIEGEL, H.

[1] Gruppenalgebren mit zentralem Radikal, Arch. Math. Basel *23* (1972), 380–384.

SPIRCU, T.

[1] Radicalul strict regulat al unui inel (Strongly regular radical of a ring), Stud. Cerc. Mat. *26* (1974), 751–754.

STEINFELD, O.

[1] Verbandstheoretische Betrachtung gewisser idealtheoretischer Fragen, Acta Sci. Math. Szeged *22* (1961), 136–149.

[2] Über die Quasiideale von Ringen, Acta Sci. Math. Szeged *17* (1956), 170–180.

[3] On ideal quotients and prime ideals, Acta Math. Acad. Sci. Hung. *4* (1953), 289–298.

[4] Ein Beweis des Wedderburn-Artinschen Struktursatzes, Magy. Tud. Akad. Mat. Kut. Int. Közl. *3* (1958), 63–65.

[5] Negativan rendezett struktúrák primelemeiről (On prime elements of negatively ordered structures), Magy. Tud. Akad. Mat. Fiz. Oszt. Közl. *17* (1967), 467–472.

[6] Primelemente und Primradikale in gewissen verbandsgeordneten algebraischen Strukturen, Acta Math. Acad. Sci. Hung. *19* (1968), 246–262.

[7] Eine Charakterisierung der primitiven Ideale eines Ringes, Acta Math. Acad. Sci. Hung. *19* (1968), 219–220.

[8] Über die Struktursätze der Semiringe, Acta Math. Acad. Sci. Hung. *10* (1959), 149–155.

[9] On semigroups which are unions of completely 0-simple semigroups, Czechoslovak Math. J. *54* (1966), 23–41.

[10] On a generalization of completely 0-simple semigroups, Acta Sci. Math. Szeged *28* (1967), 135–145.
[11] On Litoff's theorem, Ann. Univ. Sci. Budapest R. Eötvös, Sect. Math., *13* (1970), 101–102.
[12] An analogue of Litoff's theorem, Semigroup Forum *2* (1971), 86–88.

STEINFELD, O.–WIEGANDT, R.

[1] Über die Verallgemeinerungen und Analoga der Wedderburn-Artinschen und Noetherschen Struktursätze, Math. Nachr. *34* (1967), 143–156.

STENSTRÖM, B.

[1] Radicals and socles of lattices, Arch. Math. Basel *20* (1969), 258–261.

STEVENSON, R. A.

[1] Jacobson structure theory for Hestenes ternary rings, Trans. Amer. Math. Soc. *177* (1973), 91–98.

STEWART, I.

[1] Baer and Fitting radicals in groups and Lie algebras, Arch. Math. Basel *23* (1972), 385–386.

STEWART, P. N.

[1] Semisimple radical classes, Pacific J. Math. *32* (1970), 249–254.
[2] Local radical and semisimple classes of rings, Ph. D. Thesis, Univ. British Columbia, Vancouver 1969.
[3] On the locally antisimple radical, Glasgow Math. J. *13* (1972), 42–46.
[4] Divinsky's radical, Proc. Amer. Math. Soc. *31* (1972), 347–353.
[5] Strongly hereditary radical classes, J. London Math. Soc. (2) *4* (1972), 499–509.
[6] Strict radical classes of associative rings, Proc. Amer. Math. Soc. *39* (1973), 273–278

STONEHEWER, S. E.

[1] Group algebras of some torsion-free groups, J. Algebra *13* (1969), 143–147.

STRATIGOPOULOS, D.

[1] Hyperanneaux non commutatifs: Le radical d'un hyperanneau, somme sous-directe des hyperanneaux, hyperanneaux artiniens et théorie des éléments idempotents, C. R. Acad. Sci. Paris, Sér. A–B, *269* (1969), A627–A629.

STREB, W.

[1] Über nil Algebren mit nilpotenten Kommutatoridealen, die keine von Null verschiedenen Zentren besitzen, Rend. Sem. Mat. Univ. Padova *46* (1971), 391–395.
[2] Über Ringe, die von ihren Einheitengruppen erzeugt werden, Rend. Sem. Mat. Univ. Padova *47* (1972), 313–329.

SUBRAMANIAN, H.

[1] Von Neumann regularity in semirings, Math. Nachr. *45* (1970), 73–79.

SULIŃSKI, A.

[1] Some characterizations of the Brown–McCoy radical, Bull. Acad. Polonaise Sci. Série math. astr. phys. *5* (1957), 357–359.
[2] Radicals in associative algebras of finite order, Bull. Acad. Polonaise Sci. Série math. astr. phys. *5* (1957), 361–363.
[3] Некоторые вопросы общей теории радикалов, Мат. Сборник *44* (1958), 273–286.
[4] On subdirect sums of simple rings with unity, Bull. Acad. Polonaise Sci. Série math. astr. phys. *8* (1960), 223–228.
[5] A classification of semisimple rings, Bull. Acad. Polonaise Sci. Série math. astr. phys. *9* (1961), 1–6.
[6] The Brown–McCoy radical in categories, Fund. Math. *59* (1966), 23–41.

SULIŃSKI, A.–ANDERSON, T.–DIVINSKY, N.

[1] Lower radical properties for associative and alternative rings, J. London Math. Soc. *4* (1966), 417–424.

ŠULKA, R.

[1] Note on the Sevrin radical in semigroups, Mat. Fyz. Časopis *18* (1968), 57–58.
[2] О нильпотентных элементах, идеалах и радикалах полугруппы, Mat. Fyz. Časopis *13* (1963), 209–222.
[3] Заметка о радикалах в фактор-полугруппах, Mat. Fyz. Časopis *14* (1964), 297–300.
[4] Радикалы и топология в полугруппах, Mat. Fyz. Časopis *15* (1965), 3–14.
[5] On the nilpotency in semigroups, Mat. Časopis *18* (1968), 148–157.
[6] The maximal semilattice decomposition of a semigroup, radicals and nilpotency, Mat. Časopis *20* (1970), 172–180.

SUSSMANN, I.–FOSTER, A. I.

[1] On rings in which $a^{n(a)}=a$, Math. Ann. *140* (1960), 324–333.

SZÁSZ, F. A.

[1] Über Ringe mit Minimalbedingung für Hauptrechtsideale I, Publ. Math. Debrecen *7* (1960), 54–64.
[2] Über Ringe mit Minimalbedingung für Hauptrechtsideale II, Acta Math. Acad. Sci. Hung. *12* (1961), 417–439.
[3] Über Ringe mit Minimalbedingung für Hauptrechtsideale III, Acta Math. Acad. Sci. Hung. *14* (1963), 447–461.
[4] An observation on the Brown–McCoy radical, Proc. Japan Acad. *37* (1961), 413–416.
[5] Verbandstheoretische Bemerkungen zum Fuchsschen Zeroidradikal der nichtassoziativen Ringe, Arch. Math. *12* (1961), 282–289.
[6] Bemerkungen über Rechtssockel und Nilringe, Monatsh. Math. *67* (1963), 359–362.
[7] A topologikus algebrákról és gyűrűkről (On topological algebras and rings) I, Mat. Lapok *13* (1962), 356–378.
[8] A topologikus algebrákról és gyűrűkről (On topological algebras and rings) II, Mat. Lapok *14* (1963), 74–87.
[9] Az operátormodulusok Kertész-féle radikáljáról (On the Kertész radical of operator modules), Magy. Tud. Akad. Mat. Fiz. Oszt. Közl. *10* (1960), 35–38.

[10] A teljesen reducibilis operátormodulusokról (On completely irreducible operator modules), Magy. Tud. Akad. Mat. Fiz. Oszt. Közl. *11* (1961), 417–425.

[11] Ringe, deren endlich erzeugbare echte Unterringe Hauptrechtsideale sind, Acta Math. Acad. Sci. Hung. *13* (1962), 115–132.

[12] Bemerkungen zu assoziativen Hauptidealringen, Indag. Math. *64* (1961), 577–583.

[13] Reduktion eines gruppentheoretischen Problems von O. Ju. Schmidt, Bull. Acad. Polonaise Sci. Série math. astr. phys. *7* (1959), 369–372.

[14] Radikalbegriffe für Halbgruppen mit Nullelement, die dem Jacobsonschen ringtheoretischen Radikal ähnlich sind, Math. Nachr. *34* (1967), 157–161.

[15] Einige Kriterien für die Existenz des Einselementes in einem Ring, Acta Sci. Math. Szeged *28* (1967), 31–37.

[16] Eine Charakterisierung des Jacobsonschen Radikales eines Ringes, Bull. Acad. Polonaise Sci. Série math. astr. phys. *15* (1967), 53–56.

[17] Lösung eines Problems bezüglich einer Charakterisierung des Jacobsonschen Radikals, Acta Math. Acad. Sci. Hung. *18* (1967), 261–272.

[18] The sharpening of a result concerning the primitive ideals of an associative ring, Proc. Amer. Math. Soc. *18* (1967), 910–912.

[19] Die Lösung eines Problems bezüglich des Durchschnittes zweier modularer Rechtsideale in einem Ring, Acta Math. Acad. Sci. Hung. *20* (1969), 211–216.

[20] Simultane Lösung eines halbgruppentheoretischen und eines ringtheoretischen Problems, Acta Sci. Math. Szeged *30* (1969), 289–294.

[21] Reduktion eines Problems bezüglich der Brown–McCoyschen Radikalringe, Acta Sci. Math. Szeged *31* (1970), 167–172.

[22] Unabhängigkeitsfragen des Kuroŝschen Axiomensystems der Radikale der Ringe, Publ. Math. Debrecen *15* (1968), 287–292.

[23] Beiträge zur Radikaltheorie der Ringe, Publ. Math. Debrecen *17* (1970), 267–272.

[24] Ein radikaltheoretischer Vereinigungsendomorphismus des Idealverbandes der Ringe, Ann. Univ. Sci. Budapest R. Eötvös, Sect. Math. *12* (1969), 73–75.

[25] Gyűrűk maximális jobbideáljairól (On maximal right ideals of rings), Magy. Tud. Akad. Mat. Fiz. Oszt. Közl. *17* (1967), 473–476.

[26] Gyűrűk radikáljairól (On radicals of rings) I–III, Mat. Lapok *19* (1968), 259–301; *20* (1969), 99–116, 311–345.

[27] On Frattini one-sided ideals and subgroups, Math. Nachr. *46* (1970), 235–242.

[28] Notes on modules, I–III, Proc. Japan Acad. *46* (1970), 349–350, 351–353, 354–357.

[29] Rings with an almost nil adjoint semigroup, Proc. Japan Acad. *46* (1970), 593–594.

[30] On radicals of semigroups with zero, I, Proc. Japan Acad. *46* (1970), 595–598.

[31] On the adjoint semigroups of rings, Proc. Japan Acad. *46* (1970), 773–775.

[32] Generalized bi-ideals of rings, I, II, Math. Nachr. *47* (1971), 355–360, 361–364.

[33] On minimal bi-ideals of rings, Acta Sci. Math. Szeged *32* (1971), 333–336.

[34] On the idealizer of a subring, Monatsh. Math. *75* (1971), 65–68.

[35] Almost right quasiregular adjoint semigroups of rings, Math. Nachr. *48* (1971), 309–314.

[36] The radical property of rings such that every homomorphic image has no nonzero left annihilators, Math. Nachr. *48* (1971), 371–375.

[37] A class of regular rings, Monatsh. Math. *75* (1971), 168–172.

[38] Further characterizations for the Jacobson radical of a ring, Proc. Japan Acad. *47* (1971), 810–811.

[39] Certain subdirect sums of finite prime fields, Colloquium Math. *25* (1972), 43–48.

[40] Über artinsche Ringe, Bull. Acad. Polonaise Sci. Série math. astr. phys. *11* (1963), 351–354.

[41] Eine Äquivalenzrelation für eine Charakterisierung des Jacobsonschen Radikals, Acta Math. Acad. Sci. Hung. *22* (1971), 85–86.

[42] Ideals of a ring with modular intersection, Revue Roum. Math. Pures Appl. *16* (1971), 609–616.

[43] Bemerkungen zu meiner Arbeit: "Beiträge zur Radikaltheorie der Ringe" Publ. Math. Debrecen *23* (1976), 49–51.

[44] Rings, which are radical modules, Math. Japon. *16* (1971), 103–104.

[45] Steinfeld Ottó egy gyűrűelméleti eredményének moduluselméleti analogonja (The module-theoretic analogue of a ring-theoretic result by O. Steinfeld), Magy. Tud. Akad. Mat. Fiz. Tud. Oszt. Közl. *21* (1972), 443–447.

[46] An almost subidempotent radical property of rings, Colloquia Math. Soc. J. Bolyai, 6. Rings, Modules and Radicals, Bolyai J. Mat. Társulat, Budapest 1973, 483–499.

[47] Rings with radical maximal submodules, Monatsh. Math. *77* (1973), 354–356.

[48] On some weakly supernilpotent radicals of rings, Colloq. Math. *28* (1973), 195–201.

[49] Über das im Operatorring enthaltene allgemeine Radikal eines Untermoduls, Acta Sci. Math. Szeged *34* (1973), 371–376.

[50] On left magnifying elements and quasimodular right ideals of rings, Math. Japon. *18* (1973), 221–224.

[51] On some simple Jacobson radical rings, Math. Japon. *18* (1973), 225–228.

[52] On Hashimoto universal algebras with some properties of Hopf, Math. Japon. *18* (1973), 229–234.

[53] An almost subidempotent radical property of rings, Math. Nachr. *48* (1971), 371–375.

[54] Remarks on my paper "The radical property of rings such that every homomorphic image has no nonzero left annihilators", Math. Nachr. *64* (1974), 391–395.

[55] On hereditary radicals of semigroups with zero, Math. Japon. *20* (1975), 89–91.

[56] The join-representation of some intersections in complete lattices, Math. Japon. *19* (1974), 17–19.

[57] On strong semisimplicities of semigroups with zero, Periodica Math. Hung. *5* (1974), 145–148.

[58] On generalized subcommutative regular rings, Monatsh. Math. *77* (1973), 67–71.

[59] The solution of a problem on upper radicals, Math. Nachr. *66* (1975), 101.

[60] A second almost subidempotent radical for rings, Math. Nachr. *66* (1975), 283–289.

[61] A criterion for the existence of the two-sided unity element of semigroups, Proc. Japan Acad. *47* (1971), 82–814.

[62] Short elementary proof of a ringtheoretical result, Math. Japon. *17/2* (1972), 113–114.

[62] Some generalizations of stongly regular rings I–III, Math. Japon. *17* (1972), 115–118; *18* (1973), 87–90; *18* (1973), 91–94.

[64] Ein kurzer elementarer Beweis eines Satzes von O. Steinfeld über die einstufig nichtprimen Ringe, Periodica Math. Hungar. *7* (1976), 9–10.

[65] Bemerkungen zu meiner Arbeit: "An almost subidempotent radical property of rings" Estudos de Matematica em homenagem a Prof. A. Almeida Costa, Lisboa 1974.

[66] Vizsgálatok algebrai strukturák radikálelméletében (Investigations in the radical theory of algebraic structures), I, II, Magy. Tud. Akad. Mat. Fiz. Oszt. Köz. *22* (1973), 215–255.

SZÁSZ F. A.–WIEGANDT, R.

[1] On the dualization of subdirect embeddings, Acta Math. Acad. Sci. Hung. *20* (1969), 289–302.

[2] On the duality of radical and semisimple objects in categories, Acta Math. Acad. Sci. Hung. *21* (1970), 175–182.

[3] On hereditary radicals, Periodica Math. Hung. *3* (1973), 235–241.

SZELE, T.

[1] Nilpotent Artinian rings, Publ. Math. Debrecen *4* (1955), 71–78.

[2] Simple proof of the Wedderburn–Artin structure theorem, Acta Math. Acad. Sci. Hung. *5* (1954), 101–104.

[3] Ein Zerfällungssatz für radikalfreie Ringe, C. R. du Premier Congr. Math. Hongr. 1950, 429–434.

SZÉLPÁL, I.

[1] Die unendlichen abelschen Gruppen mit lauter endlichen echten Untergruppen, Publ. Math. Debrecen *1* (1949), 63–65.

SZENDREI, J.

[1] On the Jacobson radical of a ring, Publ. Math. Debrecen *4* (1955), 93–97.

[2] Asszociativ félgyűrűk bi-ideáljairól (On bi-ideals of associative semirings), Szegedi Tanárképző Főiskola Tud. Közl. *2* (1971), 169–172.

SZWED, P.

[1] A scale of a radical in commutative B_0-algebras, Bull. Acad. Polonaise Sci. *20* (1972), 745–752.

TAMAŞ, V.

[1] Weakly nilpotent, nilpotent subalgebras, Annales Stii. Univ. Alexandru Jon Cuza, Iaşi, Sec. I. a. Mat. *13* (1967), 263–266.

TANGEMAN, R. L.

[1] Strong heredity in radical classes, Pacific Math. *J. 42* (1972), 259–265.

TANGEMAN, R. L.–KREILING, D.

[1] Lower radicals in nonassociative rings, J. Austral. Math. Soc. *14* (1972), 411–423.

TELEMAN, S.

[1] Sur les anneaux réguliers, Revue Roum. Math. Pures Appl. *15* (1970), 407–434.

[2] On the regular rings of John von Neumann, Revue Roum. Math. Pures Appl. *15* (1970), 735–742.

TEPLY, M. L.

[1] Some aspects of Goldie's torsion theory, Pacific J. Math. *29* (1969), 447–459.

[2] Homological dimension and splitting torsion theories, Pacific J. Math. *34* (1970), 193–206.

[3] Direct decomposition of modules, Math. Japon. *15* (1970), 85–90.
[4] Torsionfree projective modules, Proc. Amer. Math. Soc. *27* (1971), 29–34.
[5] A class of divisible modules, Pacific J. Math. *45* (1973), 653–668.

TEPLY, M. L.–FUELBERTH, J. D.
[1] The torsion submodule splits off, Math. Ann. *188* (1970), 270–284.

THIERRIN, G.
[1] Sur les idéaux complètement premiers d'un anneau quelconque, Acad. Royale Belgique Bull. *43* (1957), 124–132.
[2] Sur le radical corpoïdal d'un anneau, Canad. J. Math. *12* (1960), 101–106.
[3] Duo rings, Canad. Math. Bull. *3* (1960), 167–172.
[4] Extensions radicales et quasi-radicales dans les anneaux, Canad. Math. Bull. *5* (1962), 29–35.
[5] Anneaux métaprimitifs, Canad. J. Math. *17* (1965), 119–205.
[6] Quelques caractérisations du radical d'un anneau, Canad. Math. Bull. *10* (1967), 643–647.

TIAGO DE OLIVIERA, J.
[1] Residuals of systems and radicals of rings, Univ. Lisboa, Rev. Fac. Ci. A (2), *5* (1956), 177–248.

TIRNOVEANU, M.
[1] Über die Nilpotenz und das Radikal, Bul. Inst. Politechn. Bucureşti *24* (1962), 27–35.

TÔGÔ, S.
[1] Radicals of infinite dimensional Lie algebras, Hiroshima Math. J. *2* (1972), 179–203.
[2] Characterisations of radicals of infinite dimensional Lie algebras, Hiroshima Math. J. *3* (1973), 25–26.

TÔGÔ, S.–KAWAMOTO, N.
[1] Ascendantly covalescent series and radicals of Lie algebras, Hiroshima Math. J. *2* (1972), 253–261.

(TOKARENKO, A. I.) Токаренко, А. И.
[1] О радикалах и стабильности в линейных группах над коммутативными кольцами, Сибирск. Мат. Жур. *9* (1968), 165–176.

TOMINAGA, H.
[1] Some remarks on radical ideals, Math. J. Okayama Univ. *3* (1954), 139–142.
[2] Galois theory of simple rings, Math. J. Okayama Univ. *6* (1956), 29–48.
[3] On *s*-unital rings, Math. J Okayama Univ. *18* (1976), 117–134.
[4] On *s*-unital rings II, ibid. *19* (1977), 171–182.

TOWERS, D. A.
[1] A Frattini theory for algebras, Proc. London Math. Soc. (3) *27* (1973), 440–462.

TSAI, C. E.

[1] The prime radical in a Jordan ring, Proc. Amer. Math. Soc. *19* (1968), 1171–1175.
[2] The Levitzki radical in Jordan rings, Proc. Amer. Math. Soc. *24* (1970), 119–123.
[3] An internal characterization of the prime radical of a Jordan algebra, Proc. Amer. Math. Soc. *36* (1972), 361–364.

(TSALENKO, M. SH.) Цаленко, М. Ш.

[1] Полугруппа рефлективных подкатегорий, Мат. Сборник *81* (1970), 62–78.

(TSEITLIN, D. I.) Цейтлин, Д. И.

[1] О радикалах в классах *ZA*-, *N*- и N_0-групп, Сакартв. ССР Мецн. Акад. Моамбе *66* (1972), 29–32.

TSUSHIMA, Y.

[1] Radicals of group algebras, Osaka J. Math. *4* (1967), 179–182.
[2] On the annihilator ideals of the radical of a group algebra, Osaka J. Math. *8* (1971), 91–97.

TURNIDGE, D. R.

[1] Torsion theories and semihereditary rings, Proc. Amer. Math. Soc. *24* (1970), 137–143.
[2] Torsion theories and rings of quotients of Morita equivalent rings, Pacific J. Math. *37* (1971), 225–234.

UTUMI, Y.

[1] On quotient rings, Osaka Math. J. *8* (1956), 1–8.
[2] Self-injective rings, J. Algebra *6* (1967), 56–64.

VASILESCU, F. H.

[1] Radical d'une algèbre de Lie de dimension infinie, C. R. Acad. Sci. Paris, Sér. A–B, *274* (1972), A536–A538.

VERHOEFF, J.

[1] Recent investigations about the radical of a ring, Math. Centrum Amsterdam, Rapport ZW 1953, 1–7.

VIDAL, R.

[1] Sur les complétés d'anneaux pour la filtration du radical de Jacobson, C. R. Acad. Sci. Paris, Sér. A–B, *272* (1971), A1638–A1641.

(VINBERG, E. B.) Винберг, Э. Б.

[1] К теореме о бесконечности ассоциативной алгебры, Изв. Акад. Наук СССР *29* (1965), 209–214.

(VODINCHAR, M. I.) Водинчар, М. И.

[1] О минимальном наднильпотентном радикале в топологических кольцах, Мат. Исслед. *3* (1968), № 4, 29–50.

[2] Наследственные и специальные радикалы, Мат. Исслед. *4* (1969), № 2, 17–31.

(VOVSI, S. M.) Вовси, С. М.

[1] Абсолютная свобода нестрогих радикалов, Уч. Зап. Латв. Ун-та *151* (1971), 14–18.

[2] О бесконечных произведениях классов групп, Сибирск. Мат. Жур. *13* (1972), 272–285.

[3] Радикальные классы $\mathfrak{M}$-наднильпотентных групп, Изв. Высш. Учеб. Завед. Мат. (1975), № 1, 21–26.

WADDELL, M. C.

[1] Properties of regular rings, Duke Math. J. *9* (1952), 623–627.

WAERDEN, B. L. VAN DER

[1] Algebra, I, 8. Aufl.; II. 5. Aufl., Springer-Verlag, Berlin–Heidelberg–New York 1971.

WALLACE, D. A. R.

[1] On the radical of a group algebra, Proc. Amer. Math. Soc. *12* (1961), 133–137.

[2] The Jacobson radicals of the group algebras of a group and of certain normal subgroups, Math. Z. *100* (1967), 282–294.

[3] Lower bounds for the radical of the group algebra of a finite *p*-soluble group, Proc. Edinburgh Math. Soc. (2) *16* (1968/69), 127–134.

[4] On commutative and central conditions on the Jacobson radical of the group algebra of a group I, Proc. London Math. Soc. (3) *19* (1969), 385–402.

[5] On commutative and central conditions on the Jacobson radical of the group algebra of a group II, J. London Math. Soc. (2) *4* (1971), 91–99.

[6] The radical of the group algebra of a subgroup of a polycyclic group and of a restricted *SN*-group, Proc. Edinburgh Math. Soc. (2) *17* (1970/71), 165–171.

WALLIS, W. D.

[1] A reduction of the problem of semisimplicity, J. Algebra *10* (1968), 501–502.

WALT, A. P. J. VAN DER

[1] On the Levitzki radical, Archiv Math. *15* (1965), 22–24.

[2] Prime ideals and nil radicals in near rings, Archiv Math. *14* (1964), 408–414.

WARE, R.–ZELMANOWITZ, J.

[1] The Jacobson radical of the endomorphism ring of a projective module, Proc. Amer. Math. Soc. *26* (1970), 15–20.

WARNER, S.

[1] Some rings having minimal left ideals, Trans. Amer. Math. Soc. *125* (1966), 395–405.

[2] Linearly compact Noetherian rings, Math. Ann. *178* (1968), 53–61.

WATTERS, J. F.

[1] On the adjoint group of a radical ring, J. London Math. Soc. *43* (1968), 725–729.
[2] On internal constructions of radicals in rings, J. London Math. Soc. (2) *1* (1969), 500–504.
[3] Lower radicals in associative rings, Canad. J. Math. *21* (1969), 466–476.

WEHLEN, J. A.

[1] Algebras over Dedekind domains, Canad. J. Math. *25* (1973), 842–855.

WEI, D. Y.

[1] On the concept of torsion and divisibility for general rings, Illinois J. Math. *13* (1969), 414–431.

WEINERT, H. J.

[1] Halbgruppen ohne Frattinische Unterhalbgruppe, Acta Math. Acad. Sci. Hung. *15* (1964), 309–323.
[2] Bemerkung zu einem von F. Szász angegebenen Ring, Acta Sci. Math. Szeged *32* (1971), 223–224.
[3] Zur Theorie Levitzkischer Radikale in Halbringen, Math. Z. *128* (1972), 325–341.

WEISSGLASS, J.

[1] Radicals of semigroup rings, Glasgow Math. J. *10* (1969), 85–93.
[2] Regularity of semigroup rings, Proc. Amer. Math. Soc. *25* (1970), 499–503.
[3] Semigroup rings and semilattice sums of rings, Proc. Amer. Math. Soc. *39* (1973), 471–478.

WICKLESS, W. J.

[1] A characterization of the nil radical of a ring, Pacific J. Math. *35* (1970), 255–258.
[2] Rings with the contraction property, Proc. Amer. Math. Soc. *27* (1971), 51–60.

WIDIGER, A.

[1] Über Ringe mit eingeschränkter Minimalbedingung für Rechtsideale, Beitr. Alg. Geom. *1* (1971), 141–153.

WIEGANDT, R.

[1] Über halbeinfache linear kompakte Ringe, Studia Sci. Math. Hung. *1* (1966), 31–38.
[2] Über linear kompakte reguläre Ringe, Bull. Acad. Polonaise Sci. Série math. astr. phys. *13* (1965), 445–446.
[3] Über transfinit nilpotente Ringe, Acta Math. Acad. Sci. Hung. *17* (1966), 101–114.
[4] Über lokal linear kompakte Ringe, Acta Sci. Math. Szeged *28* (1967), 255–260.
[5] Radical and semisimplicity in categories, Acta Math. Acad. Sci. Hung. *19* (1968), 345–364.
[6] On compact objects in categories, Publ. Math. Debrecen *15* (1968), 267–281.
[7] A note on Kurosh–Amitsur radical properties, Ann. Univ. Sci. Budapest R. Eötvös, Sect. Math., *12* (1969), 63–65.
[8] On the socle of an object in categories, Acta Sci. Math. Szeged *31* (1970), 245–251.
[9] A note on lower radicals, Ann. Univ. Sci. Budapest R. Eötvös, Sect. Math. *13* (1970), 165–166.

[10] Über die Struktursätze der Halbringe, Ann. Univ. Sci. Budapest R. Eötvös, Sect. Math. *5* (1962), 51–58.

[11] On subdirect embeddings in categories, Fund. Math. *68* (1970), 7–12.

[12] Local and residual properties in bicategories, Acta Sci. Math. Szeged *32* (1971), 195–205.

[13] Radicals coinciding with the Jacobson radical on linearly compact rinsg, Beitr. Alg. Geom. *1* (1971), 195–199.

[14] A note on compact objects, Publ. Math. Debrecen *18* (1971), 99–102.

[15] Some problems in universal algebra with ring-theoretical background, Proc. Mini-Conf. Univ. Algebra, Szeged 1971, 23.

[16] On the structure of lower radical semigroups, Czechoslovak Math. J. *22* (1972), 1–6.

[17] Homomorphically closed semisimple classes, Stud. Univ. Babeş-Bolyai, Ser. Math.-Mech. *17* (1972), № 2, 17–20.

[18] Structure theorems for objects, Publ. Math. Debrecen *19* (1972), 191–198.

[19] Lectures on radical and semisimple classes of rings, Lecture Notes, Univ. Islamabad, Islamabad 1972.

[20] Semisimple properties preserved by surjections, Colloquia Math. Soc. J. Bolyai, 6. Rings, Modules and Radicals, Bolyai J. Mat. Társulat, Budapest 1973, 507–514.

[21] Radical-semisimple classes, Periodica Math. Hung. *3* (1973), 243–245.

WITTHOFT, W. G.

[1] A class of nilstable algebras, Trans. Amer. Math. Soc. *111* (1962), 413–422.

WONG, E. T.

[1] Regular rings and integral extension of a regular ring, Proc. Amer. Math. Soc. *33* (1972), 313–315.

WRIGHT, F.–DIEM, J.

[1] Real characters and the radical of an abelian group, Trans. Amer. Math. Soc. *129* (1968), 517–529.

WÜRFEL, T.

[1] Über absolut reine Ringe, J. Reine Angew. Math. *262/263* (1973), 381–391.

YOOD, B.

[1] Closed prime ideals in topological rings, Proc. London Math. Soc. (3) *24* (1972), 307–323.

(YUDANINA, A. B.) Юданина, А. Б.

[1] О тензорных обобщениях радикальных кодов, Вычисл. Мат. и Вычисл. Техн. № 3, Харьков 1972, 126–128.

YUSUF, S. M.

[1] The classical radical of an additively inverse semiring, J. Natur. Sci. Math. *5* (1965), 57–63.

[2] The Krull–McCoy radical of additively inverse semirings, J. Natur. Sci. Math. *12* (1972), 397–412.

ZAKS, A.

[1] Simple modules and hereditary rings, Pacific J. Math. *29* (1968), 627–630.
[2] Injective dimension of semi-primary rinsg, J. Algebra *13* (1969), 73–86.

(ZALESSKI, A. B.) Залесский, А. Б.

[1] Условие полупростоты модулярной групповой алгебры разрешимой группы, Докл. Акад. Наук СССР *208* (1973), 516–519.

ŻELAZKO, W.

[1] On the radicals of p-normed algebras, Studia Math. *21* (1962), 203–206.

ZELINSKY, D.

[1] Linearly compact modules and rings, Amer. J. Math. *76* (1953), 79–90.

ZELMANOWITZ, J.

[1] Endomorphism rings of torsionless modules, J. Algebra *5* (1967), 325–341.
[2] A shorter proof of Goldie's theorem, Canad. Math. Bull. *12* (1969), 597–602.
[3] Regular modules, Trans. Amer. Math. Soc. *163* (1972), 341–355.

(ZHEVLAKOV, K. A.) Жевлаков, К. А.

[1] Ниль-радикал алгебры Мальцева, Алгебра и Логика *4* (1965), 67.
[2] О радикальных идеалах альтернативного кольца, Алгебра и Логика *4* (1965), № 4, 87–102.
[3] Альтернативные артиновые кольца, Алгебра и Логика *5* (1966), № 3, 11–36.
[4] Нижний нильрадикал альтернативных колец, Алгебра и Логика *6* (1967), № 4 11–17.
[5] Нильидеалы альтернативного кольца удовлетворяющего условию максимальности, Алгебра и Логика *6* (1967), № 4, 19–26.
[6] О радикалах Клейнфелда и Смайли в альтернативных кольцах, Алгебра и Логика *8* (1969), 176–180.
[7] Совпадение радикалов Смайли и Клейнфелда в альтернативных кольцах, Алгебра и Логика *8* (1969), 309–319.
[8] О радикалах и неймановских идеалах, Алгебра и Логика *8* (1969), 425–439.
[9] Квазирегулярные идеалы в конечно пороженных альтернативных кольцах, Алгебра и Логика *11* (1972), 140–161.
[10] Радикал и представления альтернативных колец, Алгебра и Логика *11* (1972), 162–173.
[11] Замечания о локально нильпотентных кольцах с условиями обрыва, Мат. Заметки *12* (1972), 121–126.

(ZHITOMIRSKI, V. G.) Житомирский, В. Г.

[1] О радикалах треугольного типа, Уч. Зап. Свердл. Гос. Пед. Ин-та *124* (1970), 176–179.

ZORN, M.

[1] Alternative rings and related questions, I. Existence of the radical, Ann. of Math. (2) *42* (1941), 676–686.

AUTHOR INDEX

SUBJECT INDEX

Page numbers refer only to where the concepts listed are defined.